PHILOSOPHY OF MEDICINE AND BIOETHICS

Philosophy and Medicine

VOLUME 50

PHILOSOPHY OF MEDICINE AND BIOETHICS

A TWENTY-YEAR RETROSPECTIVE AND CRITICAL APPRAISAL

Edited by

RONALD A. CARSON
Institute for the Medical Humanities,
University of Texas Medical Branch, Galveston, Texas, U.S.A.

CHESTER R. BURNS
Institute for the Medical Humanities,
University of Texas Medical Branch, Galveston, Texas, U.S.A.

KLUWER ACADEMIC PUBLISHERS
DORDRECHT / BOSTON / LONDON

A C.I.P. Catalogue record for this book is available from the Library of Congress

ISBN 0-7923-3545-7

Published by Kluwer Academic Publishers,
P.O. Box 17, 3300 AA Dordrecht, The Netherlands.

Sold and distributed in the U.S.A. and Canada
by Kluwer Academic Publishers,
101 Philip Drive, Norwell, MA 02061, U.S.A.

In all other countries, sold and distributed
by Kluwer Academic Publishers Group,
P.O. Box 322, 3300 AH Dordrecht, The Netherlands.

Printed on acid-free paper

Printed in the Netherlands

TABLE OF CONTENTS

SECTION II PRACTICE AND THEORY

SECTION III POLICY

IN MEMORIAM

MARX W. WARTOFSKY
1928 - 1997

With the death of Marx Wartofsky, American philosophy has lost one of its most noble, generous, and energetic members.

The *Philosophy and Medicine* series would not have come into existence without the support of Marx Wartofsky and Robert Cohen, who, in 1974, approached D. Reidel Publishing Company prior to the First Trans-Disciplinary Symposium on Philosophy and Medicine.

The Editors are very fortunate to be able to publish Professor Wartofsky's contribution in this celebratory volume. The philosophy of medicine has its character due to his creative work. We deeply regret that he did not live to see its publication.

He will be sorely missed.

RONALD A. CARSON AND CHESTER R. BURNS

INTRODUCTION

The Institute for the Medical Humanities at the University of Texas Medical Branch co-sponsored the first Trans-Disciplinary Symposium on Philosophy and Medicine which was held in Galveston in May, 1974. The papers presented at that meeting inaugurated a philosophy and medicine book series edited by H. Tristram Engelhardt, Jr., and Stuart F. Spicker.

In their introduction to Volume 1 of this series, Engelhardt and Spicker emphasized that there are "few, if any areas of social concerns so pervasive as medicine and yet as underexamined by philosophy."[1] Participants in the first symposium contributed historical and philosophical assessments of concepts of health and disease, explorations of epistemological questions in the philosophy of science and medicine, and phenomenological analyses of the relation of body and self. There was also a lively exchange on "virtue ethics." The editors envisioned a continuing series of symposia that would provide opportunities for sustained transdisciplinary dialogues about philosophy and medicine.

To help celebrate more than twenty years of extraordinary success with this series, we convened another symposium in Galveston in February, 1995. We asked the participants (some of whom had attended the first symposium) to address the following questions: In what ways and to what ends have academic humanists and medical scientists and practitioners become serious conversation partners in recent decades? How have their dialogues been shaped by prevailing social views, political philosophies, academic habits, professional mores, and public pressures? What have been the key concepts and questions of these dialogues? Have they come and gone or remained with us, and why? Have the dialogues made any appreciable intellectual or social differences? Have they improved the care of the sick?

Philosophy of Medicine and Bioethics, Volume 50 in the series, contains a variety of distinctive responses to our questions. We have arranged these responses into three sections: History and Theory, Practice and Theory, and Policy.

The authors of the essays and commentaries in Section I develop a variety of theoretical perspectives from the humanities, and from philosophy of science and philosophy of medicine in particular.

Edmund Pellegrino perceives in the current evolution of bioethics a growing and troubling methodological imbalance between "disciplines of particularity" and philosophical ethics that reflects "the current trend from objectivity to subjectivity in ethical analysis . . . ." Against this trend Pellegrino argues for a place of prominence for philosophical ethics among the disciplines in dialogue about the proper shape of bioethics as an interdisciplinary enterprise. According to the ecumenical model he develops, philosophy is one discipline among others, but deserves pride of place because it remains "the fundamental discipline for analytic and normative ethics." Other disciplines can teach how it was and is with us but "only ethics as ethics seeks to discern what in the moral life is morally right and good."

In his conceptual survey of the recent involvement of the humanities with medicine, K. Danner Clouser articulates the view that the primary purpose of the medical humanities is service. Reflecting on his years of experience as a philosopher at work in the world of medicine, Clouser finds that world permeated by the kinds of considerations characteristic of humanistic deliberation. Furthermore, he sees a parallel between the relation of the medical humanities to the humanities and the relation of medicine to science. The former use the latter to accomplish a common goal by applying knowledge from their disciplines of origin for "the benefit and well-being of patients." Medicine teaches a narrowing of focus (as does education for the professions generally) with the result that, for all the depth of knowledge this approach makes possible, there is a distinct danger of becoming locked into only one perspective. An exclusive focus on a patient's pain may distract attention from his or her suffering, which may or may not be associated with the pain. "The ability to quickly and easily shift perspectives when appropriate, to see with 'new eyes,' is a valuable clinical skill," a trained capacity that, in Clouser's view, the medical humanities are particularly adept at providing.

In "The Primacy of Practice," Stephen Toulmin provides a historical framework for understanding the recent shift in the focus of philosophical ethics from definitional to substantive issues. By the end of the last century, he observes, a reaction had set in against historicism, motivated by "the hope of underpinning contingent knowledge by necessary principles." By the mid-twentieth century philosophical ethics, confronted, not least by medicine, with requests for help in sorting out practical questions, discovered the wisdom of collectively accumulated and tested experience and "argumentation as a form of 'dialogical' exchange of opinion" to be more useful than propositional knowledge. In Toulmin's view, such discoveries are bringing about a welcome reassessment of rhetoric and a revival of practical philosophy in the wake of the demise of foundationalism.

Marx Wartofsky is interested in the ways in which cognitive practices develop and change over time and in different circumstances. He is skeptical of epistemological claims about knowledge in general, and even of general claims regarding knowledge in particular domains of inquiry or practice, such as ethics or medicine. On Wartofsky's constructivist account of historical epistemology, "Knowledge is not simply reflection upon our experiences and our actions, but is constituent of and ingredient in our practices, and therefore is as various in its character and distinctive features as are these practices, and cannot be understood or appropriately studied apart from them." For purposes of analysis or comparison we may usefully abstract features held in common by particular forms of knowledge. But to mistake the abstraction for knowledge is to fall into the trap of reification. There is no such thing as knowledge in general. This is what the practitioners (Wartofsky's "endocrinologists") know and what the epistemologists can learn from them.

"Even if what 'works' is what 'counts,' how do we know that what works is good?" is the central question Pellegrino puts to Toulmin and Wartofsky in his commentary. Pellegrino rejects the idea that foundationalist accounts have been discredited in principle in the aftermath of the collapse of Cartesian foundationalism. He welcomes the turn from theory to practice taken by Toulmin and Wartofsky, but he sees no way in which practices can provide normative guidance. In his view, practice without metaphysical underpinnings is condemned to circularity and self-justification. "Even if praxis becomes the keystone, it must be set in a supporting arch."

H. Tristram Engelhardt, Jr., takes a different tack in his proposal for grounding secular bioethics. Appearances notwithstanding, Engelhardt doubts that secular bioethics has provided those whose lives are characterized by a radical pluralism "a neutral, but still content-full, vision of proper moral deportment that all could endorse . . . ." Apparent consensus on divisive issues (such as those dealt with by the National Commission for the Protection of Human Subjects) turns out, on closer examination, to have been "manufactured" by limiting participation to those of similar mind on the issues. Secular bioethics, Engelhardt argues, has been no more successful in achieving moral agreement than has any other master narrative. But, although it cannot deliver consensus, it can insist on the principle of permission as the bedrock of relations among moral strangers. "The authority to which persons can appeal when they do not share a common ideology, moral vision, religion, or content-full political understanding is the authority of the consent of collaborators."

Bioethics seems largely to have displaced philosophy of medicine in recent decades, observes Henk ten Have. Or has it subsumed philosophy of medicine? For well over a century, prior to the emergence of contemporary bioethics, philosophy had occupied itself with medicine, but relations between the two disciplines were antagonistic. Around the turn of the century, this antagonism was ameliorated by a division of labor. Henceforth, the two disciplines would no longer compete in analyzing the human condition; rather, medicine would concentrate on accounting for and treating disease, and philosophy would analyze medicine's conceptual methods for doing that. In the course of pursuing its metamedical work, philosophy has discovered that "medicine is fundamentally embedded in culture . . . [and] cannot be understood without attention to cultural values." Bioethics, along with a broader-gauged philosophy of medicine that encompasses and extends earlier epistemological, anthropological, and ethical traditions are the upshot of this discovery.

Noting ten Have's regret that the chorus of concern over ethical issues in medicine threatens to drown out other voices speaking to important philosophical questions, Stuart Spicker asks what the likely continued hegemony of bioethics portends for the future of philosophy of medicine. And in his commentary on Engelhardt's paper, Spicker acknowledges the inadequacy of past attempts to ground morality in the claims of reason, but he is not persuaded by Engelhardt's argument that all such attempts must necessarily fail. "Is there nothing 'sacred' in the secular?" he asks. ". . . [P]erhaps moral authority can be located . . . in the integrity of the lived-body."

The authors in Section II bring various "practice" perspectives to bear in their reflections on philosophy of medicine and bioethics.

Larry Churchill critiques a methodological parochialism in bioethics that results from the urge to theorize moral experience in universal terms, an exercise that distorts experience by reducing its variety. He argues for reconceiving bioethics in the context of historically and culturally informed social inquiry that takes the particularity and complexity of moral experience as seriously as it takes the decisional aspects of moral reasoning. "What our experiences require is not *the* theory (or no theories), but several theories."

In her reflections on a week in the life of a practicing bioethicist, Judith Andre identifies three elements constitutive of bioethics as a practice—its interdisciplinarity, its practical orientation, and its eclectic approach to the acquisition and dissemination of knowledge. In that the practice of bioethics is in large measure about helping health-care professionals and the public "think more deeply and act more wisely about matters of health," bioethicists have a responsibility to be cognizant of the ways in which their work may contribute to the valorization of health and medicine in our society, thereby drawing attention away from circumstances of social inequality that not only damage health but destroy hope.

Thomas Cole discerns in the papers by Churchill and Andre modes of reflective engagement reminiscent of the rhetorical humanism of the Renaissance. Cole believes that, in its deliberative form, rhetoric is well-suited to the task of entertaining and advancing multiple perspectives on varieties of moral experience, and to sustaining a vital connection between both feeling and thought, and personal integrity and social justice.

Remarking on the "poverty of proceduralism" characteristic of bioethics, Ronald Carson urges closer attention to the rich moral languages that patients, families, doctors, and nurses bring to their encounters. The practice of medical ethics, in the view he espouses, consists of reflective, interpretive work aimed at bridging the communication gap between conversation partners who often speak to each other in foreign moral accents. Carson looks to maxims, understood as illuminative of experience rather than prescriptive, as malleable but "steady guides" in the exercise of conscientious judgment.

In her commentary on Carson's paper, Anne Hudson Jones expresses doubt that the substitution of maxims for principles in moral reasoning will provide a true alternative to an impoverished proceduralism because it reinforces the privileged position of the doctor (or the ethicist). Jones identifies similarities between Carson's position and a radical strand of narrative ethics which she believes to be more promising in that it encourages a dialogue in which "everyone involved in a particular case . . . become[s] part of the chorus of voices that seeks its best resolution."

To make the case for a more culturally sophisticated bioethics, Carl Elliott scrutinizes the distinction between identity-altering therapies that aim to cure a condition, and those used to enhance the self. The former are commonly considered acceptable, whereas the latter are thought to be questionable. But, Elliott argues, because "cure" and "enhancement" derive their meaning from local understandings of identity and of life's purposes, a cultural appraisal of such concepts is necessary if the distinction is to be ethically useful. For this reason, "bioethics should look less like metaphysics and more like anthropology."

Gerald McKenny doesn't disagree but wants to revise and add a critical component to Elliott's proposal for an anthropological bioethics. McKenny insists that ideas about the self and the meaning of life not be disembodied but rather "formulated to include views about the place of the body in a morally worthy life." He argues further that a critical anthropological bioethics should not only describe and interpret the ways in which our identities are formed in relation to cultural ideas and ideals but should also expose the self-deforming power of certain societal norms.

"On what basis do we rationally establish that another culture's practices should be stopped?" asks Loretta Kopelman. Kopelman argues that one should look to the claims made by defenders of such practices to determine what they claim to be the practice's purposes and benefits. These claims can then be assessed in light of relevant cross-cultural evidence. Advocates of female circumcision/genital mutilation, for example, claim that the practice prevents disease, promotes health, and contributes to the well-being of mothers and children. Evidence from interculturally valid studies of medical practices belies this claim. It is on the basis of such shared medical values and methods that rational criticism of the morally objectionable practices of another culture can be developed.

Grant Gillett is sympathetic to Kopelman's search for commonalities of conviction and belief sufficient to warrant one culture's criticism of another's practices. But he is skeptical about the prospect of finding common ground in the values and methods of Western medicine. Moreover, Gillett doubts that abstract reasoning of any kind will make much headway against entrenched moral views and cultural practices. In his view, a phenomenological approach is likely to be more promising. "It is plausible that if we examine some of the things that are basic to human life—the actual experiences of nurturing, being welcomed, having one's hurts tended, having one's greeting snubbed, pain, loneliness, and so on—we can erect a common ground in lived experience for some of our moral intuitions."

Courtney Campbell claims that medicine often responds ineffectively to patients' suffering because its grasp exceeds its reach. In his view, an untenable assumption of modern medical practice is that suffering is "an eliminable experience," when, in fact, "pain and suffering are inevitable features of the *human* condition" and "much suffering is medically intractable." In place of medical (or metaphysical) mastery of ineliminable suffering, Campbell commends empathy—"embodied presence," characterized by the patience to listen to, and the disposition to learn from, a suffering person.

Expressing appreciation for his explication of the personal nature of suffering and the healing power of empathy, David Barnard nonetheless finds that Campbell's notion of "presence" does not assume recognizable form. This, Barnard surmises, is because Campbell misconstrues a fundamental problem facing physicians when they try to respond to patients' suffering. "If medicine is to be faulted on this subject, it is for its failure to accomplish the possible, not for its arrogance in pursuing the impossible." What doctors should do is to manage the physical aspects of pain and suffering and, in Cicely Saunders' words, "persevere with the practical."

The contributors in Section III turn their critical attention to the contemporary policy arena where they raise important questions about the evolution and direction of research ethics, and the implications for medical practice of the radical reorganization of systems of health-care provision.

Baruch Brody notes the consensus regarding the ethics of research involving human subjects that had emerged by the early 1980s and explains why the bioethics community has evinced relatively little interest in these issues since that time. Biomedical research has continued apace, but research ethics has been limited to consideration of only a few issues, such as those arising in genetic, fetal, and animal studies. Brody proposes as a remedy for this situation the creation of relevant training opportunities in research ethics and the formation of new national societies and IRB networks.

By means of a review of recent publications and activities dealing with research ethics, Harold Vanderpool questions the degree to which this subject area has been neglected by bioethicists. Noting a number of formidable practical impediments to the development of the sorts of training programs and institutional structures advocated by Brody, Vanderpool nevertheless supports the call for a "return to basic concerns that were initially, but by no means fully, explored when bioethics emerged as disciplined inquiry regarding clinical medicine and research."

Haavi Morreim makes a case for the necessary interdisciplinarity of intellectual work at the intersection of medicine, law, economics, and ethics. Economic pressures are forcing changes in the way medicine is practiced. These changes are prompting a revaluation of the scope and limits of physicians' legal duties and ethical obligations. "Traditional moral expectations of virtually unlimited resources and unstinting loyalty are impugned by [these] newer realities. . . ." The determination of appropriate use of health-care resources should centrally involve physicians' views but also judgments that extend beyond medical expertise about what medical goals and health outcomes are worth pursuing.

In his response to Morreim's analysis, William Winslade concurs in the diagnosis of the moral and legal upheaval precipitated in medical practice by the adoption of "a business model in which patients are customers and consumers." He agrees with Morreim about a need for bioethicists to be intellectual cross-dressers, but he also strongly believes that both providers and patients must learn how to understand the intersections of ethics, law, and economics as they shape the polyadic and polyvalent relationships of emerging health-care organizations and systems.

Evident in the work presented in this volume is a continuing preoccupation with methodological questions, especially those having to do with the "right" relation between theory and practice in the philosophy and ethics of medicine.

Apparent as well is the critical attention presently being paid to "policy ethics" and to the fashioning of moral judgments across cultures.

We extend our gratitude to the authors of these essays and commentaries. Their responses to the questions we asked about the dialogues of recent decades indicate a continuing (and important) preoccupation with methodological questions, especially those having to do with the "right" relation between theory and practice in the philosophy and ethics of medicine, as well as the "best" way to adapt the insights of academic humanists to the needs of medical practitioners and patients. Apparent in their responses as well is recurring attention to the problems of transcultural morality in today's global cosmopolitanism and the need for carefully fashioned policies that address the pluralism and complexities of evolving public and professional cultures.

This, the third, symposium in Galveston was also convened to honor William Bennett Bean, M.D., the first director of the Institute for the Medical Humanities. We thank Thomas N. James, M.D., for providing financial support from the President's Office so that Dr. Bean could be remembered in this way.

We are especially grateful to those who provided considerable help in completing this project: Stuart F. Spicker and H. Tristram Engelhardt, Jr., for responding positively to the idea of an "anniversary volume" and for guidance as the series editors; Sharon Goodwin for coordinating the symposium; Deborah Cummins for a critical reading of the draft papers; Diane Pfeil for preparing the volume for publication; and Eleanor Porter for proofreading the manuscript.

NOTE

[1] Engelhardt Jr., H. T., and Spicker, S.F.: 1975, *Philosophy and Medicine,* Volume 1, D. Reidel, Dordrecht, The Netherlands, p. 1.

SECTION I

HISTORY AND THEORY

EDMUND D. PELLEGRINO

BIOETHICS AS AN INTERDISCIPLINARY ENTERPRISE: WHERE DOES ETHICS FIT IN THE MOSAIC OF DISCIPLINES?*

INTRODUCTION

Bioethics is an emergent universe of moral discourse barely a quarter of a century old. Yet, even as it grows in influence, its identity, limits, and scope as a field of inquiry are increasingly debated [3], [6], [17], [62], [75]. Precisely what disciplines and modes of inquiry should characterize this new field? How much of "bioethics" is biology, how much is ethics, and how much is in the domain of the humanities or social sciences? How do the many disciplines that now claim a role in moral deliberation relate to each other? Where does philosophical ethics fit in the expansive visions of bioethics that are now fashionable? Is it merely one discipline among many, or can it make some claim to the role of *primus inter pares?* If bioethics is, indeed, an interdisciplinary exercise, how inclusive and how diversified should bioethics be? How much of bioethics should be ethics?

These questions seem particularly appropriate at this time and in this symposium which celebrates the twentieth anniversary of the Spicker and Engelhardt series in Philosophy and Medicine. The volume that initiated this impressive scholarly collation was the product of a conference at the University of Texas Medical Branch at Galveston under the aegis of the Institute for the Medical Humanities which is sponsoring today's conference as well. That institute has nurtured some of today's preeminent leaders and scholars in bioethics. It continues to be distinctive in the breadth of disciplines it brings together to examine the ethical and value questions pertinent to medicine, the health professions, and health care.

This same breadth of approach characterizes the interests, scholarly work, and philanthropy of Dr. John McGovern, in whose name I have the honor of delivering this lecture. It was also the approach of Dr. William B. Bean, at one time Director of the Institute, who, along with Dr. McGovern, is among the few who truly merit the title of medical humanist.

R. A. Carson and C. R. Burns (eds.), Philosophy of Medicine and Bioethics, 1-23.

Those questions are more than nostalgic evocations of the medieval disputes about the identity and classification of disciplines. The practical importance of bioethics is such that its definition is more than a semantic or pedantic diversion. Bioethics, today, is invoked in judicial and legislative decisions, in public debate, and in ethics committees, commissions, and clinical consultations. The outcome of those varied "bioethical" exercises imply a certain moral authority and credibility. A new profession of "bioethicists" has been born to provide technical expertise in the analysis and resolution of a wide range of "ethical dilemmas"—at the bedside, in policy formation, and in everyday life.

"Bioethics" is a bifid neologism marrying biology and ethics. One of its terms, "ethics," has traditionally been regarded as a branch of philosophy. Yet, today, many who call themselves bioethicists do not consider their work to be a branch of philosophy. Many of them regard philosophy as inadequate to encompass the complexities of the moral life; some even see it as a deterrent. They deem philosophical ethics to be overly rationalistic, abstract, and insensitive to the contextual, experiential, and pluralistic milieu of actual moral choice. Theirs is an expansive vision of bioethics embracing the widest variety of disciplines, which, taken together, are presumed to remedy the deficiencies of philosophical ethics.

There is merit and substance to some of the criticisms of philosophical ethics, and they must be taken seriously. But there is also genuine danger that the advantages of the systematic, critical, and rigorous reflection specific to philosophical ethics will be diluted or lost. To transform "ethics" into a simulacrum of the other disciplines is to risk losing that which gives it its specific character and contribution to the other disciplines.

In this essay, I will take it that bioethics is, and should be, interdisciplinary. The question I want to examine is how philosophy and the other disciplines (e.g., literature, law, history, theology, language, and linguistics) and the humanistic end of the social sciences (anthropology, economics, sociology, psychology) can relate to each other without loss of identity of ethics as a central discipline. I would agree that bioethics connotes a broad field of inquiry, but I would contend that, within that field, philosophy has a special place. Philosophical ethics certainly must be in dialogue with all the other pertinent disciplines, but it cannot and should not be subsumed or replaced by them.

In an attempt to define this position more clearly, I shall cover the following points: (1) some historical *obiter dicta* on the recent history of three "visions" of bioethics; (2) challenges to philosophical ethics (a) from outside philosophy, (b) from within philosophy in general, and (c) from within bioethics; (3) five models of bioethics and the relationships of philosophy to biology and the humanities and social sciences within each model.

I will confine my analysis to philosophical ethics within medical and bioethics. Obviously, medical ethics is not synonymous with the whole of ethics; nor is the work of philosophical ethics, even in medicine, confined to clinical decision-making, ethics consultations, or ethics committees. But, I will use those as examples of the issues in question because these are the areas in which questions of identity are most urgently asked and alternatives to philosophical ethics most often sought. It is there, too, that the disciplines are most likely to confront each other with practical and concrete ethical consequences.

I. THREE "VISIONS" OF BIOETHICS: SOME NOTES ON RECENT HISTORY

Warren Reich [56] has recently detailed the provenance of the term "bioethics." Using primary sources and interviews, he traces the term to two sources: (1) Professor Van Rensselaer Potter at the University of Wisconsin and (2) Andre Hellegers and Sargent Shriver at Georgetown University. Reich suggests that, almost simultaneously, those three coined the term "bioethics" in 1971.

Reich characterizes Potter's view as the "Wisconsin" approach, one that is "global," and based in an expansive vision of values, method, and principles ([56], pp. 321-322). This vision draws extensively on disciplines like ecology, evolutionary biology, and cybernetics. Its aim is to combine science and philosophy into a "new ethic" of bioethics. Reich describes the "Georgetown" approach as much narrower. He finds its focus in the application of normative ethics to the resolution of concrete ethical dilemmas in medicine and health care. On this view, bioethics is a new branch of a very old discipline, philosophy, which remains intact in method and content.

Reich's dichotomization describes two divergent visions of bioethics. In one, philosophy plays a central role; in the other, it is one among many disciplines, and indeed, if Potter's vision is fully developed, it is one in which philosophy must be fused with science into a "new" discipline and "new" ethics. Reich's formulation really defines the extremes of a spectrum of definitions. It leaves out an important third view to which I shall return below.

In the first edition of his *Encyclopedia,* Reich ([55], p. xxxii) defines bioethics as ". . . the systematic study of human conduct in the area of the life sciences and health care insofar as this conduct is examined in the light of moral values and principles." In the second edition, his definition is ". . . the systematic study of the moral dimensions—including moral vision, decisions, conduct, and policies—of the life sciences and health care employing a variety of ethical methodologies in an interdisciplinary setting" ([57] p. xxi). There are two significant differences between these two definitions that bear on the issue that focuses this essay—namely, the proper relationship between philosophical ethics and the other disciplines that can contribute to bioethics.

The first difference is dropping the term "principles." This change is intended to open the field for ethical theories and methods other than "principlism," the *prima facie* principles of Beauchamp and Childress [7], [8] that has so dominated biomedical ethics until very recently. This move is understandable in the light of subsequent events, although principles and "principlism" are not the same thing, as Reich, himself, points out.

The second difference is the addition of "interdisciplinary setting" and a variety of ethical methodologies. This change acknowledges the contributions of the many disciplines that now are part of the enterprise of bioethics. It expands the domain of bioethics to global proportions. This identifies bioethics much more closely to the expansive "Wisconsin" vision and much less closely to moral philosophy or ethics as in the "Georgetown" perspective. It might be added here that Hellegers' view included a close affiliation with reproductive biology and demography. These latter disciplines, within a short time, particularly after Hellegers's death, gave way to philosophical as well as theological ethics. In this sense, the Georgetown "vision" was broader than that philosophy but always grounded in ethics as the foundational discipline.

Reich did not include a third "vision," one located conceptually between the prototypical Wisconsin and Georgetown visions. I refer to the view of the same subject matter taken before 1971 and continuing into the decade of the seventies. At that time, many of the same items of discourse that now constitute bioethics were pursued under the rather clumsy title of "Medical humanities, Human values, and Ethics." No imaginative neologism defined this third vision—which is perhaps one reason it has not received the attention it deserves. It was pursued, however, by the founders of the Society for Health and Human Values. Almost simultaneously, it was the perspective that inspired the early teaching and research programs at the University of Texas at Galveston, the Hershey Medical Center of Penn State, and the State University of New York at Stony Brook.

This third vision lies conceptually, if not geographically, between Wisconsin and Georgetown. In it, philosophical ethics plays an interdisciplinary and cooperative role with the humanities and the humanistic end of the spectrum of the social sciences as well as with law and political science. On this view, ethics retains its philosophical identity. I believe that, of the three, this vision is most viable conceptually and practically. It avoids the narrowness of the one and the ambitious expansiveness of the other. Let me sketch in the provenance of this view—which may be rather loosely called the "Humanistic Vision"—one that encompasses philosophy in dialogue with all the "humanistic" disciplines, those that inquire into all aspects of human life and human values, especially as they are played out in the universal human experiences of health, disease, dying, and death.

This third vision began inconspicuously and inchoately in the early 1960s with a small group of campus ministers and medical educators who shared a common concern and a common belief. Most were not professional scholars in philosophy, theology, or the humanities. However, they shared a conviction that, if introduced into medical education, the humanities, broadly speaking, with their focus on "humanity" and "human values," could temper, if not reverse, the dehumanizing and depersonalizing effects of technology on patient care and raise the ethical sensitivities of future physicians [26], [27].

To this end, they formed a Committee on Health and Human values, which, in 1969, they formalized as the Society for Health and Human values. The Society functioned with the help of the United Ministries on Higher Education of the Presbyterian Church. In 1970, the society received a grant from the National Endowment for the Humanities to establish under its aegis the Institute on Humanities in Medicine.

The NEH funds enabled the group to convene the first meetings between representatives of campus ministries, medical educators, and scholars in the humanities [26], [27]. Those who attended represented a broad spectrum of expertise in the humanities, ethics, social science, philosophy, and theology. In the subsequent decade, the institute assisted medical schools to establish curricula under the broad titles of "Human values," "Humanities," and "Ethics." It supported summer fellowships for physicians, humanists, and social scientists. Many of today's programs in bioethics and their leaders got their start and encouragement under the aegis of the institute.

The Institute on Humanities in Medicine used the support to stimulate teaching and research in the new field [28], [29], [30], [31], [32], [50], [69], [70]. In that period, institute members made some eighty-five visits to medical schools to provide consultation to faculty members and administrators on the organization and teaching of medical ethics and the humanities in the medical curriculum. For many medical schools, this was the occasion for initiation of a teaching program. Thomas McElhinney and I [51] are in the process of preparing a detailed review of this decade of experience and accomplishment since few recent histories of bioethics [34] allude sufficiently to the contributions of the institute.

In some ways, the work of the institute was an act of faith. There was no evidence then that courses in the humanities could make future physicians more humane. Indeed, the history of humanism and the humanities, themselves, was not encouraging. Too often in the past, humanists had proven inhumane. George Steiner has pointed out how fragile the humanistic tradition proved against the bestiality of the death camps ([63], p. 73). Albert Camus ([11], p. 230) even said flatly, "The humanists are the greatest executioners of the Twentieth Century." The questions of ultimate impact on the teaching of ethics, humanities, or human values on the education and behavior of physicians and other health professionals demands to be definitively examined.

Despite early convictions about the interdisciplinary nature of bioethics, there were then, as now, certain tensions between and among the disciplines that are disquieting for the conceptual future of bioethics. Some of the tensions are of ancient lineage and arise in the different methods and temperaments with which different scholars and different disciplines study the same human phenomena. Other tensions are more specific to the settings of ethics consultations and committees where bioethics is practiced and in teaching programs where it is taught for health professionals. In a general way, these tensions reflect differences in perspective and method between disciplines of two types: the disciplines of particularity, those grounded in experiential and existential detail, e. g., literature, history, language, and the behavioral sciences; and the discipline of general and abstract moral concepts and judgment, i.e., philosophical ethics. All the involved disciplines have a theoretical base, but in the enterprise of bioethics, some contribute primarily through their richness in particulars, rather than through their particular methodology. Those differences reflect the differences between cognition and affect. Additional stresses arise out of the current trend from objectivity to subjectivity in ethical analysis, from increasing doubts about claims to moral expertise, and from the growing tendency to resolve moral conflicts by legislation, public debate, or referendum.

From the outset, bioethics has been an interdisciplinary dialogue between the abstract and the particular. The moral life, itself, involves essence and existence, the universal principle and the particular case, cognition and affect, experience and abstraction from experience. Imbalances between the realms of the abstract and the concrete can distort the enterprise of bioethics. They pose dangers to any serious effort to clarify ethical concepts or arrive at general normative statements of duty or principle or normative conceptions of the relationship between professionals and those who seek their assistance. These imbalances primarily are a challenge to the traditional conception of ethics as a philosophical enterprise.

II. CHALLENGES TO PHILOSOPHICAL ETHICS

For 2,500 years, philosophers claimed ethics as their own. For Socrates and the Stoics, philosophy was virtually synonymous with ethics. It was no surprise, therefore, that philosophers became interested in the new ethical issues that were emerging in biology and medicine twenty-five years ago. The surprise is that they had not done so previously—at least in the United States. In any case, much of the most influential work in the early years of bioethics was philosophical or theological. Theological ethics and its relationship to bioethics deserves separate consideration. I shall limit this essay to the question of philosophical ethics, taking it as a given that theology is one of the disciplines in the interdisciplinary mix.

Today, philosophy's dominance as a foundational discipline is questioned from within as well as outside its own ranks. It is now clear that the complex nature and scope of the uses of biological and medical knowledge cover a broad range of discourse among philosophers, humanists, and social and biological scientists. As a result, philosophy is forced now to re-examine its proper place in the domain of ethical reflection, which it had so long regarded as its undisputed patrimony.

The dominance of medical ethics by philosophy in the last quarter century tends to obscure the fact that medical ethics has been a truly philosophical discipline for only a short time. To be sure, medical morals have existed for 2,500 years, but medical *ethics* —i.e., the formal, systematic, and critical inquiry into medical morals—is only a few decades old [47]. Until the very late sixties and early seventies, professional philosophers had largely ignored medical ethics. When they did engage it, it was to examine medical dilemmas from the perspective of a pre-existing system of morality, e.g., Aristotle's virtue ethics, Kant's deontology, Mill's consequentialism, or, most prominently, a modernized version of W. D. Ross's [8, 59] *prima facie* principles.

Those approaches were highly successful for several reasons. They offered systematic and well-reasoned analyses of the burgeoning complex of concrete problems encountered by physicians and patients. Clinicians needed a framework against which to make difficult clinical decisions. They also wanted a philosophical alternative to what, for many centuries, had been largely an enterprise of moral theology. The shift to philosophy suited the pragmatic, positivist, secularized, and democratic temper of the times and of many physicians and patients. It also made up for the paucity, at that time, of specific guidelines applicable to the newer ethical problems created by scientific and social change.

One particular version of philosophical ethics, the so-called "four principle approach" rapidly became the standard method of analysis of clinical-ethical dilemmas. It also became the lightening rod—somewhat unfairly, in my opinion—for the mounting criticism of philosophical ethics that began to surface in the early eighties. But, it is not "principlism," alone, that is under attack; the target increasingly is any systematic, deductive, or analytical approach to ethical analysis. Aristotelian, Thomistic, Kantian, and Utilitarian ethics have all been targets of recent criticism. They have in common the idea that moral truth is ascertainable by human reason and that general, universalizable, normative principles can be applied to ascertain the rightness or goodness of human conduct. The challenge to traditional ethical systems has become a challenge to the very possibility of ethics as a rational enterprise capable of grasping moral truth.

A. Criticism from Outside Philosophy

From the outside, the most frequent charge against philosophical ethics is that of remoteness from the concrete, lived experiences of moral choice. It is argued that philosophy's abstractness ignores or belittles the richness of detail, the emotional and affective states, the value desiderata, the contextual complexity, and the limitations on rationality that are part of the moral life. The universal moral truths, justifications, and solutions traditionally sought by philosophical ethics are judged futile in a world characterized by multiple visions of the good life, many of which are incommensurable with each other. Those visions are contained in narratives and stories, the character of the agent, the agent's religious beliefs or ethnic identification—all of which philosophical ethics is alleged to have ignored.

The multiplicity of visions of reality and the moral life expressed in those criticisms are not new. They were recognized by Giovanni Battista Vico's (1668-1744) recognition of the singularity of each society's worldview as expressed in metaphors, stories, and values often incommensurable with its own (Cf. [10], pp. 8-9; 59-62). Johann Gottfried Herder (1744-1803) later pointed to the same plurality of visions among civilizations ([10], pp. 10; 57-58). While they, themselves, were not relativists in the contemporary sense, they laid some of the groundwork for skepticism of sociologists, historians, and anthropologists about philosophical ethics. The search of traditional philosophical ethics for universalizable moral principles and guidelines seemed incongruent with cultural, historical, and ethnic differences about what constituted right and wrong. As a result, interest turned from philosophical reflection to studies of the disciplines that were intimately enmeshed in the concrete particulars of the experiences of individuals, societies, and races.

As a result, a variety of substitutes for, or additions to, philosophy have been offered—e. g., literature, psychology, sociology, economics, law. Almost any discipline that engages in the particularities of some aspect of human existence has been proposed to remedy philosophy's abstractions. All have come to be endowed by one writer or another with the normative force once reserved for philosophical reflection.

The ages-old tensions between the concrete and the abstract, between essence and existence, and between the universal and the particular have become particularly acute in ethics. Ethics is the most practical of the branches of philosophy, so it cannot remain in the abstract; however, neither can it do without the abstract lest it fall into situationism. Since its beginnings, ethics has sought to define such abstract notions as the good life, the virtuous person, and the good society, and, from these, to provide a system of guidelines for right and good conduct. The distance between those abstractions and their embodiment in real persons and events is accentuated in bioethics where humans are forced to choose among complex, fearful, and threatening alternative modes of being and acting. Some contemporary ethicists deny this gap can ever be closed, leaving ethics powerless and moral agents helpless except to respond to their feelings or to make ethics entirely an exercise of praxis [68]. Others—and I am among them—insist the gap can and must be closed because the abstract and the concrete are unavoidably built into human experience and are, and always will be, the substance of moral choice and moral praxis.

B. Criticism Within Philosophy in General

The criticisms from the humanistic disciplines outside philosophy have been reinforced by trends within philosophy, itself. In the twentieth century, normative, deductive, principle-based philosophical ethics has been reformulated to various degrees in emotivist, linguistic, and positivistic terms. Today, many philosophers are uncertain about the identity of moral philosophy and about what, if anything, it brings to the dialogue. The history of philosophy has been called a record of the dialogue and dialectic between generations and systems of philosophy. Internal criticism, therefore, is not new to philosophy. But today's skeptics seem more radical. Their skepticism centers on such fundamental philosophical concepts as moral truth, the nature of reason, and the existence of an autonomous reasoning self. Thinkers like Rorty, Derrida, and their followers, for example, believe that we are already in a post-philosophic era in which the utility of philosophy is minimal, non-existent, or, at best, ephemeral.

For the less radical critics, philosophy may not be entirely dead, but its utility is certainly in doubt. On their view, its survival demands transmutation into something else—hermeneutics, historiography, rhetoric, some form of social inquiry or coherentism—all take the search for ultimate, foundational, or universal truths or concepts to be futile or illusory. Indeed, any talk of overarching theories or principles in ethics is anathema in the current anti-foundationalist temper of contemporary ethics. The result is a series of seemingly unhealable dichotomies, e.g., the *a priori* vs. the empirical, certainty vs. fallibility of knowledge, unity vs. heterogeneity, universality vs. plurality, contingency vs. rationality [5].

Late twentieth-century philosophy offers a potpourri of alternative responses to the perceived deficiencies of classical philosophy, like hermeneutics, postmodernism, critical theory, existentialism, or phenomenology. The critics of ethics within philosophy share a common theme with the criticisms from the other humanistic disciplines outside philosophy—e.g., the failure of philosophy to take into account particularities, contingencies, language, and the myriad details of concrete physical, social, and psychological existence as well as the incommensurability of diverse perceptions and visions of the world of self and other.

Beneath these criticisms is the more sweeping skepticism about the possibility of any foundation for our grasp of the real world, moral or otherwise. Antifoundationalism, as it is called, opposes anything like Aristotle's argument in the *Posterior Analytics* [2] for a sound conceptual basis for knowledge of any kind. Of course, this is not new. Antifoundationalism comes in many forms, as does foundationalism, but its common theme is a radical doubt in traditional metaphysics and epistemology, which, in turn, casts doubt on any structured justification for the propositions of moral philosophy. The practical implications for moral philosophy of the erosions of its philosophical underpinnings are significant ([71], p. 99).

C. Criticism from Within Ethics and Bioethics

It was predictable that the stresses from the disciplines outside philosophy and those within philosophy, itself, would be reflected in bioethics and medical ethics, which are subsets of the larger universe of philosophical ethics. This is not the place to detail the current debates about different suggested antidotes to principlism and to the more classical approaches to moral philosophy. In Anglo-

American ethics, we need only mention the renaissance of virtue ethics, the turn to casuistry, the ethics of caring, feminism, hermeneutics, or reflective equilibrium. For their part, European philosophers and bioethicists think Americans are overly analytical or too engrossed in practical dilemmas and puzzles. They urge a closer integration of phenomenology, postmodernist, and poststructuralist philosophies into bioethics [73]. They are also more sympathetic to the need for a philosophy of medicine linked to medical ethics.

Each alternative is inspired by the same need to accommodate the diversity and dichotomies evident in philosophy's internal criticism. Current philosophical "anti-foundationalism" to the contrary, ethics cannot escape being shaped by our conceptions of philosophy, truth, reason, and rationality. Medical ethics, in my view, must be grounded not only in ethical theory but in a philosophy of medicine. The debate about the possibility or existence of such an entity as the philosophy of medicine and its definition are adding to the present confusion about the nature of bioethics [13], [20], [49]. Engelhardt has consistently argued that the rational resolution of ethical controversy is possible only where there is a common understanding of moral values and premises. He holds that ethical discourse is possible only within communities of belief; it is doomed to failure if it attempts to take place across communities of belief [19].

Current criticism of philosophical ethics has weakened the conceptual contributions philosophy might make to the complex field of bioethics. The upshot in recent years has been to seek what philosophical ethics lacks in the other humanistic disciplines, which, being "humanistic" each in its own way, provide sources for reflection on the moral life. The result is a recasting of the way the humanistic disciplines relate to each other within bioethics and to philosophical ethics either by subsuming it, transforming it, or replacing it in the dialogue.

III. FIVE MODELS OF THE RELATIONSHIPS OF THE DISCIPLINES IN BIOETHICS

Taken together, those criticisms underscore the need for a clearer explication of the relationships that might obtain between philosophical ethics and the other humanistic disciplines now pertinent to bioethics. I would suggest five models, the advantages and shortcomings of which bear examination.

I would label these models as follows: the *traditional,* the *antiphilosophical,* the *process,* the *eclectic-syncretic,* and the *ecumenical* models. In the *traditional* model, bioethics is simply a branch of moral philosophy classically construed. This would correspond to Reich's version of the "Georgetown" model. In the *antiphilosophical* model, philosophy is considered effete and is banished in favor of one of the other disciplines or transformed into one of them. The *process* model eschews disciplinary identification in favor of bioethics as a method for deliberation and collaborative decision-making, primarily in ethics consultations and committees and in public policy. In the *eclectic-syncretic* model, something is taken from all the disciplines and fused into an entirely new discipline. This would correspond in many ways to the "Wisconsin" model. In the *ecumenical* model, philosophy is one discipline among others, each with its own identity intact, each with its specific tasks to perform. This is the model in support of which I will argue. In it, philosophy remains the foundational discipline for analytical and normative ethics.

A. The Traditional Model

In the *traditional* model, ethics is classically construed as a philosophical enterprise. Bioethics is a branch of philosophy, i.e., the application of the method and content of philosophy to ethical issues in biology and medicine. This view is caricatured by its critics as being too abstract, too removed from the context of lived worlds and the concrete experiences of the moral life. These critics contend that this model needs the insights, contextual detail, and compassionate apprehension of moral dilemmas that the other "humanistic" disciplines can provide. In its extreme (some would say, distorted) form, the traditional model can lead to legalism, formalism, and insensitivity to affective states. When this occurs, philosophy's claim to moral truth or right moral judgment is justifiably suspect.

The critics of philosophical ethics say it is reflective, but not sufficiently perceptive. There is some truth in this criticism, but it does not damn the whole enterprise of philosophical ethics. Rather, it indicates that philosophical ethics can be too narrowly conceived and that it has much to gain from dialogue with the disciplines of the particular, the imaginative, and the experiential. The question is how this dialogue is best structured without losing what philosophical ethics has to offer to the whole enterprise of bioethics.

Later in this essay, I will suggest that philosophical ethics should be retained as the foundational discipline in bioethics. But it cannot serve the enterprise of bioethics without a continuing, deep dialogue and dialectic with the disciplines of fact, particular experience, and affect in an ecumenical model of interrelationships. (See Section IIIE.)

B. The Antiphilosophical Model

Antipathy to philosophical ethics is becoming popular among practitioners of the other humanities as well as the social and biological sciences. This view results from the convergence of the criticisms from outside philosophy, within philosophy, and within bioethics already discussed in Section II of this essay. Attempts to reassemble ethics from the debris of this demolition have turned to particular experiences and praxis, dialogue, and linguistic games. But without traditional truth claims or principles, the best the skeptics can do is to fashion a species of moral gnosticism or intuitionism—if they can succeed in avoiding moral nihilism and relativism. For some, this will mean that the right and the good are what feels "comfortable," what can be agreed upon by "reasonable people," or what "works." Unfortunately, there remains the philosophical task of discerning by what means we judge the comfortable, the reasonable, or the functional. While not so openly antiphilosophical, there is a strong strain of antipathy to, or rejection of, philosophical ethics in the *process* and the *eclectic-syncretic* models.

C. The Procedural or Process Model

The procedural or *process* model evades the conceptual issues. Its emphasis is on the way moral issues or differences are to be resolved among the participants in moral decisions. Anyone involved in concrete moral decision-making in medicine—at least in individual clinical cases, ethics consultations, or ethics committees—recognizes the importance of a process for collaborative deliberation. The need for an interplay between philosophical ethics and moral and group psychology, ethnic and cultural differences, anthropology, theology, religion, etc., is obvious. But a deliberative process must have some purpose beyond deliberation. That purpose is to arrive at what is the right and good way to act is in each concrete situation, not just to help the participants in moral decisions to feel "comfortable" with the way in which they have decided to act.

The purpose of practical moral reflection is right and good conduct. It cannot result from process alone. If the process is to get anywhere, it will need principles, rules, virtues, or duties to guide it. Process without guidelines is a form of mutual psychotherapy of value to the participants. It is not the substance of ethical deliberation—unless one equates the right and the good with whatever emerges from a process of compromise or consensus. To be sure, there is an ethics of the *process* of moral decision. But this, too, must be an ethical, not a psychological project. A "good" process is a necessary requisite for a good group decision, but it is not sufficient to assure the goodness of that decision in moral terms. For that, the process, itself, must be the subject of critical, philosophical analysis.

D. The Eclectic-Syncretic Model

The *eclectic-syncretic* model has a special appeal in an age sensitive to the importance of diversity and plurality in worldviews and moral beliefs. Eclecticism is appealing when, as now, traditional systems of morality are suspect or weakened by the extent and rapidity of scientific and social change. Eclecticism was the moral philosophy of the Alexandrian School in ancient philosophy. Victor Cousin (17921862) was its exponent in his view that rival philosophies were incomplete, not false, and that the good should be extracted from each. The *eclectic* model recognizes merit in opposing positions and selects from each what it thinks valid without adopting the whole system. What is selected will depend on the principle of selection, and that question will bring us back to the need to justify that principle.

The *syncretic* model is more ambitious: it tries to resolve the differences among systems by fusing them into a new one. Eclecticism says, "Let a thousand flowers bloom." Syncreticism says, "Let's fuse sunflowers and nasturtiums and make a new flower." Both syncretists and eclectics suffer from the arbitrariness of their selections and the incoherence of a moral system built on a patchwork of pieces selected without a clear organizing principle of some kind. If there is resort to such a principle, then eclecticism and syncretism fall back into one of the other models.

The general problem with any eclectic-syncretic model is that it robs each discipline of its specific difference and, therefore, of its contribution to moral discourse. Depending on what elements of diverse systems are put together, this model can easily end up in a "new" ethics that is no ethics at all. Since this mode is so popular at the moment, I shall examine two variations in a little greater detail—one is the fusion of ethics with biology, the other with literature. The problems inherent in these two are illustrative of other syncretic-eclectic models invading law and the social sciences.

1. Ethics and Biology

Van Rensselaer Potter's quest for a fusion of science and philosophy is the first example of a syncretistic approach. On his view, bioethics should strive for global health through progress in biological science. Edward O. Wilson [74] is more specific. He sees the fusion of ethics and population biology even more ambitiously as the eradication of ethics by neurology. Others would apply Darwinian evolutionary biology more generally as an approach to moral problems and issues—again, as a remedy for the perceived deficiencies of moral philosophy [1]. David Heyd [24] sees a challenge beyond the capabilities of existing ethical systems in the creation of a "future people" through genetic manipulation. He proposes a new field of "genethics" to deal with these problems beyond the context of philosophical ethics. An increasing number of writers are interpreting "bioethics" as "biological ethics" and touting it as a "new" ethic more responsive to the needs of the times.

Whenever any new "scientific" ethic is generated, it usually ends by loosening the ethical constraints on science or reducing ethics to determinism of some kind. It is one thing to bring the latest data from population biology into the realm of moral analysis; it is another to create a new ethic out of biology. Making science the arbiter of its own ethical probity is to give it normative authority outside the perimeters of the scientific method. Science is in need of ethical justification which science, itself, cannot give. This is the case whether that science is genetics, ecology, environmentalism, or neurophysiology.

2. Literature and Ethics

In the case of literature, the dangers of displacing ethics are more subtle and, at the moment, more widely promulgated. No serious writer or reader doubts that any creditable work of literature bears significantly on good and evil. How could it be otherwise since literature deals with the human condition? It is in this sense that Nussbaum's ([43], p. 172) judgment that James's *The Ambassadors* is a major work in moral philosophy has meaning. The same is true of Annie Dillard's ([16], p. 11) comment that "living by fiction" is doing "unlicensed metaphysics in a tea cup."

But being rich in moral content does not confer special epistemological status on literature. The moral content of novels and stories cannot be self-justifying any more than science can be self-justifying. Literature greatly enriches our understanding of the moral life. It evokes sympathy and introspection. Nussbaum is right to say that we cannot assume ". . . without argument that literary works are dispensable in an inquiry that aims at ethical truth" ([42], p. 12). But this does not give sanction to the replacement of philosophy *entirely* by text or narrative as some have concluded. In the fictive situation of the protagonist, the novelist provides a moral laboratory in which we can test our own moral beliefs without danger. Fiction and poetry can provide content, but they are interpretable in different ways by different readers. The reader brings to the task his own idiosyncratic perceptual approaches. This is why multiple books continue to be written about each of Dante's poems, Shakespeare's plays, Michelangelo's sculptures, or Rembrandt's paintings. Their power to evoke is enormous. What they evoke is various, but no evocation can stand on its own as a moral model.

For all their power, fictive characters are fictions. They are as much reflective of their authors' moral lives as the reader's. Their "truth value" must also be judged outside the reader's experience of the story if they are to have more than solipsistic moral value. Whose story do we take as our guide: St. Augustine in *The Confessions* [4], Camus's Rieux [10], Santayana's *Last Puritan* [60], Thomas Mann's Hans Castorp [37], Huysman's Des Esseintes [25], or Melville's Captain Ahab [40]? These are characters and lives radically different in their moral testimony. The author's self-reflection can expand the sensitivity and responsiveness of the reader's moral introspection. She or he can enmesh us in the variegated particulars of an imagined life, but that cannot replace the hard work of normative ethics. In the end, the reader must choose whether to accept, reject, or modify his or her own way of life in light of the experience gained by the evocations of affect and thought in a work of fiction.

It is too much to expect literature to do all of what Nussbaum requires of it—to teach us "how we should live" ([43], p. 173). This was the quest of all ancient philosophical ethics. Nussbaum, whose knowledge of classical philosophy is profound, seems to put too much faith in the normative possibilities of literature. Interestingly, Iris Murdoch ([41], p. 492), who is both a novelist and a professional philosopher, returns to the Platonic notion of the Good in her recent novels. John Gardner ([21], p. 43), too, takes fiction to be a moral enterprise but warns against "the shoddy morality of most of our fiction." Indeed, the writer may be ". . . interested in knowing the world in order to make real and honest sense of it . . ." ([16], p. 151), but the reader must judge whether the writer has succeeded.

We are rational as well as affective beings. Thus, we do respond to the story's moral arguments as well as the feelings it inspires. We judge the content or message if the author has one. But to do this, not only must we enter the story but we must transcend it. The very details that give a story substance must be critically examined for the way they are assembled. Judging whether a story is good or bad is not just a literary judgment. However imperfect it may be, the task of ethics is to make that judgment. Ethics subjects the story to a metanarrative analysis.

So far as bioethics goes, one very important aspect of imaginative literature is its impact on the quality of ethical decisions through its capacity to raise the sensitivities of physicians, nurses, and other health professionals to the intricacies of the moral life. Imaginative writers can evoke experiences of empathy, revulsion, or compassion, particularly in their descriptions of the intricacies of the healing relationship. Indeed, in my view [46], philosophy of medicine begins with the existential phenomenology of that relationship. Stories can incarnate ethical experiences in the human person of doctor and patient. But to come to defensible normative conclusions, we must subject the story, itself, to ethical analysis. "Clinical stories," that is, the patient's narrative, can too easily be a *post facto* tailoring of an *a priori* theory. This is especially the case in psychoanalytical interpretations of a patient's illness, an author's themes and motives, or an autobiography [38].

To be sure, biomedical ethics and its practitioners can and should improve their skills by imaginative reading of literature. The world's literature abounds in creative evaluations of the moral life. The length and breadth of Joanne Trautmann's and Carol Pollard's [66] annotated bibliography attests to the richness of those readings. But compassion, empathy, and sympathy need reason, too. We can kill with kindness by misappropriating stories, using the wrong stories, or identifying with the wrong characters. Reason and feeling, *logos* and *patheia,* are inextricable from the healing relationship and from the ethical reflection that true healing requires. But to eradicate the distinctions between them is to lose moral balance.

Literature is an essential adjunct to philosophy. It provides detail, contextuality, complexity, description, and appreciation of values just as Nussbaum [43] suggests. But this is not to give moral hegemony to literature. Phenomenology, existentialism, postmodern philosophies, behavioral science, and biology also deal in details and particularities. Each has grasp of an essential part of the manifold phenomena that constitute human life. But none, by itself, gives a complete account of that life.

E. The Ecumenical Model

The *ecumenical* model of bioethics can best accommodate the need of philosophical ethics for concrete particulars without losing the critical capacity and search for moral truths that is the task of normative ethics. The ecumenical model differs from eclecticism and syncretism in that philosophical ethics retains its traditional identity. So do the other disciplines with which philosophy must be in dialogue. Literature, anthropology, history, and evolutionary biology are all "disciplines of the particulars" and of concrete phenomena. Those disciplines observe and study the phenomena of the moral life in all its complexity, but each does so from a different perspective.

The scholastic philosophers would say that each shares its *material* object with other disciplines, i. e., the subject each studies is the moral life. But each has its own distinctive *formal* object, its peculiar perspective in the way the moral life is studied. Those differences are what distinguish each discipline and enable each to make its own contribution to the common dialogue. Literature, for example, deals in imaginative and fictive worlds of moral decision; anthropology, in social and cultural value systems; history, in the continuity and interrelationships of moral viewpoints over time; ecology and environmental science, with the impact of human life on the biosphere; molecular biology, with the chemical mechanisms of life; etc. For all disciplines, therefore, the moral life can be a subject for moral study, but only ethics as ethics seeks to discern what in the moral life is morally right and good. This is its primary aim—i.e., it focuses on what ought to be, not what **is**, or on what is the good for human beings, not what humans do or did in their pursuit of ends which might be morally good or bad.

A complete description of a particular moral event requires an account of the agent's intent and experience of the act, its nature, and its circumstantial context [48]. It is the task of the many disciplines of the particular (literature, history, anthropology, etc.) to describe the intricacies, the contextuality, the personalized, psycho-social dimensions of an act. Those are the disciplines best suited to the task of descriptive ethics. But it is the task of philosophical ethics to sift through those details for what they tell us about the nature of the right and the good, both in theory and practice. In any viable ecumenical model of bioethics, normative and analytical ethical reflection remain philosophical enterprises. Descriptive ethics is the province of the humanities, law, biology, and the social sciences. Descriptive ethics supplies the details which analytic and normative philosophy examine for those conceptual elements and principles that transcend detail. In this dialogue with the disciplines of the particular, bioethics can acquire that compassionate sensitivity essential to its task without losing its special capacity to rise above detail.

In the end, perhaps what we are talking about is the distinction between ethics and morality as Wayne Meeks [39] has defined it. Ethics is what Meeks calls a "second order" reflection concentrating on the analytical and normative. This is an important but restricted subset of the larger concept of morality. Meeks identifies morality as "a dimension of life, a pervasive, and often only partly conscious, set of value-laden dispositions, inclinations, attitudes, and habits" ([39], p. 4). The more expansive models of bioethics are better directed to morality in this broad sense than is ethics. The "ecumenical model," as I have defined it, is closer to ethics traditionally considered but broadened by drawing on the wider world of morality. Still, if bioethics is to be ethics at all, it must retain a specific perspective within that world.

If philosophy is optimally to occupy a foundational role in the interdisciplinary exercise we call "bioethics," it will need to resolve some of the conflicts internal to philosophy in general and to moral philosophy in particular. I have not made an argument for any particular way of doing moral philosophy. Obviously, the utility and effectiveness of philosophical ethics will be a function of the validity and effectiveness of the kind and quality of philosophy it turns out to be. The current debates within biomedical ethics, and ethics generally, as to theory and method will be salubrious if they do not end up in moral nihilism, skepticism, or simply in a theory of praxis—useful as that might be. This is a topic for another day.

I have argued that bioethics should not leave ethics behind. The social sciences, humanities, and the biological sciences, rich as they are in existential detail about human existence, cannot fill the vacuum if philosophical ethics is destroyed or attenuated. If they try to do so, they, themselves, will become philosophies; that is to say, they will become metabiology, metapsychology, metasociology, etc. These neo-philosophies will have to assume philosophy's tasks, and, if they do, they will have to be judged as disciplines of the abstract and no longer as disciplines of the particular. Rather than transmuting philosophy, they will, themselves, be transmuted into philosophy. The human mind eventually gravitates to the general and the universalizable in its search for ultimacy in morals as in its other encounters with reality: philosophical ethics cannot be avoided.

CONCLUSION

Bioethics is a complex endeavor served by many disciplines and growing in influence in private, professional, and public life. There is little doubt that the multifaceted nature of the problems with which bioethics deals calls for contributions from many disciplines. In that interdisciplinary effort, there is need for clarification of what is connoted in the term "ethics" when it is joined with the prefix "bio" and where ethics fits in that mosaic of disciplines.

I have argued that ethics should remain a branch of philosophy as it always has been. I also argue that to fuse philosophical ethics with other humanistic disciplines or to transform it into some syncretic mixture of disciplines is to lose something essential to moral deliberation. Bioethics is, and should be, an interdisciplinary enterprise, but "ethics" is what it is about, and this means that in the interdisciplinary dialogue, philosophy remains the foundational discipline.

Of the several models proposed to remedy the perceived deficiencies of philosophical ethics, the one which retains the identity of philosophical ethics, yet maintains dialogue with the other humanistic disciplines, is preferable. In this model, the "disciplines of the particular" (literature, history, psychology, anthropology, etc.) perform the task of descriptive ethics. The analytical and normative components remain to philosophy, however. This is intellectual ecumenism of a sort that fosters dialogue and dialectic that can only be effective if the difference each discipline makes is sufficiently identifiable to make that difference its contribution to the still-evolving mosaic of bioethics.

Georgetown University Medical Center
Washington, D.C., U.S.A.

* A version of this chapter was delivered at the John P. McGovern Award Lecture in the Medical Humanities at The University of Texas Medical Branch at Galveston, Texas, on February 15, 1995.

REFERENCES

1. Alexander, R. D.: 1987, *The Biology of Moral Systems,* Aldine de Gruyter, New York.
2. Aristotle: 1981, *Posterior Analytics*, H. Apostle (trans.), Peripatetic Press, Grinnell, IA.
3. Arras, J. D.: 1994, 'Principles and particularity: The role of cases in bioethics', *Indiana Law Journal* 69, 983-1014.
4. Augustine: 1991, *The Confessions,* H. Chadwick (trans.), Oxford University Press, New York.
5. Baynes, K., J. Bohman, and T. McCarthy: 1989, *After Philosophy: End or Transformation?,* MIT Press, Cambridge, MA, pp. 1-18.
6. Beauchamp T. L.: 1994, 'Principles and other emerging paradigms in bioethics', *Indiana Law Journal* 69, 955-972.
7. Beauchamp, T. L. and J. F. Childress: 1979, *Principles of Biomedical Ethics,* Oxford University Press, New York.
8. Beauchamp, T. L. and J. F. Childress: 1994, *Principles of Biomedical Ethics*, 4th ed., Oxford University Press, New York.
9. Berlin, I.: 1981, *Categories and Concepts,* Penguin Books, New York.
10. Berlin, I.: 1991, *The Crooked Timber of Humanity,* Alfred Knopf, New York.

11. Camus, A.: 1961, *Resistance, Rebellion and Death,* J. O'Brien (trans.), Alfred Knopf, New York.
12. Camus, A.: 1948, *The Plague,* S. Gilbert (trans.), Alfred Knopf, New York.
13. Caplan, A. L.: 1992, 'Does the philosophy of medicine exist?', *Theoretical Medicine* 13, 67-77.
14. Carson, R. A.: 1994, 'Teaching ethics in the context of the medical humanities', *Journal of Medical Ethics* 20, 235-238.
15. Danto, A. C.: 1983, 'Presidential address: Philosophy as/and/of literature' *Proceedings of the American Philosophical Association* 1983, 5-20.
16. Dillard, A.: 1983, *Living by Fiction,* Harper-Colophon, New York.
17. Dworkin, R. B.: 1994, 'Introduction', *Indiana Law Journal* 69, 945-954.
18. Engelhardt, Jr., H. T.: 1987, *Bioethics and Secular Humanism: The Search for a Common Morality,* Trinity International Press, Philadelphia, PA.
19. Engelhardt, Jr., H. T.: 1995, *The Foundations of Bioethics,* 2nd ed., Oxford University Press, New York.
20. Engelhardt, Jr., H. T. and K. Wm. Wildes: 1995, 'Philosophy of medicine', in *Encyclopedia of Bioethics,* Volume 3, 2d ed., W. T. Reich (ed.), MacMillan, New York, pp. l680-1684.
21. Gardner, J.: 1978, *On Moral Fiction,* Basic Books, New York.
22. Gilligan, C.: 1982, *In a Different Voice: Psychological Theory and Women's Development,* Harvard University Press, Cambridge, MA.
23. Gross, P. R. and N. Levitt: 1994, *Higher Superstition: The Academic Left and Its Quarrel with Science,* Johns Hopkins University Press, Baltimore, MD.
24. Heyd, D.: 1992, *Genethics. Moral Issues in the Creation of People,* University of California Press, Berkeley, CA.
25. Husysmans, J. K.: 1969, *Against the Grain,* Dover, New York.
26. Institute on Human Values in Medicine: 1972, *Proceedings of the First Session,* Society for Health and Human Values. Philadelphia, PA.
27. Institute on Human Values in Medicine: 1972, *Proceedings of the Second Session,* Society for Health and Human Values, Philadelphia, PA.
28. Institute on Human Values in Medicine: 1976, *Human Values Teaching Programs for Health Professionals;* Report 7, Society for Health and Human Values, Philadelphia, PA.
29. Institute on Human Values in Medicine: 1976, *Reports of the Institute Fellows 1975-1976;* Report 8, Society for Health and Human Values, Philadelphia, PA.
30. Institute on Human Values in Medicine: 1977, *Reports of the Institute Fellows 1976-1977;* Report 9, Society for Health and Human Values, Philadelphia, PA.
31. Institute on Human Values in Medicine: 1977, *Reports of the Institute Fellows 1977-1978;* Report 11, Society for Health and Human Values, Philadelphia, PA.
32. Institute of Medicine: 1972, *Report of a Conference: Educating for the Health Team,* National Academy of Sciences, Washington, D.C.
33. James, H.: 1923, *The Ambassadors,* MacMillan, London, U. K.
34. Jonsen, A. R, ed.: 1993, 'The birth of bioethics: A special supplement' *The Hastings Center Report* 23, sl-s13.

35. Kalin, J.: 1992, 'Knowing novels: Nussbaum on fiction and moral theory', *Ethics* 102, 135-151.
36. Lengers, F. P.: 1994, 'The ideas of the absurd and the moral decision: Possibilities and limits of a physician's actions in the view of the absurd', *Theoretical Medicine* 15, 243-254.
37. Mann, T.: 1953, *The Magic Mountain,* Knopf, New York.
38. McHugh, P. R.: 1995, 'What's the story?' *The American Scholar* 64, 191-203.
39. Meeks, W. A.: 1993, *The Origin of Christian Morality,* Yale University Press, New Haven, CT.
40. Melville, H.: 1926, *Moby Dick,* Modern Library, New York.
41. Murdoch, I.: 1993, *The Green Knight, Allen Lane,* Penguin Books, New York.
42. Nussbaum, M. C.: 1986, *The Fragility of Goodness,* Cambridge University Press, New York and London, U. K.
43. Nussbaum, M. C.: 1990, *Love's Knowledge: Essays on Philosophy and Literature,* Oxford University Press, New York.
44. Pellegrino E. D.: 1974, 'Medicine and philosophy: Some notes on the flirtations of Minerva and Aesculapius', Annual Oration of the Society for Health and Human Values, Society for Health and Human Values, Philadelphia, PA.
45. Pellegrino E. D.: 1976, 'Philosophy of medicine: Problematic and potential', *Journal of Medicine and Philosophy* 1, 5-31.
46. Pellegrino, E. D.: 1983, 'The healing relationship: The architectonics of clinical medicine', in E. Shelp (ed.), *The Clinical Encounter, The Moral Fabric of the Patient-Physician Relationship, Philosophy and Medicine* 4, D. Reidel Publishing Company, The Netherlands, pp. 153-172.
47. Pellegrino, E. D.: 1993, 'The metamorphosis of medical ethics: A 30-Year retrospective', *Journal of the American Medical Association* 269, 1158-1163.
48. Pellegrino, E. D.: 1995, 'Toward a virtue-based normative ethics for the health professions', *Kennedy Institute of Ethics Journal* 5, 253-277.
49. Pellegrino, E. D.: 1996, 'What the philosophy *of* medicine *is*', forthcoming in a volume edited by Giovanni Russo, Rome, Italy.
50. Pellegrino, E.. D. and T. K. McElhinney: 1982, *Teaching Ethics, The Humanities, and Human Values in Medical Schools: A Ten-year Overview,* Institute on Human Values in Medicine; Society for Health and Human Values, Philadelphia, PA.
51. Pellegrino E. D. and T. K. McElhinney: in preparation, 'The role of the National Endowment for the Humanities and the Institute on Human Values in Medicine in the development of bioethics in the United States'.
52. Pellegrino, E. D. and D. C. Thomasma: 1996, *The Christian Virtues in Medical Practice.* Georgetown University Press, Washington, D.C.
53. Pellegrino, E. D. and D. C. Thomasma: 1996, *Helping and Healing,* Georgetown University Press, Washington D.C.
54. Philips, D. Z.: 1982, *Through a Darkening Glass: Philosophy, Literature, and Cultural Change,* University of Notre Dame Press, Notre Dame, IN.
55. Reich, W. T.: 1978, 'Introduction', in W. T. Reich (ed.), *Encyclopedia of Bioethics,* Free Press, New York.

56. Reich, W. T.: 1994, 'The word 'Bioethics': Its birth and the legacies of those who shaped it', *Kennedy Institute of Ethics Journal* 4, 319-335.
57. Reich, W. T.: 1995, 'Introduction', in W. T. Reich (ed.), *Encyclopedia of Bioethics,* 2d ed., MacMillan, New York.
58. Rivinus, T. M.: 1992, 'Tragedy of the commonplace', *Literature and Medicine* 11, 237-265.
59. Ross, W. D.: 1988, *The Right and the Good,* Hackett Publishing Company, Indianapolis, IN.
60. Santayana, G.: 1937, *The Last Puritan,* Scribners, New York.
61. Schaffner, K. F. and H. T. Engelhardt, Jr.: forthcoming, 'Philosophy of medicine', in *The Encyclopedia of Philosophy,* Routledge and Keegan Paul, London, U. K.
62. Schneider, C. E.: 1994, 'Bioethics with a human face', *Indiana Law Journal* 69, 1075-1104.
63. Steiner, G.: 1971, *In Blue Beard's Castle: Some Notes Towards a Redefinition of Culture,* Yale University Press, New Haven, CT.
64. Steiner, G.: 1989, *Real Presences,* University of Chicago Press, Chicago, Illinois.
65. Taylor, C. C.: 1994, '[Special Section:] Symposium: Emerging paradigms in bioethics', *Indiana Law Journal* 69, 945-1122.
66. Trautman, J. and C. Pollard: 1975, *Literature and Medicine: Topics, Titles, and Notes,* Society for Health and Human Values, Philadelphia, PA.
67. Tsouyopoulos, N.: 1994, 'Postmodernist theory and the physician-patient relationship', *Theoretical Medicine* 15, 267-276.
68. Toulmin, S.: 1996, 'The primacy of practice', in this volume, pp. 41-53.
69. University of Illinois Advisory Committee on Ethics for Health Professionals: 1974, 'Fostering ethical values during the education of health professionals,' Society for Health and Human Values, Philadelphia, PA.
70. University of Illinois Task Force on Humanistic Studies: 1977, *Exploring Humanistic Health Science Education,* University of Illinois, Chicago, IL.
71. Veatch, H. B.: 1971, *For an Ontology of Morals,* Northwestern University Press, Evanston, IL.
72. Ward, P.: 1990, 'Ethics and recent literary theory: The reader as moral agent', *Religion and Literature* 22, 21-31.
73. Welie, J. V. M. and U. Wiesing, (eds.): 1994, [Issue Dedicated to the Topic:] 'Medicine between existentialism and post-modern philosophy', *Theoretical Medicine* 15.
74. Wilson, E. O.: 1978, *Sociobiology,* Rutgers State University, New Brunswick, NJ.
75. Wolf, S.: 1994, 'Shifting paradigms in bioethics and health law: The rise of a new pragmatism', *American Journal of Law and Medicine* 20, 395-415.

K. DANNER CLOUSER

HUMANITIES IN THE SERVICE OF MEDICINE: THREE MODELS*

I. INTRODUCTION

This volume celebrates twenty years of the *Philosophy and Medicine* series. The charge to this author was to reflect on his twenty-seven years working in the medical humanities. Thus the role of this article is similar to the role of a sermon in a scholarly gathering of clergypersons and theologians. Though the heart of the conference is careful scholarship, hermeneutics, and analysis, there is usually an opening or closing sermon, loosely relating to the theme of the conference, that is more exhortation than scholarship, more overview than analysis, and more suggestive than rigorous. In keeping with such a role, I am also taking the liberty of some informality.

This article is an overview of the enterprise connecting humanistic disciplines with medicine. For the general public, it will be vaguely descriptive of what the medical humanities is about; for the professionals in the field, it will be a perspective on their daily involvement with medicine.

An overview *could* consist of enumerating the growth and accomplishments in the last twenty years: the societies, the journals, the astronomical increase in articles and in participating professionals, the growing budgets, the influence on medical qualifying exams, the degree programs combining medicine with one or more humanities fields, the many hundreds of workshops that have been and are being offered, the first time Congress has required set-aside funds for research in the ethical, legal, and social implications of medical science investigation, and so on. We could even mention the huge behavioral changes that have taken place in medicine with respect to such things as informed consent, protection of human subjects and animals in research, and allowing and encouraging patient self-determination. However, those in the field already know these things, and the general public, though maybe impressed by the accomplishments, might still wonder what it is we do.

R. A. Carson and C. R. Burns (eds.), Philosophy of Medicine and Bioethics, 25-39.

In any case, this article will be a *conceptual* overview of the involvement of the humanities with medicine rather than a review of the apparent successes. I will be suggesting three ways of thinking about what we do vis-à-vis the world of medicine. These are not mutually exclusive views; they are consistent with each other and complementary. Through the years, these are ways I have thought about what it is I do. In some strange way, they have been a help to me. But because they are at such a level of abstraction, it is not entirely clear how or why; it is hard to see immediate applications. Rather they function for me as models of a sort—models that very loosely give me some sense of direction.[1]

II. THE BABY SWISS CHEESE MODEL

Exactly where do medical humanities and medicine meet? How do they come together, and what is the nature of that contact?

The usual image is that, first and fundamentally, there exists the body of science and medicine and then, secondarily and peripherally, the humanistic disciplines are added on: bedside manners, interactive techniques, and other stylistic concerns. It is quite natural to think: "Of course the humanities are related to medicine. Medicine deals with humans, so obviously understanding something about humans might give a bit of polish to the doctor-patient relationship—and that's where the humanities come in." And thus the humanities would be equally relevant to grocers, hairstylists, and car salesmen. Indeed, this view looks like mostly a matter of style, concerning how the physician interacts with the patient on a personal level in terms of congeniality. This is the add-on view of the medical humanities. Analogously, medicine in turn would be said to influence the humanities by virtue of providing occasional subject matter for a short story, an interest for an historian, or a moral dilemma for a philosopher. These possibilities would surprise no one, nor would they be particularly interesting. This kind of disciplinary co-mingling is akin to a physician who happens to be a painter or the philosopher who happens to be a volunteer paramedic. The fields are joined by virtue of an individual who does both; yet conceptually the fields may be no closer. We just have individuals who are living in two nonintegrated worlds. On this view, it is not clear that there is any real integration. The stereotypic question: "Would you rather have a competent physician or one with a good bedside manner?" seems to presuppose this add-on, nonintegrated view of the relationship of humanities and medicine.

But can a case be made for a stronger, more integrated relationship? Can humanities relate to the *core* of medicine? This stronger relationship would seem to be the only one worth pursuing. But we need to ask: what *is* medicine's "core?"

To find it, we could start very broadly and gradually zero in on a smaller and smaller core until we find something that is pure medicine and to which humanities could *not* relate. And that, by definition, would be medicine's core, rock bottom, pure medicine. We begin, then, with the broad view that would encompass the entire practice of medicine: diagnosis, therapy, prevention, education, distribution, research, and technology. Fairly obviously the humanistic disciplines would be relevant to some of these tasks—e.g., prevention, education, distribution of health care—involving as they do questions of allocation, rationing, costs vs. quality, ranking, pedagogy, and historical parallels and insights. But maybe it is not really clear that those areas of prevention, education, and distribution are in any way central to medicine; maybe they are not its core.

So perhaps we need to narrow down further, to rid ourselves of these mushier edges; we need to close in on the real, hard core of medicine. How about just the doctor-patient interchange being considered the core of medicine? Diagnosis and treatment, at least, seem to be at the heart of the practice of medicine. But clearly humanities would be relevant to some aspects of the doctor-patient relationship. Within the total doctor-patient relationship there is, of course, a lot of plain, ordinary human-to-human interaction going on, so we need to zero in beyond all that, still looking for that hard core of medicine that would be pure medicine, devoid of the softer humanities-like content.

Could we single out just the diagnosis and treatment aspect from the whole relationship and proclaim that as the hard core? It seems reasonable until we consider that the doctor-patient relationship as a whole has an important influence on actual diagnosis and therapy. That is, a deeper understanding of the patient's goals, values, beliefs, life style, view of suffering, and so on actually influence the diagnosis and the choice of treatment. It looks like the medical core and the doctor-patient relationship cannot be cleanly peeled apart; they seem tightly intertwined, perhaps fused.

Still seeking the increasingly elusive medical core, perhaps we could circumscribe medicine even more narrowly. We might say that *real* medicine is just pathology—knowledge of disease, its causes, mechanisms, and cure. Now, with this conceptual maneuver, we ought to be able to say that humanities does not relate to the core of medicine. But not even this move will do the job. After all, what constitutes a disease arguably may be grounded in value judgments, and the nature of causation is a major philosophical topic. (Were Koch's postulates pathology or philosophy?) Furthermore, what a mechanism is and how powerfully explanatory it is are questions frequently analyzed by philosophers. Of course this does not mean that one needs explicit training in philosophy to do pathology, but it does suggest that even at this very constricted core of medicine, there are relevant humanistic disciplines impinging on it.

And, in any case, confining the core of medicine to pathological causes, mechanisms, and cures is totally unrealistic. It would exclude far too much: it would leave out such items as setting broken bones; caring for wounds, pregnancies and disabilities; dealing with alcoholism, rehabilitation, and prevention; and doing surgery.

The point of the foregoing excursion has been to get a feel for the difficulty in answering whether or not humanities relate to medicine. By progressively narrowing the scope of medicine, we were trying to find a place where there would be no humanistic possibilities present. It was just an impressionistic exercise, but the obvious difficulty in finding a segment of medicine not susceptible to humanistic impact or insight at least suggests that issues that are of interest to and within the competence of humanistic disciplines are woven in, under, around, and through the science, the practice, and the concerns of medicine.

In a sense, none of this should surprise us. As has been pointed out quite frequently over the years, the very concept of "disease" is arguably value-laden. That is, in many cases, which conditions count as diseases depend on value judgments and not just on observation; it is not simply a scientific determination. If that is true, the implications are important: namely, the key and central concept of medicine is value-laden. That would make medicine significantly different from all the natural sciences.

And, in fact, there are many other instances of value issues permeating the body of medicine, instances with which we are by now quite familiar. Not just "disease" but even life and death are concepts that are not straightforward. We have needed law and philosophy to join with medical sciences in attempting to define death, and we have needed all those and more to define life—or to show why it cannot be defined. Even the frequently heard phrase that has the form "Such and such therapy is not indicated at this time," or "Such and such therapy is clearly indicated," though sounding like an inescapable conclusion to a syllogism, is in fact loaded with value judgments. To say that a treatment is indicated or contraindicated is not simply a clear scientific matter involving no value judgments. More recently, we have seen this matter surface again in discussions of medical futility. Many would like medical futility to be a nonvalue, factual matter. But others see value even in a persistent vegetative state.

Consider also the way value issues enter even the clinical labs. For example, in a blood test for a disease, establishing the point (or range) at which a positive result is to be declared, there are value judgments about what is gained, lost, or risked by drawing the line at that place. The point selected is what determines the number of false negatives and false positives for that particular test. Among others, the judgment must be made as to whether it is better to have a false negative or a false positive for that particular disease, given the therapies available as well as the values and beliefs of most patients. There are different consequences for each place the line might be drawn, and those consequences involve value judgments.

Somewhat similarly, scientific medical knowledge is expressed in probabilities that generalize across large groups of data. But medicine must apply all this statistical knowledge to a particular person, the physician's patient. This must be part of what is meant by "the art of medicine." It is a judgment call influenced by many factors, conscious and unconscious, but at least seemingly beyond scientific determination.

So, also, is gatekeeping a judgment call. In deciding whether to refer, to run more tests, to wait and see, or whatever, there are many value judgments, instincts, and appraisals—conscious and unconscious—being made.

The point is simply that no matter where one turns in medicine, looking for that hard core of pure medicine, we seem to find the *kind* of considerations and issues about which the humanistic disciplines, in particular, deliberate. Compare that to the "add-on" view mentioned earlier in which the humanities are seen as something tacked on after all the serious work of scientific medicine is done. What we have instead is a Swiss Cheese Model of the relation of humanities to medicine. Perhaps a *Baby* Swiss Cheese Model would be even better, since in it the holes are smaller and more frequent! They permeate and they are diffuse. These are the interstices within the core of medicine wherein humanistic issues have a concern and a relevance. Humanistic issues are in, under, and through medicine. Medicine is shot through with value issues (though they are not all necessarily ethical values).

III. ANALOGOUS GOALS: MEDICINE AND MEDICAL HUMANITIES

A second model for thinking about the relationship of the humanities to medicine is suggested by the possible parallelism: medical humanities is to the humanities as medicine is to science.

Though, traditionally, medicine was considered a practical art for curing people, making them feel better, or enabling them to cope, as the life sciences have developed systematically, medicine has become more and more entwined with them. We think of medicine as science. Aware of all the research going on, all the clinical trials, and the scientific methods used to analyze and solve problems, we easily conclude that surely medicine must be a science.

And yet it is not. However much medicine uses the results of science or the methods of science, and no matter how often it joins with science, medicine still is not itself a science. That is simply because they have two different goals: for science, it is truth about the world; and for medicine, it is the healing of human beings. Obviously medicine can *use* science to accomplish its goals, but the goals are different, and that makes the enterprises themselves different. [2]

How so? Basically because medicine's main concern is with what works and not with what is true. Medicine *as such* does not have to pursue why's and wherefore's; it is enough to know that it helps, alleviates, comforts or is otherwise beneficial to a patient. Medicine has often cured, alleviated, and prevented by using certain drugs and procedures long before it understood how or why those drugs and procedures worked. Science cannot be content with that. It has nothing to say until it knows, or has good reason to think it knows. It cannot rest until all explanations have been pursued or until the appropriate experiment has been carried out beyond all possible doubt. It has all the time in the world. Its sole goal is truth. Medicine and science differ on what is an acceptable outcome or endpoint. What satisfies one will not necessarily satisfy the other. Some years ago the Congressional Office of Technology Assessment estimated that only ten to twenty percent of what medicine used as a standard had an empirical basis [1]. Medicine does what it can to alleviate or cure. It has often cured inflammation, prevented viruses, and killed germs before it even knew anything about either the pathogens or the mechanisms of the medicine. Science, on the other hand, seeks knowledge of the world and conclusive proof.

In short, medicine's mission is the good of the patient in matters of health and maladies. Whatever benefits the patient is the goal of medicine. The physician cannot pursue truth; she has a patient who comes first. Now our question is: is there a parallel with the medical humanities?

In the context of medicine, should our humanistic disciplines be aimed *not* at some kind of disciplinary "truth" (the specific description of which would no doubt differ for each of literature, philosophy, history, and religion), but at being of specific benefit to patients, either directly or through the education and training of health-care professionals? Has this been our explicit or implicit goal? It makes a difference in what we do and in how we judge the quality of what we do.

One place where it makes an obvious difference is in those *aspects* of one's field that are chosen for use in the medical context. This determines what of our discipline we choose to expose to health-care professionals (and perhaps to patients). I would not teach my medical students symbolic logic, philosophy of language, epistemology, or metaphysics—at least, *not as such.* I might use segments of those areas in order to make certain points, but those points would be determined by the ends of medicine, which will now have become the ends of medical humanities. But I would not be pursuing a true understanding of, say, philosophy of language or trying to arrive at a consistent and coherent metaphysics—I would just be using pieces of these bodies of thought for rather practical ends. Similarly and obviously, we would not chose areas of history—history of diplomacy, history of economics, history of Peoria—simply because they would not ordinarily be relevant to the goal of medical humanities, namely, the good of the patient within the context of health and maladies.

But there is a problematic aspect to the suggestion that the goal of the medical humanities is, ultimately, the good of the patient. That difficulty concerns the parallelism with the medicine/science distinction that suggests that "truth" is not so much our objective as is the good of patients. To deal with this question adequately, we would have to establish the sense of truth that the humanistic disciplines ordinarily would be thought to pursue, and that is beyond the scope of this article. Nevertheless, something more needs to be said.

What would it be for, say, philosophy to sacrifice its usual goal (whatever that is) for the more practical one of the patient's good? We certainly might cut short an analysis whose details were too fussy and too involved for the audience of health-care professionals or patients. An example might be an analysis of reductionism, since a *sense* of reductionism might really be worthwhile for physicians to know about, but not necessarily to the point where they are facile with the concept. The same could be true for determinism, or the concept of person, or even the foundations of morality. We would say that education of health-care professionals is no place to argue out all these things, if only because the arguments are interminable, and we have a limited time and a practical goal within the context of medicine. Just as medicine uses science but does not become science, so the medical humanities use the humanistic disciplines for ends closely analogous to the ends of medicine—practical and purposeful ends. This is no doubt why our colleagues back in our pure, native disciplines look askance at us in medical humanities. We are, in their eyes, corrupted; we serve practical ends. This is not unlike the way many scientists view their counterparts in medicine.

Though we might find acceptable the pragmatic short shrift given analyses and detailed discussions of concepts, the crucial question is whether we would stick with an analysis we knew to be inadequate simply because it was in some sense satisfying to the patient, or helpful to the health-care professionals in dealing with patients. We probably would, providing it helped by offering an explanation or otherwise was satisfying or comforting. But would we use a view, idea, belief, or reading that we believed to be wrong? We probably would not. That would be comparable to lying to a patient for what we perceived to be for the patient's good. Yet obviously there is a thin and tenuous line between believing something is inadequate and believing it is wrong.

Undoubtedly, perceiving our goal to be the patient's well-being represents a moral hazard for us, because our primary goal is not truth, but rather patient benefit. If we are intensely focused on benefit, it may be all the more difficult for us to become aware of wrong, misguided, or inadequate conceptual work on our part. It is a fragile balance between our proper goals and the ease of going wrong.

This discussion is itself so short as to border on the irresponsible. It needs much more investigation. No doubt each humanistic discipline would work this matter out differently. Is literature sacrificing any of its pure academic goals by having plays written to highlight moral and personal dilemmas of patients interacting with health-care professionals, or by encouraging patients to write poetry for therapeutic reasons? Dare our literature colleagues recommend readings that seem to be healing and comforting to patients when modern criticism may find those readings devoid of literary merit? Is it legitimate to use literature to inspire and motivate health-care professionals to work with the handicapped, the elderly, and the chronically ill? Similarly, we can see the same dilemma in the case of religious studies. Should their goal be simply to comfort the patient, writing or saying whatever will immediately accomplish that, or do they owe something to the integrity of the reason, experience, and interpretation that underlie religious beliefs and traditions? There is obviously a lot more to be said on this issue, and it is a balance that must be worked out for each discipline. But it is a dilemma not faced by "pure" humanities, which presumably pursues its "truth" wherever it leads and for as long as it takes.

IV. PERSPECTIVE

The last of the three models is that of perspective. In many ways this has been for me the most important. I believe that an important role of medical humanities is to provide perspective *about* medicine and *within* medicine.

Very roughly, by perspective I mean seeing the same circumstances, the same scenario, the same situation from different angles and with different foci. One perspective highlights certain features of the total circumstances, another brings aspects of the background to the fore and certain previously dominating features are diminished. Dominant goals from one perspective give way to underlying goals when the circumstances are viewed from another perspective. Recall the Baby Swiss Cheese Model wherein the humanistic issues permeated the body of medicine. Perspective is like a pathologist's stain which, in this case, would highlight or otherwise bring to the fore certain aspects of the baby swiss cheese not previously noticed.

This shifting perspective is most conveniently thought of (though perhaps slightly inaccurately) as analogous to those trick drawings that we have all seen—portraying an ascending staircase, or an old woman, or a rabbit. If one stares at the drawings long enough, or blinks, or otherwise shifts perspective, then the ascending staircase is seen as descending, the old woman as a young woman, and the rabbit as a duck. Nothing has changed in the drawing, the same data is impinging on the viewer's retina, but the perspective has changed and thus also, in effect, what is seen.

The suggestion is that a significant role of the medical humanities is to engineer a shift in perspective and even to train for the ability to shift one's perspective. Think of medical situations as one of these trick (or ambiguous) drawings that, with a shift in perspective, we might see rather differently. Where perhaps we could only see business as usual—meaning more and more futile attempts to keep a dying patient alive—with a shift in perspective, we now might see the situation differently, so that our aim becomes making the patient comfortable and his remaining life meaningful. Where we had focused only on a patient's pain, with a shift of perspective, we might come to see that it is his *suffering* that is truly awful and needs to be addressed; or perhaps our gestalt shift brings to the fore a patient's fears that we had heretofore thought insignificant. These shifts might have been brought about by forceful new information, a short story or a poem, a photograph or painting, a recollection, a moral concern, or whatever, which casts the circumstances in a new light. But first we must be made susceptible to getting out of our rut and to seeing things from a new perspective.

Before continuing with this description of perspective's role for medical humanities, we should look at what underlies this kind of experience and why the shift in perspective is so important.

I have often described professional education as a kind of conceptual ghetto. Law, Divinity, Education, Medicine—each has its own ghetto. Its denizens get locked into a certain way of seeing the world, themselves, their relationship to that world, and their role within it. That may be part of what is meant by being "professionalized." The very focus of our training leads us to view the world in special ways. Certain things and relationships are highlighted, others are of no concern and fade to the background; certain causal chains predominate, others are barely noticed. Thus one's perceptions become highly selective; that, in turn, becomes our professional perspective. This is strengthened and reinforced throughout our professional education—by the emphases in our courses (reinforced by exams), by the concepts we use daily, by the special language we use, by the technical names we have given to each of the elements of our perspective.

Thus our perceptions and lines of reasoning get channeled; they get etched into our professional thinking. Notice that this is not unlike experiences we have almost daily. Often after having focused on some feature of the world, that feature seems to jump out at us wherever we look. For example, we learn a new word, and suddenly we see and hear it every place. We are apt to think that the whole world discovered the word and started using it the same day we did. We go shopping for wallpaper, looking through book after book of samples, discussing with our spouse every color and design of the different wallpapers, and for the next week or so we notice every wallpaper in every room of every building or home that we enter, even if we have been there a hundred times before without really noticing the wall coverings. The same can happen in seriously shopping for almost anything—suits, shoes, cars, or whatever. Young people go off to college, take an art course, and upon returning home see—as if for the first time—art work on the homestead walls they had never noticed in all their years of growing up. Certain features of the world have been highlighted for them, and they now see those features everywhere.

But all of those examples are very short-term experiences. By contrast, consider medical school training (or any professional school, for that matter) during which for many years a point of view is emphasized over and over by action, by language, by example, by constant reinforcement. It is analogous to what would happen to those trick or ambiguous drawings if one were to darken or otherwise to emphasize certain lines in the drawing. The ambiguity would be destroyed. When that is done, one is locked into seeing the trick drawing as a duck, or as a young woman, or as a receding staircase. That is what professional education does. It takes the naturally rich, provocative, ambiguous world, and, as it were, darkens the lines and highlights certain features, thus locking us in to one point of view.

Consider an example from medical education. From the very first term in medical school, students are in the gross anatomy laboratory looking at the human body very mechanistically: its structures, its substructures, its interrelationships. Then they do the same on the microscopic level. The students' focus and attitude become very reductionistic. They anticipate the next underlying level of function; they know the ultimate answer lies in still a further underlying biochemical level. This mind-set is continually reinforced by exams, conversation, vocabulary, and peer expectations, and it continues into the clinic, where again the focus is on the quantifiable. "What are the numbers today?" they are asked about a patient. What are the levels, the inputs, the outputs, the rates?

Contrast that with being taught to see the human body through the eyes of painters and poets. This causes a significant shift in perspective. There is a different focus, a different goal, and a different vocabulary; and these differences effectively change what we see—what we see as important, stimulating, and insightful. One's thought processes do not automatically move from the part to the subpart, but from the part out to the whole, and beyond that, to the setting. We talk about line, dynamics, juxtaposition, movement, mood, expressiveness, and composition. We are seeing the human body through "new eyes."

I want it to be clear that I am not denying the need and fruitfulness of the reductionist mind-set or any other mind-set inculcated by specialized education. We need recipes, formulas, and algorithms; we need mechanism and reductionism; we need quantifiable tests and special vocabularies. The point, rather, is that we *also* need to avoid being locked up in a single mind-set. We need other perspectives; we need imaginative flexibility, enabling and encouraging us to shift perspectives. Of course, the point is not that one perspective is right and the other one is wrong. *The correctness of the mind-set or perspective is a function of the goal to be achieved.* What is needed is *flexibility* in perspectives, to avoid getting locked into one or another. Then when purposes change, perspectives can change accordingly. There are clearly times both in a physician's personal life and in his or her professional life when the situation calls for a different perspective, a shift in mind-set—perhaps where a cure is not possible, or when the numbers are not relevant, or when that mind-set is screening for certain clues not appropriate to the case at hand, or in short, doing what it knows how to do in a situation where that is not what is called for.

"Flexibility of perspectives" is one of the most important contributions of the humanistic disciplines to medicine and particularly to medical education. The ability to quickly and easily shift perspectives when appropriate, to see with "new eyes," is a valuable clinical skill. Seeing the human body through the eyes of the artist; studying the concept of disease through the contexts of an historian; understanding suffering through the views of a theologian; analyzing knowledge claims through the conceptual tools of a philosopher—these break the shackles of a single vision, goal, method, and focus. These enrich and enhance.

But notice: there is a relativism about which is the appropriate perspective. It is just another way of seeing, in some sense, the same situation and circumstances. Are all perspectives equally valid? No, of course not. The validity of a perspective is determined by what it accomplishes, how well it helps to achieve the ends—in this case, the ends of medicine, which is the well-being of the patient. So some perspectives are better than others; some are irrelevant. Some are crucial for the purposes of morality, that is, to see the situation from the moral point of view. Other perspectives have different purposes having to do with the welfare of the patient. But it is the ability to *shift* perspectives that is important, the ability to avoid being locked into one fixed perspective. And a major contribution of the medical humanities is to make one flexible of perspectives. A physician colleague had complied with the family's urging to keep the terminal diagnosis from the patient. Subsequently that physician read Tolstoy's *The Death of Ivan Ilych.* It was a revelation to him; he saw the situation from a new perspective in which the medical details receded to the background and Ivan Ilych's feelings of abandonment and being deceived by the conspiracy of the family came to the fore. It was the power of the literary artistry that drew the physician into this different way of seeing the circumstances [2].

V. IN CONCLUSION: THE MODELS AND SOME IMPLICATIONS

My charge for this volume was to reflect on my years of trying to integrate humanities with medicine. I have described three "models" of that interaction that have emerged for me through the years. They may even be more personal and subjective than I had imagined, thus having no dependable use or objective validity. They certainly function at a level of abstraction that makes them questionable. But for me they have been richly suggestive and have kept me focused by sounding alarms and posing problems.

Space does not allow for a detailed account of how these models have been a provocation and a guide. But by way of summary, several brief observations about the practical implications of each are in order.

The Baby Swiss Cheese Model continually reminds me that humanistic issues pervade the body of medicine, and thus it is a graphic reminder that wherever one looks in medicine, there will be humanistic issues involved. Just keep looking. Some, of course, are more interesting than others. Furthermore, it is a challenge to uncover them because we are unwittingly manipulated by hidden assumptions and values that we do not even know are there. So there is a reason for bringing them to the surface for examination. This model, exemplifying the tight interweaving of humanistic issues with medicine, also reminds us how crucial it is for medical humanities to know a lot about medicine, if our job is to be done thoroughly. We should be doing our thing in the midst of medicine. It reminds us that humanities is not an addendum, not a simple add on, not just some additional knowledge at the peripheries of medicine, but part and parcel of the very functioning of medicine, in all the interstices from diagnosis, to laboratory, to therapy, and the incremental steps along the way.

The Baby Swiss Cheese Model also keeps before us the need for interdisciplinary teamwork if the tight interweaving of the humanistic disciplines and medicine is to be explored. It provides the grist for courses and for individual seminars. Clinicians respond enthusiastically, probably because they have always suspected that something of the sort is true about medicine, namely, that it is permeated by a broad spectrum of human concerns and values. Furthermore, for them it creates a whole new dimension of interest in medicine.

The model of Analogous Goals has both helped and annoyed me through the years. Recall that this model emphasizes the very practical end of medical humanities, just as medicine's practical end is diagnosing and treating the patient's disease, as distinguished from science whose mission is truth. This model exists as a continuing reminder that our goal in the medical humanities is a service and not an endless academic pursuit in which we involve ourselves only with the esoteric research and in discussions with our own kind in our own discipline. This is tough and, no doubt, largely unacceptable. We like to think of ourselves (as scientists and many of the physicians like to think of themselves) as following truth wherever it leads. We do not fashion truth and make it palatable, rather we say it like it is. Perhaps so. But this is why it is important to raise this issue explicitly. Maybe we in medical humanities *should* be more like physicians than like scientists. Maybe we *should* be aiming at the well-fare and good of the patient rather than at making some esoteric point within our discipline. We are in that sense an applied discipline. We must have in mind to make sense to those whose welfare is our goal. Naturally we are often in both camps, and then we speak, research, and write for fellow professionals in the field, but when we "translate" this for our medical humanities, we should be willing to lessen the rigor, if necessary, to have the appropriate effect. This may seem almost a silly point, and yet I have found it helpful through the years to keep a perspective on what I was doing in my own discipline and what I should be aiming for in medical humanities. It helped me keep in check the urge to lay too much detail on my medical students, rather than merely pursuing some point just enough to be helpful to them, and no more. Fifteen to thirty years ago, I think this point was clear to us. Our mission was to contribute to the ends of medicine. But as time has gone on, our fascination with our subject matter tempts us to turn our applied disciplines into typical academic fields, to be pursued in typically academic ways. Thus application would no longer be primary, having been replaced by the search for some kind of "truth." We would be like those researcher-clinicians who get so involved in the research that the application of the research and patients themselves no longer are of interest. In this context it may be fruitful for us to think in terms of "the art of the medical humanities" as analogous to "the art of medicine." Just as a part of the meaning of the "art of medicine" is the skill necessary to fashion the universals and incomplete truths of science to the needs of the particular patient, so we must fashion the concepts, methods, and incomplete truths of our "pure" disciplines so as to be beneficial to patients directly or through the health professionals we help to train. Making extensive arguments, clever rebuttals, and nice debating points is not what the medical humanities are about when they are fulfilling their role.

The Perspective Model has had many practical implications for teaching medical students, only one of which I will mention in this concluding section. The goal of flexibility of perspectives suggests to me that I am not or should not be burdening the students with even more information, but ideally only using what they already must know about medicine. My job is to lead them to see their standard material from a different perspective. To be sure, a bit more information will be necessary in order to get them to shift perspectives and for me to defend why the new perspective is valuable. Nevertheless, I am basically using their information, things they already know or will be required to know. I might want them, for example, to see the concept of disease in a new light. They already know a lot about disease. What I might do is give them some historical examples that they might not have known about (that would be additional information) but it is not as though they must memorize a lot of new information. Once they see their familiar material in a new light, that is all I ask. If I loaded them with information, they would forget it by the time they could use it and the very idea of a gestalt shift would be lost in the details. That is, by loading the students up with details, arguments, emphases, and a whole body of knowledge, in effect, I would be more likely locking them into another perspective than liberating them from a particular perspective. Emphasizing flexibility of perspective *ipso facto* emphasizes *integration* of humanistic concerns in health care. Thus the humanistic concerns seem part and parcel of medicine, just different perspectives on it. And that's different from seeing medicine as a collection of disciplines in each of which one must become an expert.

Nothing has been said in this article concerning the contributions of medicine to the humanities. That is unfortunate, since those contributions are significant, and an article of equal length could easily be written detailing them. As I have pointed out in another place, the world of medicine—teeming with human turmoil, anguish, pathos, exuberance, hope, tenderness, and characters and machines of all descriptions—is enormously rich and fertile ground in which the humanistic disciplines can rediscover their own taproot of concern for the human condition. As the "pure" humanistic disciplines have retreated more and more into esoteric issues of their own making, becoming a technology unto themselves, it takes something like the reinvolvement in concrete experience, such as in the world of medicine, to revitalize their ancient humanistic concerns. Research and rigor are fun, fulfilling, and important. But let us in medical humanities not forget our real goal, which is to use these conceptual tools and insights to the benefit and well-being of patients—both directly and through the training of health-care professionals.

University Professor of Humanities
The Pennsylvania State University College of Medicine
Hershey, Pennsylvania, U.S.A.

* A version of this chapter was delivered at the Marcel Patterson Memorial Lecture in the Medical Humanities at The University of Texas Medical Branch at Galveston, Texas, on February 17, 1995.

NOTES

[1] Some of the ideas in this article have been expressed elsewhere by me, though in a different context: 'The Body of Medicine: Riddled with Values' in L. W. Hodges (ed), *Social Responsibility: Business, Journalism, Law, Medicine,* Vol X, Washington and Lee University, Lexington, Virginia, pp. 41-54, 1984; 'Medicine, Humanities, and Integrating Perspectives', *Journal of Medical Education* 52, pp. 930-932, 1977; 'Humanities in Medical Education: Some Contributions', *The Journal of Medicine and Philosophy* 15: 289-301, 1990. In the current article, Section IV "Perspectives," particularly borrows from this last reference.

[2] Ronald Munson makes this same point, though not for the same purposes, in a fine article, 'Why Medicine Cannot Be A Science', *The Journal of Medicine and Philosophy* 6, pp. 183-208, 1981.

REFERENCES

1. The Congressional Office of Technology Assessment: 1978, 'Assessing the efficacy and safety of medical technology', Washington, D.C.
2. Charon, R., Brody, H., Clark, M. W., Davis, D., and Martinez, R.: 1996, 'Literature and ethical practice: Five cases from common practice', *The Journal of Medicine and Philosophy* 21, 243-265.

STEPHEN TOULMIN

THE PRIMACY OF PRACTICE: MEDICINE AND POSTMODERNISM

I

A dozen years ago, in my essay, 'How medicine saved the life of ethics'[3], I argued that the debate in ethics among analytical moral philosophers, in particular the approach to the subject known as "meta-ethics," ran into the sand in the 1960s or 1970s, and that the coincidental revival of interest in the moral problems of medicine took place just in time to redirect moral philosophy into substantive new fields of discussion. Even then, I might have done more to place this shift from definitional issues about the meaning of terms like "right" and "good" to substantive ones about the termination of artificial life support and the like within a broader historical frame, as it was already becoming clear that at a deeper level, also, the philosophical tide was turning. More recently, however, the philosophical shift from formal definitional issues to matters of substance has been unmistakable.

In one respect, the change of focus in philosophical ethics between the 1950s and the 1980s had an obvious context. From early in the twentieth century, one central strand in European philosophy had been a series of critiques of Cartesian rationalism. This was launched in Germany by Edmund Husserl and Gottlob Frege, in England by Bertrand Russell and G. E. Moore, and in the United States by William James and John Dewey. Other powerful figures involved in the critique were Lev Semenyovich Vygotsky and Mikhail Bakhtin in Russia, Ludwig Wittgenstein in Vienna, Cambridge, and Norway; while, at one stage, Martin Heidegger had a significant part to play.

More recently, Richard Rorty reformulated the central themes and slogans of this critique to good effect; while, in the wings, a parallel movement in literary theory was developing under the slogan of "deconstruction." This movement shared, at least, the conviction that all earlier quests for a comprehensive system of knowledge, based on permanent, universal systems of overarching principles, were misguided from the start, and are by now discredited. Claims to philo-

R. A. Carson and C. R. Burns (eds.), Philosophy of Medicine and Bioethics, 41-53.

sophical universality and permanence can be ignored: their only interest lay in the ways that they could serve as a "cover" for the collective interests of the nations, social groups, or genders with which their authors—novelists, philosophers, or playwrights—were affiliated.

Instead of rehearsing the points at issue in the current debate once again, let me here begin by concentrating on some underlying continuities that were present in this critique from the start: these will point our attention back to critical themes in European philosophy right up to 1600, but were swept aside from the 1640s on, as one side-effect of the intellectual movements that we have known for fifty years and more as "modern" science and "modern" philosophy.

II

Go back to the beginning of our century, and recall the terms in which the general philosophical debate was framed. In several countries and languages, this was a time of reformation. From the early-nineteenth century, the central issues under discussion had rested on one or another attitude to history—in Germany from Herder, via Fichte, to Hegel and Marx, among English speakers from Coleridge and Carlyle to Whewell and Peirce, in France from Auguste Comte to Émile Durkheim. By the late-nineteenth century, this approach was open to suspicion for subordinating an understanding of our basic framework of ideas to the contingencies of history, and a reaction set in. Frege, for instance, condemned historicism as a fallacy; Husserl and Russell devised methods of giving the foundations of epistemology a formal stiffening—or "logical grouting"—but they all regarded their nineteenth-century inheritance as corrupt. The separate streams of phenomenology and logical analysis thus sprang, alike, from a desire to go behind the flux of historical change to the unchanging foundations supposedly underlying them.

So, the attack on Cartesian rationalism at the beginning of the century was initiated in the name of *logic:* this is true both of Frege and Husserl in Germany, and of Moore and Russell in England. The objects of philosophical analysis were *propositions:* its method was to show explicitly the "necessary connections" implicit in the meanings of propositions and, consequently, of the *terms* that figure in them. (As Frege insisted, "Only in the context of a proposition does a term have meaning.") At the outset, most philosophers chose to disregard utterances by particular writers or speakers to particular readers or audiences on particular occasions, above all, those framed in the colloquial language of everyday life. Everyday language, Russell argued, is a "rough and ready" instrument devoid of any real exactitude, and open to different interpretations as we go from situation to situation. The first step (it was generally assumed) was to rewrite the meaning–content of any utterance in a way that brought to the surface all the different relations to context and occasion that its colloquial use left implicit and unstated.

This could be done in several ways. One was to recast this meaning-content in the formalism of Russell and Whitehead's *Principia Mathematica,* as in Russell's theory of definite descriptions. (Willard van Orman Quine still retains this method, adding spatial and temporal markers to turn temporary, local "utterances" into explicit "propositions.") Alternatively, one might use G. E. Moore's refined lexicography to relate the meanings of complex, ambiguous terms and propositions back to their primitive, "unanalysable"—and presumably unambiguous—foundations. Either way, the focus was on the formal entailments embodied in the "logical structure" of knowledge and language. At times, this exclusive emphasis on necessity was challenged, as in Quine's well-known paper, "Two Dogmas of Empiricism"[1]; but the reasons for this were epistemological, and did not redirect attention toward rhetoric or the general theory of communication. As in Aristotle's *Organon,* the initial emphasis was on *analytics:* so, even the philosophy of science—renamed "inductive logic"—became an annex to formal logic.

Historically, this move was familiar. Parmenides as against Heraclitus, Descartes as against Montaigne: the hope of underpinning contingent knowledge by necessary principles has always had its appeal. Either way, historically conditioned nineteenth-century theories gave way to "desituated" axioms or principles. Plato's initiates returned from the Cave with an ability to recognize Eternal Forms; Descartes' "reflective thinkers" found their experience shaped by the "clear and distinct ideas" guaranteed by God's Benevolence; and, in 1900, clear-headed phenomenologists and formal logicians claimed to identify the "timeless" axioms of their philosophical systems as self-validating.

Different philosophers arrived at these timeless propositions by different routes: Russell grounded mathematics in logic, G. E. Moore used sophisticated definitions, Husserl a subtler mode of introspection. In each case the unit of philosophical analysis—the unit of *meaning*—was not a local and timebound utterance, made at one place and time or another, in one *situation* or another, but a "proposition:" an abstract, timeless entity expressing in eternal terms the *intellectual content* of a statement or utterance. In any comprehensive philosophical system, such timeless propositions would be linked by equally timeless logical relations, to yield an account of the world whose validity did not depend on when or where it was presented. The outcome was as *theoretical* a view of philosophy as one could want, whose validity was independent of all historical times and places. Historicism was thus transcended, and philosophical arguments were transferred to a desituated, ahistorical plane.

I use the inelegant neologism "desituated" for a reason. When I first wrote about this subject, I used another inelegant, but more misleading term—"decontextualized." We have all met arguments that speak of experience as a *text,* and relations within it as *inter*textual, *con*textual or whatever. But the subordination of ideas about life to ideas about language finally comes to an end. The relation of events to particular situations is not that of a *text* to its *con*text: in crucial respects, the ways in which—and purposes for which—an "utterance" fits the broader patterns of *human activity* are prior to, and independent of the ways it fits into longer written or spoken texts or verbal exchanges. A philosophical "text" (e.g., *cogito ergo sum*) can of course be construed with an eye to its *literary* context; but it is also made at some particular time and place—on a particular *occasion*—and this too throws indispensable light on its meaning. As Aristotle puts it, such utterances must also be read *pros ton kairon*—"with an eye to the occasion."

This attempt to detach philosophy from history did not last. From the 1900s on, among analytical philosophers, the study of *desituated propositions* widened, bit by bit, to embrace *situated utterances*—"speech acts," "language games"—and thence into reflections on different kinds of human life and activity: *Lebensformen.* Writers of phenomenological tastes meanwhile focused more and more on a life world (*Lebenswelt*) that turned out to be just as variable. The significant relationships among utterances (e.g., a jury's *verdict* and the judge's *sentence*) were no longer formal matters for logic alone: instead, they were matters for, say, sociology, history of law—even public policy—and must be decided in substantive terms.

Before the mid-twentieth century, philosophers believed that they had an independent standpoint from which to discuss the underlying virtues of, say, physics or law. Since 1960, the argument in the mainstreams of European and American philosophy has moved back into the historical arena from which it had fled around 1900. As a result, philosophers have not just been led back into the world of history; they have also been compelled to focus anew on questions of rhetoric, and issues of practice. Unlike timeless propositions, *utterances* are always the utterances of a particular writer or speaker, presented to a particular audience, delivered in a particular style, and with a particular spin. This is not to imply that all *utterances* are somehow less than candid—even dishonest: only, that the shift from "propositions" to "utterances" takes language out of eternity, and *resituates* it in a world of practice and action.

III

Viewing the development of philosophy in Europe and North America after 1945 through a wide-angle lens, we find a slow but inexorable change. As the program of clarification proceeded, it was less and less possible to substitute explicit propositional "markers" for the contextual or situational features that determine the meanings of actual utterances. Instead, philosophical attention shifted to the great variety of ways in which language enters life, so generating a taxonomy of "utterances" or "speech acts."

We see this shift taking place in many different writers. Wittgenstein's focus in the *Tractatus Logico-Philosophicus* on "propositions" (*Sätze*) shifted, in the *Philosophical Investigations,* to *Sprachspielen* ("language games") and ultimately to *Lebensformen* ("forms of life") as the occasions for language games. What this shift made clear is that words, sentences, and other lexical items are not connected to their occasions of use by formal, logically necessary relations. The words "ball" and "strike," used by an umpire behind the baseball plate, are understood as his role in the "language games" of baseball, and these, in turn, gain definite meanings from the parts they play in the "form of life" we know as baseball; but neither relation is strictly formal or logical. For his part, J. L. Austin reminded us how many utterances operate more as *performances* than as representations of facts (*Bilder der Tatsachen*). John Searle similarly argued that they gain a meaning not in the way mathematical formulae do—by formal definition—but as other human acts do, from their roles in larger constellations of *human action.* Linguistic utterances differ from human gestures and songs, that is to say, only in being "speech" acts, rather than acts of other kinds.

By now, the focus is not on logical structure but on collective function: we have come closer to sociology or ethnography than to, say, geometry. At this point, a door opens from formal logic into more pragmatic fields, even into rhetoric. Given his devotion to Aristotle, John Austin surely understood that he was moving philosophy away from the formal inferences of the *Analytics* to the practical concerns of such later parts of the *Organon,* e.g., the *Topics, Nicomachean Ethics,* and *Art of Rhetoric,* but, writing in the 1950s, he saw little reason to labor the point.

Other, less historically sensitive philosophers were not so prepared to recognize the destination at which they had inadvertently arrived. They were taught for so long to despise rhetoric that they could not recognize it, even if it bit them in the hand. Only in late 1992 did the Center for Philosophy of Science at the University of Pittsburgh open its doors to the rhetoric of science. Yet, the first thing a young research scientist will learn is to "write up" his results in the style and manner needed if his essays are to have any chance of being published in the professional journals of his subdiscipline, or taken seriously by other members of the profession that he is hoping to enter. That task is, surely, a *rhetorical* task, or it is nothing.

Elsewhere, this move was made more easily. In Russia between the World Wars, Vygotsky and Bakhtin had already moved the study of language away from the analytic and formal, and toward the pragmatic and rhetorical. Starting with a literary thesis on *Hamlet,* Vygotsky worked his way, via psychology of learning, toward neurology—an odyssey completed by Alexander Romanovich Luria after Vygotsky's death. In the 1930s, Vygotsky had already given an account of how we master language games, long before Wittgenstein had given currency to the idea of a "language game" itself. Meanwhile, Bakhtin argued a case for giving rhetoric a place in the theory of meaning, and recognized that timely, local utterances cannot be raised to the level of timeless, universal propositions without destroying their timely, practical significance. Instead of analyzing all *arguments* as formal chains of propositions—like those in Euclid—and dismissing rhetoric as a technique for undermining formal validity, we should think of *argumentation* as a "dialogical" exchange of opinion, and rhetoric as a way of focussing on the interests that are at stake for the parties to such an exchange.

This was not the first time in the history of philosophy that logic and rhetoric had coexisted peaceably. Thanks to Aquinas' appreciation of Aristotle, this was also the case in the High Middle Ages, and remained so throughout the Renaissance, as late as 1600. All the varieties and circumstances of language use (tropes and figures, the general and special topics, presumptions and rebuttals) were open to debate among philosophers as much as among literary scholars, and no objection was seen to the idea that the parties to an argument (theoretical or intellectual, practical or political) have legitimate interests to present. Interests need not *distort* an argument: rather, they can serve to *structure* it, defining what kind of argument it is, and thus what "topics" will be appropriate to its formulation.

In this respect, as in many others, the emergence of "modern" philosophy served as a retreat from medieval and Renaissance Aristotelianism, and a revival of Platonism. The ambitions stated in René Descartes' two great essays of the 1630s—the *Meditations* and the *Discourse*—were not unprecedented. Rather, they followed Galileo's example, inviting natural philosophers to revive the epistemological program taught in antiquity, when Plato presented rigorous geometrical proofs as an ideal to be aimed at in serious argumentation of all kinds, and held out "theoretical grasp" (*episteme*) as a standard to be demanded of authentic knowledge in all fields of inquiry.

Why did this message appeal to forward-looking young intellectuals from 1640 on? Why did this renewed commitment to Platonic *episteme* lead to an eclipse of traditional Aristotelianism—at least, outside the ranks of orthodox Catholic ecclesiastics? (This is no place to address those questions, which are central issues in my book, *Cosmopolis* [4].) What we should notice here is that, after this shift in philosophical priorities, several fields of debate that had been perfectly respectable in the late-sixteenth century lost intellectual prestige, as compared with those favored by the Cartesian rationalists. Rhetoric was dismissed in favor of formal logic, casuistry was rejected in favor of moral theory, clinical medicine and jurisprudence yielded place to physiology and political theory, history to physics. In each case, local, short-term demands of practice gave way to the idealized dreams of theory: substantive to formal, particular to universal, concrete to abstract, timely to timeless. Theoretical system building (*episteme*) was held out as the main task of philosophy; practical wisdom about the circumstances of urgent decisions (*phronesis,* as Aristotle called it) was downgraded.

From 1650 on, the dream of a comprehensive, self-justifying, overarching, abstract rational theory, as the foundation of human knowledge, seized the minds of European intellectuals—above all, Francophiles. From the start there were skeptics, who found a voice in the delicate wit of Lawrence Sterne's *Sentimental Journey.* But the dream of Descartes kept a charm and power for three centuries. See, for example, what Bertrand Russell says in his *Autobiography* about his early attraction to philosophy. Even now, French arguments *against* any such overarching theory (like those of Lyotard) are colored by a sense of loss. They do not acknowledge that Descartes' dream was the outcome of a misconception, a misplaced hope for universal *episteme,* from the start. So, instead of accepting Aristotle's warning that the certainties and necessities of formal geometry are inappropriate to more substantive fields (above all, ethics) they strike a pose, implying that "We wuz robbed:" in failing to meet irrelevant Platonist standards, the world demonstrates its absurdity. Yet, if anything in the story is absurd, it is not the world we have to deal with, but the manner in which we have attempted to deal with it.

IV

Since 1950, then, the intellectual shift that transformed European modes of thought from 1590 and 1650 has been reversed. Few philosophers today still seek the system of "clear and distinct ideas" that Descartes hoped for. But the death of foundationalism did not leave the field empty. The twentieth-century critique of rationalism undercut only its abstract, theoretical aspects; all of its concrete, practical aspects that flourished under Aristotle's wing up until 1600 can now return to the center of the picture. When, for example, Wittgenstein argued that "philosophical" issues are vacuous—in failing to engage the world in a way that "makes a difference" in actual experience, and gaining plausibility only because of our misunderstandings about logical grammar—his critique was aimed at an abstract theoretical conception of philosophy. It did not apply as powerfully to the concrete, substantive arguments of casuists or bioethicists, as they arise in actual practical situations.

In traditional epistemology, we can perhaps argue that language is "on holiday;" but this is not true of the ways in which we use language to deal with the problems that face us in the intensive care unit or in preventive medicine. There, the thrust of our practical problems is particular rather than universal, local and timely rather than general and timeless, and the solutions to these problems are devised *pros ton kairon*—"as the occasion requires." Clinical medical ethics is only one example of this change: the same is true (*mutatis mutandis*) of rhetoric and casuistry, environmental philosophy and ecology, jurisprudence and the philosophy of law.

To put the point harshly: the philosophers of the early modern period were the first "pure intellectuals" of Europe—the first to believe that conclusive answers to the crucial questions about nature and humanity can be reached by abstract, general reflection, in isolation from particular, practical enterprises, and established with a certainty hitherto confined to theorems in formal geometry or dogmas in theology. For us in 1995, the Pyrrhonism of Montaigne's *Essais* (written in the 1570s and 1580s) is more congenial than the geometrical dogmatism of Cartesian rationalism sixty years later. The most fruitful questions in philosophy arise out of specific, substantive inquiries like clinical medicine, civil law, public administration, and ecological protection. Because something substantive is at stake in arguing about the liability of the mentally disturbed, about the care of the terminally ill, or the control of drift-net fishermen, the conceptual issues that they involve "make the difference" that is required if they are to be given a determinate meaning.

This revival of practical philosophy also reverses the relation of theory to practice. In *Consequences of Pragmatism,*[2] Richard Rorty distinguishes the abstract, general uses of terms like theory, truth, proof—which, for Wittgensteinian reasons, can too easily become vacuous—from concrete, substantive uses of those same terms, which avoid this outcome. This distinction he marks by using upper case initial letters for vacuous, grandiose theoretical uses, and lower case for more modest, pragmatically fruitful uses. The pursuit of elusive universal timeless *Truth*, he argues, is something we can avoid; but this in no way undercuts our ability to understand local and timely *truths* about the particular features of everyday situations. Nor should a rejection of the demand that all intellectually serious arguments conform to the standards of geometrical *Proof* lead us to undervalue the other, less formal standards by which practical arguments and beliefs are put to the test—even "*proved.*" Above all, giving up the dream of an overarching comprehensive *Theory*, having the formal necessity of some geometrical system, does not stop us appealing to particular (small–t) *theories* in one or another substantive field of concrete practical activity. On the contrary, it helps us recognize that, in substantive enterprises, appeals to theory are one kind of argumentation (*topos*) among others.

As with the value of case analysis in health-care ethics, this is borne out in the practical experience of clinical physicians. Medical diagnostics and therapeutics are illuminated by appealing to theories from biochemistry, genetics, or physiology; but, in clinical experience, doctors must never allow appeals to theory to distract them from the plain facts of their patients' histories. If the general theories of biomedical science explain why particular procedures succeed or fail in clinical practice as and when they do, this is intellectually rewarding; but, practically speaking, both failures and successes are facts of clinical experience, and depend on biological theory inferentially, at a remove. So it is a mistake to speak of clinical medicine as "applied" biomedical science—as was often done before the 1970s. A theoretical explanation legitimately strengthens our confidence in a given clinical procedure, but there is always more to timely, well-judged clinical practice than scientific theory can explain. What directly demonstrates its value is not its scientific foundation, but the record of its clinical performance. We may understand *why* it works in theory; but what matters is *that* it works in practice.

V

To return to our starting point: Philosophers can learn something worthwhile from medicine, not just in ethics, but in general philosophy. Medical ethics, specifically medical anthropology, undercut the shallow moral subjectivism and relativism that was popular in the years just before and just after World War II. For broken limbs, bleeding wounds, obstetric crises, pneumonia and malaria, say, use the same language in any country or culture. Similarly, the treatments they call for work (or fail to work) in similar ways and for similar reasons with humans from all nations and backgrounds. If a bull's horn impales a toreador, he bleeds in the same way as a Rwandan villager or O. J. Simpson's multiply stabbed wife. For clinical medicine and its moral practice, California, Central Africa, and Spain are all the same.

So, we can generalize about the *right* procedures in treating medical conditions, not because the basic truths of medical ethics are "universalizable" *as a matter of definition,* or for other *a priori* reasons. We can do so, rather, because the clinical demands and the maxims of equity evoked by human suffering transcend all the differences among societies and cultures. The central insights of physiology are universal—we may say—not axiomatically, but *experientially.* This is true also of the moral insights that respond to the phenomena of physiological function and malfunction.

In saying this, we are at the same time acknowledging the limits of the traditional fact/value distinction. Not for nothing Claude Bernard referred to clinical research as belonging to the field of "experimental medicine." The central issues of physiology, like those of clinical medicine, are naturally framed (as here) by contrasting *function* with *mal*function—the desired modes of activity of different bodily organs with those that fall short, and threaten an organism's health. Unless we depend on this contrast, indeed, we cannot discuss physiology in a way that nourishes our understanding of medicine proper. The biochemistry and biophysics of carcinomas, say, may simply represent alternative chemical processes and physical mechanisms on a "factual" level; but on that level we are not yet addressing the issues of health and disease with which medicine and physiology are both concerned. However "factual" and "objective" the methods physiologists employ, the core questions on which their research throws light have to do—evaluatively—with "good" or "bad" bodily functioning.

Does this mean that these core questions have moral or ethical implications? Some would reply that, even if these issues guide our actions, they are *technical* not *moral.* Yet this distinction is one that can be insisted upon only from inside a "modern" point of view: it would have made no sense to Aristotle. In the *Nicomachean Ethics,* for instance, Aristotle contrasts *theoretical judgments* that are universal and timeless, with *practical judgments* that are relevant to one particular place and one particular time. Theory is the realm of *episteme* (conceptual grasp); practice relies rather on *phronesis* (local perception). As illustrations of *phronesis* he points to steersmanship and clinical medicine: steering your way through the shoals of sickness calls, in his eyes, for *phronesis* as much as *episteme.* Even though the general discoveries of physiology may be as "scientific" and "objective" as those of any other discipline, their bearing on the clinical condition of any individual patient will be a matter of judgment: a matter of discerning the particular set of circumstances that hold in this case, and the particular course followed in the history of the patient's current illness.

Nor is the contrast of *episteme* with *phronesis* relevant in practical disciplines like clinical medicine alone. Even in seventeenth-century natural philosophy, concrete particular judgments could not be eliminated from physics. Within any physical theory, there are "demonstrative" or "deductive" inferences, that explore the conceptual ramifications of the theory in question: to that extent, the conclusions of a physical argument follow *necessarily* from the premises. Consider, for instance, the argument by which, from his three laws of motion, Newton proved that a physical body moving in empty space, under a single inverse-square attraction toward a distant, much more massive body, must trace an orbit with the form of a conic section: an ellipse, parabola, or hyperbola. That argument was as much a deduction as those in Euclid's geometry. But it was *only* a deduction. It explored the implications of the terms "force," "mass," etc.; but it did not apply the terms to any concrete situation. For that, one must also assume that the space around the sun and between different planets was free of matter that significantly stood in the way of planetary motions. This enabled Newton to infer that, since planets are *presumably* "moving in empty space" around the "much more massive" sun, their orbits *presumably* have the form of conic sections—at least, insofar as his theory of gravitation fits the facts. But he was in no position to check this empirical assumption directly: for the time being this conclusion was hypothetical—an empirical presumption based on plausible but unproved empirical assumptions.

After the appearance of Newton's *Principia,* people forgot the difference between (necessary) inferences among abstract statements *within* a theory, and (presumptive) inferences such as are involved in applying such a theory to *outside* concrete situations. After fifty years, Newton's theory was so solid that its external applications had the same seeming inevitability as its internal deductions. Still, epistemologically, the difference between the two sorts of inference is crucial: even in physics, the formal necessity of abstract inferences *within* a theory does not rub off on to the substantive presumptions of the best established applications. Even there, intellectual grasp of deductions within a theory may be *episteme;* but our ability to recognize those situations to which these deductions apply rests, as much, on the practical wisdom that Aristotle called *phronesis*—that familiarity with the way the world works *in actual practice* that allows scientists to recognize particular concrete situations to which their theories are relevant.

In the long run, then, seventeenth-century rationalism was a viable option in neither epistemology nor philosophy of nature. But, even after the loss of Descartes' dream, philosophical reflection on the methods and arguments of *practical* enterprises remains a live option for philosophy. In this respect, the air of absurdity that clings to attempts to revive the rationalist program for philosophical theory is irrelevant to those programs that seek, rather, to reappropriate the "practical philosophy" of the Aristotelian tradition. Between them, our intellectual and practical experience during the twentieth century may have discredited the rationalist program for *theoretical* philosophy, but they have left the claims of *practical* philosophy untouched.

VI

At this point, we can carry this reinterpretation further. Some of the most powerful critiques of the tradition of modern philosophy—obviously, but not exclusively, those by John Dewey and Richard Rorty—are by writers who have chosen to call themselves *pragmatists*. What we may now ask is, whether pragmatism (so understood) is one more philosophical theory *comme les autres,* or whether, rather, it prefigures a radical reordering of priorities in the field of philosophy: a final abandonment of rationalism, with its focus on *a priori* arguments, in favor of reflection on the methods of inquiry of practical disciplines, that are grounded explicitly in *phronesis,* and do not concede the traditional claims of *episteme* to intellectual supremacy?

Certainly, one main intellectual task in philosophy of medicine is reflective analysis of the different kinds of established medical practices—classification (as in nosology), perceptual discernment (as in diagnostics), legitimation (as in therapeutics), or half a dozen others. The ways in which these tasks relate to one another, and the modes of analysis to which they lend themselves, are something about which Descartes and his tribe give little help, since they saw philosophy as committed above all to constructing a comprehensive theory of nature, and paid scant attention to the pedestrian elements in clinical practice. We shall do better, rather, to turn back to Descartes' predecessors: to such Pyrrhonist skeptics as Montaigne, or to the classical rhetoricians whose tradition led from Aristotle, by way of Cicero and Quintilian, to Ramus.

Seen from this point of view, the late-twentieth-century collapse of foundationalism—celebrated by literary theorists today as projecting us into a "post–modern" world that is both unstructured and potentially absurd—looks very different. Loosen the constraints imposed on philosophy by the rationalists' exclusive emphasis on theory, and restore practice to its proper place in the philosophers' field of attention, and it becomes clear that we can reappropriate intellectual structures that were available in philosophy *before* the seventeenth century. The *post*-modern world is thus, in part, *neo-pre*-modern.

Some philosophers who are otherwise perceptive, for example, Bourdieu and Habermas, see the need to do justice to practice, but assume that this requires a theory of practice. The present argument ends with a very different conclusion. To call for a theory of practice itself perpetuates the Cartesian tradition of *subordinating* practice *to* theory: of assuming that practice is intellectually serious only if it has a foundation in theory. But, once we recognize that, in substantive intellectual enterprises, "appeals to theory" are one *topos*—or style of argumentation—among others, theory takes its place *within* the world of practice. The *Lebenswelt,* made up of *Lebensformen*—which has alone survived unscathed the critique by phenomenologists and analytical philosophers—is a world not of propositions but of practices. The intellectual core of that world contains those practices in different substantive enterprises that have survived the pragmatic tests to which they were subjected in the evolution of those enterprises. None of them are perfect by Platonist standards. All of them will, no doubt, be refined and improved upon. But—at least "for the time being"—they are the best we have, and they are the starting point for any subsequent rational refinements. So, we cannot afford to undervalue them.

The Center for Multiethnic & Transnational Studies
University of Southern California
Los Angeles, California, U.S.A.

REFERENCES

1. Quine, W.: 1963, *From a Logical Point of View,* Harper, New York.
2. Rorty, R.: 1982, *Consequences of Pragmatism: Essays 1972-1980,* University of Minnesota Press, Minneapolis, MN.
3. Toulmin, S.: 1982, 'How medicine saved the life of ethics', *Perspectives in Biology and Medicine* 25, 736-750.
4. Toulmin, S.: 1990, *Cosmopolis,* Free Press, New York.

MARX W. WARTOFSKY

WHAT CAN THE EPISTEMOLOGISTS LEARN FROM THE ENDOCRINOLOGISTS? OR IS THE PHILOSOPHY OF MEDICINE BASED ON A MISTAKE?

The thesis of this paper is that philosophy needs to do an internship in medicine in order adequately to pursue one of its own central tasks, namely, epistemology or the theory of knowledge; and further, that the proper pursuit of the philosophy of medicine is a critical reflection on the theory and practice of medicine, and not the bringing of philosophically ready-made concepts to the rude precincts of the barbarian medical mind. Note that my point here is not that philosophers need to study medicine in order to do epistemology of *medicine*—that's obvious and needs no argument—but rather, in order to do epistemology in general.

I begin from the premise that medicine is one of a set of fundamental, cognitive practices [2]—like law, the arts, the sciences, the various technologies of production—that, with the advent of tool making, pictorial representation, and language, have historically shaped the human mind, literally brought it into being as an emergent and evolving historical artifact constituted by these practices, or as the Germans say, *in und durch* ("in and through") such practices themselves. Medicine thus constitutes one of the basic and earliest forms of human knowledge, sharing with the other forms certain features and constraints having to do with human learning and the development of skills. Yet it is, at the same time, a distinctive mode, different in its aims and practices from other ways of knowing. Medicine therefore has its own history, though not a strictly autonomous one. It has been affected by, and in turn has helped to shape, other modes of thinking, other forms of practice.

The epistemologist, seeking to understand human knowledge in general, ignores the distinctive characteristics and history of medical knowledge at her peril. What the epistemologist *qua* epistemologist can learn from the endocrinologist, then, is something not only about the nature of *medical* knowledge, but about human knowledge in general, which is, after all, the subject matter of

R. A. Carson and C. R. Burns (eds.), Philosophy of Medicine and Bioethics, 55-68.

epistemology in general. The question then is, what is "epistemology in general," and is there such a thing?

Epistemology, broadly speaking, is that subdiscipline of philosophy that inquires into the nature of human knowledge: what it is; how it is possible (or *whether* it is possible) to acquire it; and how we can know that we have it. Like many of the pursuits of philosophy, it seems to be like a snake swallowing its own tail: in raising the question of how we can *know* that we know something (or as we are prone to say these days, what are the truth conditions under which we are justified or warranted in asserting certain propositions), we seem to be presupposing just what it is we set out to investigate. An even hairier account of epistemology sees it as a fancy name for the study of Pontius Pilate's question, "What is Truth?" Thus, epistemology is not about what we happen to believe, nor about how we come to believe what we believe. That's the task of a descriptive social psychology, or perhaps a branch of cognitive epidemiology—how beliefs or propositional attitudes spread, or infect, a given population.

Epistemology, by contrast, is *normative* rather than descriptive. It purports to set out the conditions for determining whether what we believe is true or whether it is false. It therefore establishes the foundations for knowledge, or the criteria for judging whether our knowledge claims are justified or warranted. This foundationalist arrogance has given epistemology a bad name recently, especially among those philosophers who already know that our claims to truth either have no foundations whatever, or that the sciences—paradigmatically, the mathematized natural sciences—are our best bet for finding out the truth about anything (and everything), and that therefore, some philosophers' claims to epistemological privilege in investigating or determining the conditions of knowledge are without foundation.

Putting the epistemological battles aside for the moment, we may still ask "What would an epistemologist *want* to learn from an endocrinologist?" (or a cardiologist? or an otolaryngologist or an oncologist?) Certainly not endocrinology as such, since philosophy is not taken up with the professional study of medicine. An epistemologist would presumably be concerned with the nature of medical knowledge in general. In that case, however, it would seem to matter little whether the epistemologist studied the knowledge of endocrinologists or of urologists. They are both instances of medical knowledge and insofar as a philosophical theory of knowledge is concerned with universal principles—in this case, the universal principles of medical knowledge—it would be as appropriate to study one instance as another, and perhaps best to study a variety of specialties so as to discover what is common to all these modes of medical knowing. Just as the philosopher of science attempts to reconstruct the nature of scientific knowledge on the basis of paradigmatic cases in the history of science, or in contemporary science, so too—at least according to one standard story—the philosopher of medicine abstracts and reconstructs the typical and universal features of medical knowledge, how it is acquired, how it is transmitted, how it is achieved, established, tested, changed, how it is used, or abused.

Thus, epistemology *in general* learns from medicine *in general* about medical knowledge *in general*, and sees how this relates to knowledge *in general*, albeit by studying particular instances to discover their universal epistemic features. What a nice story!

If there has begun to appear a somewhat ironic cast to the last several paragraphs (or perhaps even in everything I have said thus far), it is not entirely accidental. I have been mixing together things that I believe, with things I used to believe but don't any longer, with things I never believed, with things I would like to believe but can't. Irony is a way of self-distancing, of looking at oneself as if one were someone else, of hearing oneself say things at once assertively and skeptically, of meaning something and taking it back, all in the same breath. What I am skeptical about here is the notion of epistemology in general, or of medical knowledge in general, just as I am skeptical about *the* scientific method, in general, or about ethics in general. What I am not skeptical about is that there are scientific methods, and there is medical knowledge, and there are ethical principles, and some that I would take to be universal. But they are all rooted in concrete practices from which they have emerged, by critical reflection.

Thus, there is a positive thesis that I want to develop here, and about which I have no doubts, or very little doubt, namely, that human knowledge grows out of a variety of practices; and although it has universal, species-specific characteristics because we have genetically common cognitive structures and capacities, and also because there are transhistorical and transcultural features of our social practices as humans, knowledge is an artifact and as such bears the marks of its variety and of the differences of its conditions of origin and its development. Because knowledge is a historical and historically developing artifact, its characteristics change, not simply in terms of what it is we come to know, but in the very modes of this cognitive acquisition itself, and in the criteria of what counts *as* knowledge. Knowledge is not simply reflection upon our experiences and our actions, but is constituent of and ingredient in our practices, and therefore is as various in its character and distinctive features as are these practices, and cannot be understood or appropriately studied apart from them, even if there are abstractible features which emerge as objects of sheer reflection—what Aristotle vividly called "mind thinking itself." The reduction of epistemology to the study of the universally abstract features of perception or thought, disentangled from the reticulation of concrete practices, leaves us with a flat and colorless model of human cognition—ahistorical, largely asocial, isolated from its affective contexts, and simplified beyond recognition from the rich life of mentation that human beings typically lead.

For some time, I have counterposed to this abstract and essentialist *a*historical epistemology a model of a historical epistemology, which recognizes both the variety of modes of cognitive practice, and their proneness to change. The reason that a traditional epistemology—"epistemology in general"—won't do, is that there is no such thing as knowledge in general, just as there is no such thing as medicine in general, just as there is no such thing as philosophy in general.[1] Whatever the particular forms of knowledge have in common can, of course, be abstracted, but to represent such abstractions as what knowledge is "essentially" or in general is to commit a fallacy of reification, of taking the abstraction for the richly individuated and concrete mode of cognition it is skimmed from.

But if there is no medical knowledge in general, then what is an epistemology of medicine about? Must we then devolve upon, e.g., "the epistemology of endocrinology in the years 1990-95"? Isn't there some mediation between the concrete specificity of a (time and place indexed) practice and the generalizable features of medical thinking? Doesn't the very teleology of medical practice—roughly, the aim of maintaining or restoring good health, of alleviating the pain and suffering of ill-health or injury, of caring for the incapacitated—impose a certain general demand upon the range of medical specialties, such that they all subserve some common criterion of the good of the practice? Even where the relations are among highly disparate levels or types of medical knowledge—between, say, the biochemistry of endocrine activity and the therapeutic measures to be taken in the case of parathyroid dysfunction—isn't it obvious that there are general considerations which relate these two sorts of knowledge to each other in very systematic ways? Isn't it therefore a desideratum of medical education and training to be able to interrelate these disparate bits of knowledge, and to see them within even the personal contexts of a patient's lifestyle and character?

These are, of course, rhetorical questions, intended to show that despite the earlier assertion that there is no medicine in general, or medical practice in general, that there are, in fact, highly general and generalizable features that systematically link the specialized and differentiated aspects of medical practice to each other. That, of course, is no great discovery. The human organism, as the object of medical treatment and care, has discriminable parts, or levels of organization, from the biochemical to the anatomic/physiological to the level of psychopersonal identity. Indeed, if we take into account the fact of kinship and sexual and social existence, as we must in considerations of hereditary illnesses, or infectious diseases, or the diseases of war, of poverty, of dietary insufficiency, of substance abuse, of environmental insult, etc., then the systemic generality of medical practice that joins the medical specialties in the cooperation necessary to the ends of medicine can be seen as an essential—general—feature of medical knowledge. It is, in fact, part of the general training of physicians to know when to call upon the assistance of one or another specialist in dealing with the medically multidimensional character of health and disease.

If you like, this suggests a metaphysical model of medicine, in terms of what one might call the nature of medical reality, of the relation of whole to part, and of part to part, and of one onto-biological level to another. Such a reconstruction also affects the epistemology of medicine, since the normative requirements of medical knowledge would seem to map these ontological features. For example, the knowledge of the physiological/anatomical characteristics of endocrine diseases—e.g., the hyperfunction of pituitary growth hormones—would be systematically related to knowledge of the protein chemistry of human growth hormone, to its effects at cellular sites, and these levels in turn related to the gross bodily symptoms of, say, acromegaly or gigantism.

What the epistemologist can learn from the endocrinologist, however, is not some half-baked knowledge of endocrine function or of endocrine diseases (though some such half-baked understanding would be at least a necessary condition for anything else). The epistemologist would learn instead how the endocrinologists think about these disease entities, how they understand them, what the relationship is between empirical observation and history-taking, or between the symptoms and signs, the tests and measurements, on the one hand, and the diagnostic process itself, on the other; and finally, between the diagnosis and the therapeutic choices. But isn't this precisely the same set of questions concerning medical knowledge that the epistemologist could address to any other clinical/diagnostic branch of medical practice? Isn't this "medicine in general," despite all the protestations that I entered earlier? Of course, the endocrinologist's subject is distinctive, in terms of the focus on the regulatory functions of the glands upon various bodily processes. But this doesn't seem to make the *kind* of knowledge different from that of other medical specialties. What seems to differ is only what it is knowledge about. If my earlier suggestion is correct, I would have to show somehow that there are epistemologically relevant differences in the modes of cognition among these specialties—or at least among groups of them—as against the mere difference in the content or the object of the knowledge.

I want to suggest, however, that there are in fact epistemologically relevant differences in cognitive style or in ways of knowing among the specialties. In effect, I am claiming that what the epistemologist *qua epistemologist* can learn from the endocrinologist will differ from what can be learned from the radiologist or the pediatric surgeon. To what extent, however, are these differences rooted in the differences in the practice itself, rather than in contingencies that are epistemologically irrelevant? For it is certainly true that there are different individual temperaments, characters, even traditions and styles *within* the practice of a specialty; and there seem to be historic differences among national styles of treatment within the same specialties. Medicine speaks many languages. But what I am suggesting is that the differences in kinds of knowledge or cognitive mode that would be of interest to an epistemologist show themselves here as differences among the medical specialties themselves.

One might hypothesize that these differences are the results of the speciation of medical practices in the course of cultural evolution (by analogy to biological speciation); or perhaps that the differences among kinds of knowledge derive from objective differences in the objects of such knowledge, or their organic characteristics, or the distinctive nature of the illnesses or injuries that require treatment in the various specialties. Perhaps it is the case that gastrointestinal function and its medical problems demand a different mode of perception, a different sort of clinical judgment or conceptual gestalt than does, say, neurology. Perhaps radiology attracts a different sort of knower than does family medicine. The infamous distinction in the history of medicine between upper-class gentlemen-observers and lower-class sawbones and bloodletters, i.e., between physicians and surgeons, has its origins in radically discrete medical and social practices. Wouldn't it be fair to entertain the possibility of an epistemological distinction among these two trades, certainly in their earlier disparate histories? I remember a colleague at the medical school of the university at which I once taught recounting to me the dismay of his medical school professor, who asked him incredulously, "You haven't failed anything, and you'd make a good surgeon. Why do you want to go into psychiatry then?"

It seems plausible that such differences arise also because of the alternative construction of the medical subject as a result of different medical technologies, and institutional traditions. The "kidney in Room 410" is an institutional abstraction of the organ to be treated from the person who happens to be its bearer at the moment. The immunochemical assay to be read and interpreted within the ontological framework that the laboratory analysis induces, reduces disease to a certain configuration or profile of quantitative results, judged "pathological" or "abnormal." In the spare and reductive universe of such laboratory reports, the appropriate response would reasonably be to try to change that configuration, in the direction of an acceptable norm, abstracting from consideration, for the moment, the concrete circumstance of a patient with a disease. We might say that this makes possible the practice of (certain kinds of) medicine without a human subject. This is a familiar anti-reductionistic complaint, but it may tend to obscure something else of a positive nature, namely that reduction and abstraction is the heuristic of choice among a range of research practices in the sciences more generally, and it has, more often than not, led to the kind of deep specialization and focus that has resulted in important theoretical and experimental advances, as well as to the honing of specialized skills that demand such concentrated and abstractive attention. At the same time, such reductionist focus, however effective it may be for some purposes, loses sight of the larger, or more holistic contexts of the patient as person, or of medicine as a caring profession. And yet, the reductionist could argue that nothing is more effectively caring than the technical medical competence and concentration that such intensive and abstractive focus make possible.

In summary, what I am suggesting as an hypothesis in the historical epistemology of medicine is that differences in ways of knowing may arise as a result of several factors: a) ontological differences in the object of knowledge, or of the medical practice, demanding different epistemological approaches or modes of access (as, say, between a patient's mind, his liver, and his hormonal balance, i.e., between a mental, a physiological, and a biochemical object of cognition); b) historical or cultural differences in the development of the traditions of one practice as against another (e.g., between physicians and surgeons, or between family practice and, say, dermatology, where not only the ontological but the ideological differences play a role); and c) different technological constructions of the medical reality by virtue of the differences in the means of representing or imaging the object of inquiry (e.g., the patient or the patient's disease or condition as reconstructed by the latest means of measurement or analysis, that is, NMR, CAT, multiphasic laboratory tests, computer-assisted diagnosis—what one might call the technologization of the medical subject).

Now comes the hard part. Having laid all this out in broad heuristic strokes, two tasks emerge. First: How are we to understand the differences in modes of cognition that are alleged to characterize the different specialties (or groups of specialties)? What does it mean, epistemologically, to claim that differences in the object of treatment, or in the historical or cultural tradition of a medical specialty, or in the technological construction of the medical subject, are correlated with, or even lead to different kinds of medical knowledge? How shall we characterize the cognitive diversity? By what epistemological categories? Second, and closely related: what would an epistemological analysis of a given specialty look like? What would one look for? How would one proceed? In short, having created the hydra-headed monster of a pluralistically conceived theory of knowledge, how do I propose to tame it?

Obviously, I can attempt only a brief answer to these questions here. If worthy, this constitutes a research program and that means nitty-gritty details and specific case studies in place of the programmatic promising and armwaving I have been doing thus far. Let's try a few specifics.

There appears to be an obvious, perhaps a gross difference between what are called the medical sciences and clinical practice. It is recognized in the traditional divisions of the medical curriculum, and in the institutional distinctions between "research" and "medical practice," or more generally, and still more crudely, between "theory" and "practice." The medical scientist, e.g., the biochemist, the researcher in nuclear medicine, the geneticist, the physiologist does not have "patients" as such, but rather "experimental subjects." However, the research problems posed are likely to derive from clinical contexts, and the knowledge gained is aimed at applications that are ultimately clinical, however removed the site of clinical application of the science may be. (Of course, in actual cases, many medical researchers combine this activity with related, if limited, clinical practice; and the clinical trial, involving human subjects, is not simply experimental but putatively involves the anticipation of a therapeutic result for some of the subjects.) As basic research, however, it has a quasi-independent status as science, contributing to the growth of knowledge, apart from any specific applied or mission-oriented contexts.

Do such differences between the focus and proximal aims of medical science on the one hand, and clinical practice on the other, yield different cognitive modes of the sort at issue here? Is scientific knowledge in general different from clinical knowledge? Or, instead, is *medical* science a hybrid, that differs in some significant way from "pure" or "basic" science (whether theoretical or experimental)? Is it, instead, "applied" or "mission-oriented" and does this make it more akin , epistemologically, to clinical practice? Clearly, there are common (albeit problematic) epistemic criteria that both scientific and clinical practices share: roughly, criteria of rationality, in relating hypotheses to observational or experimental test; of systematicity and consistency, in attempting to understand how one thing relates to another; or in discovering anomalies or contradictions in some concatenation of statements (observational, diagnostic, theoretical, etc.). On the other hand, the clinical focus on diagnosis and therapy in the case of the individual patient seems to be very different from the emphasis on law-like regularities or universals of a scientific sort, which are presumably the aim of scientific inquiry proper; and the concern and care of the health professional seems to be at odds with the criterion of disinterested research, which is ostensibly the hallmark of science.

I put these questions concerning the differences between the medical sciences and clinical practice aside here, with the bare remark that my proffered answer would be that clinical practice is in great measure "scientific" in all the ways that the medical sciences are; and that the sciences in general (not just "medical" science) have a significant component of what may well be called clinical judgment, a feature totally ignored in the older, received view in traditional (logical empiricist) philosophies of science. I have argued this elsewhere [1] and allude to it here only briefly in order to make a point: although there are distinctive differences between the clinical and the scientific aspects of medicine, and although these differences are epistemically significant, they are not separable in the *practices* of medicine, and certainly not in the neat ways that epistemological analysis often suggests. Whatever conceptual analysis may characterize as differences between, e.g., "knowing how . . . " and "knowing that . . . ", or between *Verstehen* and *Erklaren* or between clinical judgment and hypothetico-deductive inference, my point is that both modes of thinking interpenetrate in *all* modes of medical practice,[2] whether "clinical"- or "research"-oriented, whether "applied" or "theoretical." I take these to be differences in focus or in the proximate aims of the practice, rather than real distinctions in the modes of cognition.

We are back at the beginning again. If the differences between "clinical" and "scientific" aspects of medicine do not account for different modes of medical cognition, what else could? What distinguishes one medical specialty from another, epistemologically? What (if anything) marks off the cognitive style of the endocrinologist from that of, say, the otolaryngologist, the osteopath, the radiologist? My suggestion, earlier, was that the object of medical concern is differently constructed in each. In effect, the medical ontology differs, and correlatively, so does the medical epistemology, the specific mode of cognition appropriate to the knowledge of these distinctive entities The construction of the medical object is a complex feature of the practice of medicine. It involves, for example, a construal of the organic or biological features of a human subject that are relevant to the aims of the practice; but also, thereby, a determination of the modes of cognitive (and technical) access that are available to the practitioner, and of the resources of other practices of medicine that may be called into play; and all of this within the mediating framework of the good that the practice pursues, e.g., the prevailing norms of health and disease, the ethical requirements and imperatives of the practice, and so on and on. In short, the distinctiveness of the cognitive mode of a given medical specialty depends upon what the practitioners *take themselves to be doing,* what they take their practice to be a practice *of,* what it is that is practiced *on,* and what they take to be the ends or the good of their practice. In effect, what I am calling cognitive mode here is a matter of the self-construction of the practitioner in and through the activity itself. The medical object is not given, it is constructed in the course of the activity of medicine, and in this sense it is a historically constructed object, the product or artifact of the relevant history of medicine. And so, too, is the practitioner, or the medical specialist, who comes to be *that* concrete individual by engaging in *that* practice. This seems obvious, but it is easily lost sight of in the ahistorical assumption that the human organism, as the object of medical practice, and the human being, as its subject, are somehow given once and for all in all their essentials, and that the different medical practices or specialties are, like natural kinds, essentially fixed by the independent ontology of the human organism. There is no doubt a biological reality which is given, here, but it is itself the product of an emergent evolution, and in turn becomes transformed as an artifact of the historical and social practices of the species, of its cultural evolution. In short, what this digression to the larger picture is meant to suggest is that the medical object—say, the endocrine system—that defines the domain and the character of a specialty—say, endocrinology—becomes a cognitive object, the object of the cognitive practice of medicine in terms of a model of that system—say, of hormonal function.

Such a constructivist view does not imply that the endocrine functions of the human organism are themselves "nothing but" constructions of the mind, or medical artifacts, or that they are not "real" independently of medical practice and human cognition. Rather, such a constructivism argues that insofar as the organism does become an object of medical practice, it becomes transformed *as a cognitive object,* in terms of some model. Since the practice of medicine is a historical matter, the representation of the medical object changes as medical knowledge changes, and even progresses. The historical epistemology of medicine is the history of such changes, not simply in the content of what is known, but in the very ways of knowing. What the epistemologist can learn from, e.g., the endocrinologist, is the way in which the distinctive modes of cognitive practice, represented in the models of that practice, come to develop and change. But this argument that I have just offered begs the question, since it *assumes* that the specialist's practice is epistemically distinctive—for example, that the endocrinologist's model of the endocrine system and its functioning is an endocrinological model different in epistemologically significant ways from the models of other specialties. That it is different merely by having a different object—hormonal function rather than, for example, immunological function or optical function—does not mark it off yet in the requisite way. How then could one characterize a given specialty as epistemically distinctive, in such a context? What, indeed, can the epistemologist learn from the endocrinologist that he couldn't learn just as well from any other medical practice? Let's look at the case in point.

What is the "medical object" taken to be in this case? We may say that endocrinology deals fundamentally with the complex and delicate balance—or the disruption of this balance—in the effects of hormones which chemically regulate a wide variety of bodily processes (growth, development of secondary sexual characteristics, metabolism, etc.) While endocrine anatomy and physiology is localizable (in terms of the various glands of internal secretion that produce the regulatory hormones), the action of these secretions is global, circulating in the blood and attaching to the different cellular sites where the specific chemical reaction takes place.

Several conceptual issues arise here. One is whether this elaborated endocrine activity does or does not constitute *a system;* a second is, what counts as *an endocrine gland?* (For example, the liver has been referred to as a paraendocrine organ because of its regulative effects, cancer has been characterized as an "autonomous, i.e., uncontrolled gland," etc.) A third global feature of the endocrine system (if it is one), or of hormonal activity, is its remarkable *homeostatic character.* For example, the thyroid gland secretes about 90 mg of thyroxin (T_4) and about 10 mg of triodothyromine (T_3) daily. When this is depleted, the pituitary kicks in, increasing the TSH, and when the "normal" level of replenishment is reached, this acts by feedback to cut off further secretion.

Of course, there are other homeostatic mechanisms in the body, so that what makes endocrinology distinctive, as a cognitive practice, cannot be this feature of endocrine function alone. But if we were to group all the regulatory mechanisms of the body that are homeostatic in this way, we could then characterize this *modus operandi* of the organism in terms of a *model*, that is, an abstract representation of a structure or process that can be interpreted or mapped onto any number of specific homeostatic processes. This representation or model then becomes a cognitive artifact or a general sort by means of which the domain of such auto-regulatory process of the organism can be understood. The crude model or first approximation can then be refined by adjustment to the specific modes of self-regulation that are studied more closely, e.g., in terms of their biochemical action or their particular physiological function, so that the model becomes more ramified, more sophisticated and, at the same time, better understood in its concrete determination, and perhaps, at the limit, rejected or radically replaced by a more adequate model.

What I am describing is the genesis of a cognitive mode, a way of coming to understand a given process or bodily function by means of a model. Am I then defining a mode of cognition in terms of a model? At first blush, it would seem not. The model of homeostatic autoregulation may simply be an abstract structure, an aid to the imagination, a sort of extraneous shorthand to keep note of the features of a medical phenomenon that we know or understand apart from the model itself. Or it may serve, in fact, as the means by which we organize or come to understand the systematic relations among the functions of the endocrine glands, and what states of the organism their action effects. However, the way in which we organize or structure our knowledge, or the ways in which we represent it is not something extraneous to that knowledge; rather, I would argue that it constitutes it as the kind of knowledge it is; and more than this, the model orients our inquiry and determines what will count as an explanation, or as the understanding of a particular function or process: in effect, it says what knowledge *is*, in the relevant context.

An objection may easily be entered here, to the thesis that we are somehow accounting for something epistemologically specific, in this story about endocrinology and its homeostatic model. If we take such a view of the endocrine system, it seems clear that there is in fact no *one* distinctive "way of knowing" that somehow attached to this entity as its exclusively appropriate cognitive characterization. First of all, the anatomy and physiology of the endocrine glands is not a distinctive one; and the biochemistry of the action of the various hormonal secretions is plain old biochemistry, not "endocrinological biochemistry." The understanding of the various bodily processes regulated by endocrine glands, i.e., by the pituitary, the adrenals, the thyroid and parathyroid, the testes and ovaries, the pancreas, etc., is like the understanding of other bodily processes, with the one difference that these are hormone-related. It would certainly seem that the endocrinologist simply applies the standard modes of medical understanding to a particular range of structures and functions, and what distinguishes this mode of knowledge is only its focus on the endocrine system. Perhaps the case had better be made not in terms of epistemically distinctive specialties, but rather in terms of the constituent elements of scientific and clinical inquiry in general, i.e., anatomy, physiology, biochemistry, adding to these the supervenient clinical relations of doctor to patient, etc. Medical epistemology would then be nothing but a combination of the epistemology of the sciences in general (perhaps leaving room for special emphasis on biology) together with knowledge of pathology and a strong dose of diagnostic and therapeutic heuristics. This is, in fact, a view more in keeping with traditional views both in the philosophy of science and in the epistemology of medicine. Indeed, the greatest of the innovators in medical epistemology, of the last century, Claude Bernard, effected a radical change in medical education and in medical cognitive practices by showing how the scientific methods and knowledge of physics and chemistry were relevant to the study and understanding of the biological organism, and at the same time, insisting on the distinctiveness of the human organism, biologically and pathologically.

Wouldn't it be sounder to adopt some more universal, even if complex ontology of the human organism and of its states of health and disease, and to base our epistemology upon such a more unified foundation, rather than to pluralize it, specialty by specialty, so that medical epistemology becomes a patchwork quilt of isolated duchies and principalities of cognitive style?

In answer to these very serious objections, I cannot offer more, at the end of this essay, than the suggestions I have already given. I am convinced of this much: distinctiveness in epistemic style is not a desideratum of fashion or a matter of temperament. Nor do I think one's conclusions on the epistemological issue simply express differences between contructivist and essentialist-realist philosophical or methodological convictions. I have argued that a closer examination of the actual cognitive practices of the different medical specialties will yield a very different epistemological picture than the received view of a unified method of medical thinking modeled on a universal scientific rationality combined with a plurality of clinical foci. But this position is fraught with open questions. I propose that it would be fruitful to pursue them.

City University of New York
New York City, New York, U.S.A.

NOTES

1. Cf. Karl Marx, 'Whenever we speak of production, then, what is meant is always production at a definite stage of social development—production by social individuals. . . . *Production in general* is an abstraction, but a rational abstraction insofar as it really brings out and fixes the common element and thus saves us repetition. . . . If there is no production in general, there is also no general production. Production is always a *particular* branch of production—e.g., cattle raising, manufactures, etc.—or it is a totality.' 1973, *Grundrisse, Foundations of the Critique of Political Economy (Rough Draft),* tr. with a foreword by M. Nicolaus, Penguin Books, Harmondsworth, Middlesex, pp. 85-86.
2. Indeed, in all forms of scientific practice generally. This is a larger issue, which I address in 'Scientific Judgment' (see Bibliography), in 'Medical knowledge as a social product: Rights, risks and responsibilities' (see Bibliography), and in 'Science and art: Heuristic and aesthetic dimensions of scientific discovery,' in *Philosophic Exchange,* 1993-1994, nos. 24 & 25, pp. 5-11.

REFERENCES

1. Wartofsky, M. W.: 1980, 'Scientific judgment: Creativity and discovery in scientific thought', in T. Nickles, *Scientific Discovery,* Boston Studies in the Philosophy of Science 60, D. Reidel, Dordrecht, The Netherlands, and Boston, MA.
2. Wartofsky, M. W.: 1981, 'Medical knowledge as a social product: Rights, risks and responsibilities', in W. B. Bondeson, H. T. Engelhardt, Jr., S. F. Spicker and J. M. White Jr. (eds.), *New Knowledge in the Biomedical Sciences,* D. Reidel, Dordrecht, The Netherlands, and Boston, MA.

EDMUND D. PELLEGRINO

PRAXIS AS A KEYSTONE FOR THE PHILOSOPHY AND PROFESSIONAL ETHICS OF MEDICINE: THE NEED FOR AN ARCH-SUPPORT: COMMENTARY ON TOULMIN AND WARTOFSKY

INTRODUCTION

Wartofsky and Toulmin hold several theses in common. Both authors are skeptical about theory in philosophy, epistemology, or ethics. Both reject all essentialist formulations. Both seek salvation from the failings of theory in context, detail, practices, and socio-cultural and historical matrices. Both are decidedly anti-Foundationalist. Toulmin's *bête noire* is Cartesian rationalism, and Wartofsky's is "foundationalist arrogance" regarding epistemology. Both seem to tip the balance Aristotle and Plato would have recommended between theory and praxis in favor of praxis. Toulmin is closer to American pragmatism than Wartofsky, whose inclinations are more toward socio-cultural and historical determinism.

Many of the points I make in my commentary in Part I on Toulmin would apply to Wartofsky, particularly his anti-Foundationalism and the way he balances the theory-praxis equation. There are, however, several additional points specific to Wartofsky's paper I think worth adding in Part II.

As a clinician, and as someone who has proposed that the philosophy and ethics of medicine should be grounded in the realities of clinical practice, I find the move to praxis appropriate. To this extent, I agree with both Toulmin and Wartofsky. My difficulties arise with their anti-essentialism and anti-foundationalism—adamant in Toulmin's case and somewhat more moderate in Wartofsky's. I agree that Descartes's brand of Foundationalism (big "F") is a grand illusion, but this does not per se condemn all foundations. Small "f" foundations are still essential, as both Toulmin and Wartofsky suggest here and there in their own propositions. In short, I can accept praxis as a keystone, but keystones need arches, too, if they are to be at all useful.

R. A. Carson and C. R. Burns (eds.), Philosophy of Medicine and Bioethics, 69-84.

PART I: RESPONSE TO TOULMIN

In his paper, Stephen Toulmin develops further the project he has pursued for some time: moving ethics and philosophy from theory to praxis. He has argued that, by engaging the moral problems of medicine, philosophy saved itself from meta-ethical desuetude and aridity ([33], pp. 736-750). Later, with Jonsen, he recast the direction of ethics from deductive reasoning to casuistic induction from paradigm cases [13]. His present paper takes us a step further, asserting the primacy of practice not just in ethics, but in the philosophy of medicine and in philosophy in general.

Out of the background of his thoroughgoing grasp of the history of philosophy, ancient and modern, Toulmin locates the difficulties of subsequent philosophy in Cartesian rationalism and the many a priori theories it has spawned. If philosophy is to survive the collapse of Cartesian-type Foundationalism, Toulmin claims it must engage practices like medicine, law, or ecology. In this way, it can return to pre-Cartesian, Aristotelian modes of thought. In that return, philosophy will be able to generalize again, not about theories of universal truth but about practices.

Toulmin's main line of argumentation—the turn from grand theories to concrete moral situations and dilemmas—is a welcome antidote to the presumptions of Cartesian intellectualism, analytical technicism, and existential and phenomenological solipsism. However, I do not think Toulmin's pragmatic turn is a sufficient cloth with which to refashion the whole of medical ethics, the philosophy of medicine, or philosophy itself, out of the remnants of his anti-foundationalist critique. Even if praxis becomes the keystone, it must be set in a supporting arch.

For one thing, all foundationalism is not "big F" Foundationalism, nor is all foundationalism defeasible by the same arguments Toulmin uses so effectively against Descartes. For another, the practice/theory dichotomy is not cleanly continuous. Praxis without theory usually ends up conceptually impoverished and verges on empiricism; theory without practice has no anchor in reality and verges on flights of fancy. Finally, praxis encompasses the ethics of the professional *qua* professional, but this is not the whole of biomedical ethics. Decisions about abortion, euthanasia, "vegetative" states, resource allocation, etc., cannot be resolved within practices. Their metaphysical and epistemological dilemmas can be finessed or ignored, but not eradicated.

I agree with Toulmin that praxis and the concrete realities of clinical medicine should be the starting point for biomedical ethics and the philosophy of medicine. I also agree that philosophy, in general, can benefit from such an approach. However, I do not agree that praxis is sufficient for the whole of biomedical ethics—too many ethical issues transcend professional ethics per se. Moreover, the issue is not a choice between theoria and praxis, but of the proper balance between them: how do we encompass pragmatism's emphasis on action without succumbing to its deficiencies? This balancing seems central to Toulmin's own project of recapturing Aristotelian modes of philosophizing in which theory and praxis, ethics and metaphysics, were mutually reinforcing enterprises.

I. MEDICAL ETHICS AND PRAXIS

A. Praxis and Professional Ethics

Medical and clinical ethics are pre-eminently practical exercises in which the concrete details and contexts of moral decision are quite decisive. We can readily agree on this score, therefore, with Toulmin's turn to praxis. But this does not, in itself, justify casuistry as the preferred ethical theory for clinical ethics. This is not the place to enter into a critique of the advantages and shortcomings of casuistry ([37], pp. 33-39). Manifestly, the casuists' identification of "paradigm" cases requires the kind of deep engagement with praxis that Toulmin urges. But, to be effective, casuistry also depends on a body of commonly held values out of which agreement can be reached on what constitutes a paradigm case and what features of that case should determine when like cases present themselves. In the late Christian Middle Ages, casuistry developed out of just such a common tradition. It also had recourse to a moral authority to judge cases. Today, the number of "communities" with shared values is becoming ever smaller. Moreover, any notion of an authority on the moral life is anathema in secular bioethics. Today, what is a paradigm case for one community or person may not be for another. Something beyond casuistry and praxis seems necessary for ethics to survive in today's highly pluralistic, and increasingly diverse and individualistic, moral universe.

To be sure, that something is not more meta-ethics. Urgent practical moral problems like those we encounter in clinical medicine are left unaddressed when we spend all our energy straightening out the language and logic of ethics. Meta-ethics is not to be ignored, but it is far from being the whole of ethics. At the very least, pressing problems like euthanasia, assisted suicide, and the moral status of persons in permanent "vegetative" states, for example, cannot be held in abeyance until the meta-ethical issues are resolved. Clinical ethics demands that moral choices be made in a particular case at a particular time and place. Notwithstanding the uncertainties, complexities, and disagreements inherent in particular cases, choices cannot be escaped.

Clearly, then, medical ethics must be linked to medical praxis. Moral analysis must begin with the facts of cases—with the predicaments of patients, families, and caregivers, with the circumstances and contexts in which they exist and experience their moral choices, and with diagnoses and prognoses. The *case*, in all its uniqueness, is the proper arena for choice. But, by the very fact of its particularities, the case cannot be its own norm. Nor is a "paradigm" case, per se, its own norm. Cases help us to understand the complexity, conflicting obligations, and qualifying circumstances of moral choices. But, the degree to which any case can be independent of some standard of the right and good outside the case itself is always open to serious question. In professional praxis itself, some theory of the healing relationship is needed to generate the norms of morally good or bad professional praxis.

In the end, classical or paradigm cases must be designated as "classical" or "paradigm." To make that designation, we go beyond the case to something else—intuition, reflective equilibrium, consensus, coherence, principles, rules, maxims, authority, etc. Call it what we will, some standard outside the case tells us where *this* case fits on the moral scale, and whether it qualifies as a "paradigm" by which other cases may be judged. To be sure, beginning with cases grounds ethics in real problems, but to judge cases morally it is also necessary to stand outside the cases.

Toulmin's aversion to theory stands in the tradition of the pragmatic philosophies of C. S. Peirce, John Dewey, and William James. Those philosophers deny the existence of any universal system of moral truths and focus on the empiric and experiential, and on action. (Here, I take "praxis" in its Greek and Aristotelian meaning as "action" in the ethical and political life, not in the sense it was later taken by Hegel and Marx ([3], pp. ix, 11-83; [18]). Like Peirce, they extend scientific modes of thought to most domains of human cognition and action, including ethics. They are all antipathetic to a moral life guided by rules or general or abstract principles.

For Dewey, the problem of moral choice is a problem of deliberation and valuation of alternate actions dependent upon a capacity to project their consequences into the future [32]. By determining those consequences intelligently and objectively, we can decide what ought to be done. The basis for decisions is in education and "institutions that foster the growth of intelligence in the life of the community" ([2], p. 123).

William James's pragmatic ethic also sees truth not in abstract reason but in action. For him the " . . . essence of good is simply to satisfy demands" ([12], p. 201). We ought to desire or value that which satisfies the most demands. This is the "guiding principle for ethical philosophy" ([12], p. 205). Determining this end is a matter of practical experience, not a philosophical task. It is marked by tentativeness and ambiguity but is nonetheless more reliable than abstract rules of conduct or any of the other proposed definitions of good.

The thought and, consequently, the ethics of James and Dewey, like that of the pragmatic revivalists like Quine and Rorty, is anti-foundationalist, that is to say, it rejects grand or overarching theories. To be sure, praxis-based ethics is a logical counter to the Cartesian dream—or perhaps more accurately, the Cartesian illusion—that physics, ethics, and medicine could all be deduced in toto from a few doubt-resistant principles if only we could make man an "angel full of clear ideas" ([17], pp. 11-30, 184). But praxis-based ethics, itself, falters when it comes to justifying its moral conclusions by no other criterion than the fact that it derives from an "intelligent" calculation of foreseen circumstances. Even if what "works" is what "counts," how do we know that what works is good? As Smith points out with reference to Dewey, there is a difference between knowing the relations and conditions of things and knowing what we ought to desire ([32], p. 146). This is especially the case with those profound ethical and moral issues in medical ethics that go beyond praxis per se.

B. Praxis and Ethics in Medicine

Medical ethics embraces at least two broad categories of moral questions. One has to do with the conduct of the healing relationship, with the obligations of physicians-*qua*-physicians in their relationships with those who seek their help. This is the realm of the ethics of medicine as a practice or *techné* with specific goals and purposes that define it as a certain kind of human activity. In this realm, Toulmin's turn to praxis has its most valid application.

There is, however, another entire realm of much more profound questions that must be decided in the course of medical praxis, problems of greater profundity than praxis per se can ever encompass. This is the realm of ethics *in* medicine. I refer, here, to the whole vexing range of "bioethical" and "biomedical" ethical issues from abortion, euthanasia, embryo research and eugenically motivated genetics, to care of the environment, allocation of scarce resources, and the corporate responsibilities of managed-care organizations. Those questions are beyond medical praxis as such, even though physicians play a role in their implementation.

Toulmin does not distinguish between these two categories. His emphasis on praxis may be applicable, in part, to the category of professional ethics. But this is certainly not the case with the second category. Here, we must wrestle with ultimately metaphysical questions—the nature and moral status of persons, fetuses, brain-damaged humans, or higher primates. We cannot expect to find answers to these questions by a contemplation of practices. As Burleigh shows, the "practice" of German physicians before and during the Holocaust was inspired by the belief that everything was relative, including codes of medical ethics, and that there were lives unworthy of life, which were owed a merciful death [4]. A pragmatist with a utilitarian end-point might project the consequences of acts of genocide and infanticide using the ethical methodology of James or Dewey and be convinced these were the right things to do. Dewey and James, and most pragmatists, would certainly shrink from such an atrocious distortion of their methodology. But without metaphysical underpinnings, it is difficult to avoid such distortion. The same holds for such utility-driven possibilities as the use of humans in permanent "vegetative" states, rather than healthy primates, for experimental purposes [6].

One need not succumb to the temptations of Descartes's brand of Foundationalism to argue for the necessity of certain universal, metaphysically-based concepts like the dignity of each human person, or the inalienable right to freedom and to life. To be sure, the critical examination of these issues must also be based in concrete reality, but the examination must also transcend the particular and the concrete to get at the more general moral truths without which everything eventually becomes relative.

Proponents of pragmatism, of course, object that metaphysics has no place in medical ethics, that metaphysical differences do not necessarily result in different ways of acting in practical moral situations [15]. Parkin refutes this view by showing that ethical theories and metaphysical presuppositions are relevant and inescapable [20]. Indeed, for the second category of medical-ethical issues mentioned above, if ethics is to be at all philosophical, it cannot avoid confrontation with concepts, arguments, justifications, principles, and values as well as the notions of person, life, death, etc.

This is foundationalism with a small "f". It is compatible with a respect for, and engagement with, practices and particularities. None of this implies acceptance of the Cartesian form of foundationalism which Toulmin rightly decries. There is a place for the "philosophic turn" in all practices. Amelie Oksenberg Rorty suggests, for example, that philosophers should enter all aspects of institutions and public affairs—government, corporations, and schools of education, medicine, and law ([30], p. 273). Rorty hopes philosophers engaged in such institutions and practices can help to determine how the epistemological and metaphysical alternatives would affect those institutions and practices. Like

Toulmin, Rorty wants philosophy to engage practices, but she seems to have more faith than Toulmin that understanding the "foundational" problems of practices might help to resolve them.

II. PRAXIS AND THE ETHICS AND PHILOSOPHY OF MEDICINE

The only way out of the circularity of an ethic with its origin and terminus in praxis is to resort to some criterion of what works well other than the criterion of praxis itself. I would argue that, for medical ethics at least, the way out of this circularity is to ground medical ethics in a philosophy *of* medicine, one which begins with medical praxis but looks for understanding of what medicine *is*, that is to say, what sets it apart as the specific kind of human activity it manifestly is. This would be a philosophy *of* medicine in the classical sense of that term ([23], [24], [26], [28]).

Toulmin also seeks a philosophy of medicine in some way related to medical praxis. He correctly recognizes the failure of Descartes's hope to deduce medicine from universal incorrigible truths. In this he is at one with the great post-Cartesian, Hippocratic physicians of the seventeenth and eighteenth centuries—Thomas Sydenham (1624-1689), Giorgio Baglivi (1668-1706), and Herman Boerhaave (1668-1738). They called for an end to speculation about disease just as the Hippocratic authors had done. In its place, they insisted on reliance on observation and experience subjected to reason [14].

They did not repudiate theory entirely. In fact, they were proposing a theory of medical praxis, one based in the inductive science of Bacon and the experimental method of Galileo. They took Aristotle's *Posterior Analytics* for granted as the method of reason to be applied to careful observation so as to make medicine a verifiable, and not a speculative, activity. Their theory of medicine was the direct antithesis of the prevalent Galenic philosophy. Like Descartes, Galen proposed to ground medicine firmly in philosophical premises ([9], [31]). In a sense, the Galenists and Hippocratics were re-debating the issues the Hippocratic treatise "Ancient Medicine" was presumed to have settled centuries before ([10], pp. 13-64).

Both the Galenists and Hippocratics, however, were in agreement on the ethics of medicine. They saw no need to link a philosophy *of* medicine, with the central moral reality of medical praxis, the physician-patient relationship. They were agreed on the viability and validity of the Hippocratic, patient-centered ethos. This is decidedly not the case for our time. The Hippocratic Oath has been reinterpreted, altered, even repudiated, and each of its precepts, dismantled [25].

In many ways, the Hippocratic ethic, like philosophy, has sustained the same kind of deconstruction and anti-foundationalist attack as philosophy. The task of medical ethics today is very like the one Toulmin posits for philosophy—rebuilding something from the shards of a shattered ancient tradition. I do not think that praxis, itself, can provide the requisite building material or the grouting that can hold it all together.

Ethics based solely in practices suffers certain disabilities, i.e., overemphasis on methodology, conflation of scientific method and philosophic reflection, and circular reasoning since both the origin and terminus of ethics are located in the practices themselves. It is impossible to avoid self-justification, which is just as dangerous in praxis-based ethics as it is in principle or rule-based ethics. Any recourse to theory beyond praxis as a way out of the dilemmas is precluded by the current fascination with the anti-foundationalist credo.

A limited set of choices seems available: (1) abandonment of any universally binding precepts of professional ethics in favor of whatever precepts are institutionalized at any historical moment or place; (2) replacement of professional ethics by positive law removing any claim to the integrity of professional ethics; or (3) a reconstruction of medical ethics out of some "foundation" (small "f") other than praxis per se.

This last option is the most viable. To be successful, it will have to begin in the realities of praxis as Toulmin suggests. But it cannot end there. It must also be grounded in a philosophy of medicine, one which also starts with the realities of praxis but seeks to discern from those realities what makes medicine the kind of activity it is, and what obligations follow from those realities for physicians, patients, and society.

Thomasma and I have pursued the linkages between a philosophy of medicine and its ethics in a series of works ([27], [28], [29]). In that effort, we believe we are following the Aristotelian mode of thought, something which Toulmin also seeks to restore. Our philosophy of medicine, and hence the ethics we derive from it, is teleologically structured. It is derived a posteriori from the universal realities of the clinical encounter, i.e., healing, helping, caring, health. Toulmin seems to start where we start with the universality of the phenomenon of illness when he speaks of the fact that broken bones, bleeding, pneumonia, etc., and their treatments are "for clinical medicine and its moral practice the same in California, Central Africa, and Spain" ([34], p.). The "right" procedures are generalizable, Toulmin says, not because they are so a priori, but because clinical demands and equity transcend cultural and societal differences.

Toulmin raises the possibility of an a posteriori as against its a priori derivation in a rationalist schema. This is consistent with the "Aristotelian hunger for concreteness" ([18, p. 25). Aristotle's search also went beyond the concrete, beyond these aspects of the concrete that are so fortuitous and accidental. " . . . [T]hat there is no science of the accidental is obvious; for all science is either of that which is always, or of that which is for the most part. For how else is one to learn or teach another?" ([1], 1027a19-22, p. 782). We can only learn what "is for the most part" by examining the concrete realities of whatever world we are interested in. In the case of medicine, this is the world of the clinical encounter.

Toulmin is right to insist that this is not *epistemé* in the sense of a complete, incorrigible, theoretical understanding of reality. Medicine is, however, not theoryfree. As Plato, and Aristotle after him, repeatedly asserted, medicine is a *techné*. It is knowledge realized only when put into practice. It is a practice based not just on experience (the realm of the empiricist), but on understanding of the real nature of its object, a capacity to explain its procedures and serve the good of its object. A *techné* needs general rules and fixed knowledge. It differs from theory in always being linked to concrete practices. To the extent that there are generalizable rules for a *techné*, it has a theory, but not a theory in the sense of *epistemé* or first philosophy [18].

Those generalizable rules of a *techné* provide the basis for an ethic of medicine because they define the conditions without which the ends of medicine cannot be achieved. In this way, the ethics of medicine connects with the concreteness of the ends of the clinical encounter. These ends in turn spell out the ethical obligations internal to the practice of medicine. Thus, a theory of medicine links the *techné* of medicine with the practice and ethics of medicine in a continuous mutual and reciprocal relationship. This is a foundation (small "f") without the disabilities of a priori Cartesian Foundationalism. I share Toulmin's skepticism about "grand" theories in medicine but I would not exclude theory and, thus, a foundation for medical practice and ethics.

Given that there are rules to a praxis or *techné*, these do not constitute a complete moral *epistemé* [5]. No rule can encompass all the particularities and concrete details under which moral actions are taken. Nor, on the other hand, is a knowledge of the particulars sufficient. It is the task of practical reason, of *phronesis* to link the proper rule to the particular action.

But *phronesis* is not the same as praxis. *Phronesis* is a virtue possessed by *phronimos*, the person of practical wisdom who through character and experience can judge "with an eye to the occasion"—what is the precise fit between rule and particular action. The practitioner is not automatically the *phronimos*. He or she may lack the virtue of *phronesis* or fail to develop the moral experience and character necessary to make right and good judgments. After all, there are "good" and "bad" practitioners, technically and morally.

But *phronimos* cannot exercise the virtue of *phronesis* without rules and principles which he or she fits to the occasion of moral action. Without rules or norms, the practitioner becomes the norm, and herein lies the danger of the hegemony of praxis without general rules. Incomplete though such rules may be, as moral *epistemé*, they are essential to moral praxis.

Thus, medical ethics, like ethics more generally, will need a proper balance of norms and principles, of *phronesis* with praxis, of moral theory with application to particular moral acts and agents. Toulmin is right to assert the importance of praxis, which is surely Aristotelian in spirit, but he seems here at least to slight theory, which also had its significant place in Aristotle's ethics and metaphysics. A medical ethic faithful to the Aristotelian mode of reasoning would aim at proper ordering of the relationship between theoria and praxis and between general and particular moral *epistemé*. Here, theory would not have merely the instrumental function it has been allocated in modern science ([7], p. 69). In its purest form, theory is knowledge for its own sake, abstracted from things. As Aristotlian "first philosophy," it is " . . . the active possession of thought as thought" ([18], p. 15). Again, to grant this is not to succumb to Cartesian rationalism but to return to the Aristotelian mode of philosophy, which ranges from experience to art to wisdom—each level with its own kind of knowledge and degree of abstraction ([1], 981b25-982a2, p. 691).

In the long run, we cannot escape the "desire to know," as Aristotle put it, and that desire will not be satisfied without a reach for ultimacy, a reach to attain " . . . knowledge about certain principles and causes," to know "the first principles and causes of things" ([1], 980, 981b:27-28; 982a1, pp. 689-692). This was the aim of philosophy for Aristotle as it was for Plato. Whereas Plato grounded his thinking in eternal forms, Aristotle sought to extract the knowledge of the ultimate from the immediate, the real and the practical in ethics, metaphysics, rhetoric, biology, and, especially, medicine. Praxis and theory come together in a human effort to know in the fullest sense of that term.

Medicine is Aristotle's "paradigm" case. Aristotle emphasizes that the physician treats individuals not "man." If a person has a theory without a grounding experience and recognizes the universal but does not know the individual included in it, that person will often fail to cure, for it is the *individual* that is to be cured ([1], 980a15-24, p. 689).

Aristotle and Plato agree that the highest life would be the life of pure *theoria*. Both recognized—Aristotle more convincingly than Plato—that human beings are not gods and that the best life for humans is one which combines *theoria* and praxis [8]. Toulmin reminds us again that a life (and medicine) based in Cartesian *theoria* is unrealistic and unbalanced. His turn to praxis—to action and doing in medicine—is welcome and needs to be heeded.

If Toulmin is to apprehend Aristotle, his next turn might well be to a better balance between theory and practice. For this, medicine is, as it was for Aristotle, a study of the special place of a *techné* between empiricism, on the one hand, and speculation, on the other. Medicine is a *techné*—a practice, grounded in experience, guided by rules and by a knowledge of what *is*, what ought to be done, the way it ought to be done, and the reasons for which it should be done in that way. In these realities lie the substance of a genuine philosophy and ethics of medicine. For Aristotle, medicine, like ethics, is a science of the practical, and that is its special point of view [11].

PART II: RESPONSE TO WARTOFSKY

Marx Wartofsky takes a similar pathway as Toulmin, moving from theory to praxis and taking philosophy towards a deeper engagement with clinical medicine. Wartofsky is, if anything, a bit more precipitate than Toulmin. He believes that epistemology specifically should do an "internship" not just with clinical practice, but with a particular form of clinical practice, namely, endocrinology. He aims at improving the philosopher's capacity not to do epistemology of medicine, but to do epistemology in general. In a sense, Wartofsky's move is from theory of knowledge to a praxis of knowledge as a way to enhance the theory of knowledge itself.

Wartofsky's move does not seem so radical if we recall that he set forth these theses in outline form some twenty years ago in a provocative editorial in the *Journal of Medicine and Philosophy* [35]. In that editorial, he asked essentially two questions: what is distinctive about medical knowledge? and, what might it contribute to human knowledge generally ([35], p. 267)? Wartofsky replied to his own questions by asserting that medical knowledge is, indeed, distinctive, that its distinctiveness resides in its cognitive modes, and that it is historically, socially, and culturally variable—the "heir to its own history." If medical knowledge has universality, that universality arises not in any a priori theory of knowledge, but in uniformities developing in actual historical practices ([35], p. 270).

Wartofsky asks what an epistemologist would want to learn from an endocrinologist—or any other medical specialist for that matter. This is a question he answers only in the most general way. His approach to an answer is an interesting mixture of skepticism and irony. But, despite this, he believes that there are scientific methods, if not *the* scientific method. There are also ethical principles, some of which he takes to be universal, but that universality, he insists, is rooted in practices. Here, Wartofsky is very close to Toulmin's assertion that philosophy can still generalize but only about right practice. Neither, however, is clear about how one makes the leap from the particular to the universal—a problem as old as philosophy itself.

Neither Toulmin nor Wartofsky really engages what praxis can offer to philosophy or epistemology. Indeed, there are epistemological lessons to be learned from a praxis that is in actuality only in part scientific. Medicine is often without the kind of evidence for its usages of tests, procedures, or treatments that would qualify as rigorously scientific. Indeed, much of the adoption and ejection of practices is based on rules of thumb, recently labeled by one author "heuristics" [19]. These are rough rules by which clinical decision are made when the scientific data are wanting, indeterminate, or conflicting. These rules come only in part from experience since they contain a rich admixture of styles, traditions, hunches, and personal theories. They are the low-level generalizations common to many practices beside medicine. Their epistemological status is somewhere between principles and raw empirical experience.

Wartofsky argues that, since knowledge is a constituent of practices, it is as various as those practices and cannot be understood apart form them. To ignore this "embeddedness" of knowledge is to leave us with a colorless, flat model of cognition. There is no such thing, he asserts, as "knowledge in general" or "medicine in general," or, for that matter, "philosophy in general." It is difficult to know, then, what Wartofsky can allow philosophy to generalize about—even when it starts in practices.

According to Wartofsky, medical knowledge is shaped by the *telos* of medicine—health, alleviation of suffering, prevention of illness. This *telos* is what links basic and clinical science with the context of *this* patient's illness. This is a position with which I heartily agree. Indeed, I have posited the *telos* of the clinical encounter to be the architectonic of clinical medicine and the *terminus ad quem* of medical activity [24].

Just as he begins to show us the value of particular *praxes* and the kind of knowledge they require, Wartofsky moves back to a more general level, asserting that there is also a more general epistemology of medicine-*qua*-medicine that imposes knowledge demands on each specialty. Here, too, he moves to what he calls a "metaphysical model" defined in terms of the nature of medical reality, the ontology of which is mapped by normative requirements of medical knowledge. Here, Wartofsky links his two theses: philosophy needs to go to medicine to pursue its study of knowledge and that study, itself, will rest in a metaphysics of medicine—i.e., in a philosophy *of* medicine—drawn not from outside medicine—i.e., from philosophy *in* medicine or philosophy *and* medicine—but from the activity of medicine itself, in a series of articles dating back to the first issue of the *Journal of Medicine and Philosophy* [22]. I have also argued, (later in collaboration with Thomasma), for a philosophy of medicine, a medical ethics, and a medical praxis rooted in the phenomena of clinical medicine ([23], [24], [26], [27], [28], [29]).

I agree that there are epistemologically-different cognitive styles among the specialties, and (I would add) in intellectual temperament. [Different kinds of minds are attracted to different specialties. One of the insufficiently noted tragedies of the medical life is what happens when the physician's cognitive style is at variance with the style of the specialty that he or she has chosen. This matching of personal cognitive styles with specialties is ignored in medical schools and, almost as frequently, in residency training.]

An internist, for example, must be attuned to ambiguity, delayed satisfaction, multiplicity of treatments; he or she must like puzzles, take joy in the intellectual dialectic of diagnosis, etc., and, at the same time, be willing to give of himself or herself to the chronically or incurably ill. A surgeon must be able to confront immediacy, be prepared to act definitively in the face of uncertainty, have great physical stamina, be able to visualize in three dimensions, see the possibilities of mechanical solutions, etc. Radiologists need the gift of dimensionality, of correlating shadow and substance with distance from the intimate drama of illness. Different specialties, indeed, have different ways of knowing, but also different psychological approaches to knowledge, which is a part of the epistemology of the specialties, which, in turn, intersect with the ontological differences in the object of knowledge, historical and sociological difference, and technologic construction.

A more thorough study of the knowledge requirements of the general internist and family doctor would take epistemology to another level of abstraction—one that would be applicable to generalists of all kinds—even to the epistemological stance of the liberally educated person. We can start with medicine, with a particular praxis, and paradoxically enhance our grasp of the universal [21].

I am wholly in agreement, therefore, with Wartofsky's emphasis on the *telos* of medicine and the need for an ontology and metaphysics of medicine. This has been the project I have followed in my own writing on the philosophy of medicine and its foundational function for an ethics of medicine. This is foundationalism with a small "f"—surely not a Cartesian Foundationalism, perhaps more an empirical phenomenology.

Wartofsky is in some danger of contradiction in his major assertion that there is no such thing as medicine in general or philosophy in general. To be sure, there are no Platonic forms; no crystalline, self-subsisting universals for medicine or philosophy. But, there *are* general statements; there is a concept of what medicine *is*—even if that concept is that it can be many things at different times and places. Wartofsky emphasizes the importance of the telos of medicine. As I have argued above, the *telos* of medicine is the *terminus ad quem,* the statement of ends and goals. The facts and details of clinical practice are the *terminus a quo.* To put all of this into a coherent whole requires an ontology and metaphysics of medicine.

In urging this, Wartofsky seems to vitiate his strong denial that there can be medicine in general, or philosophy in general. What are metaphysics and ontology if not disciplines of the most general and abstract kind? What is philosophy of *me*dicine if not an inquiry into the nature of medicine in general [26]?

CONCLUSION

I applaud both Toulmin's and Wartofsky's moves to praxis, to engage intimately with the phenomena and the concrete practicalities of medicine. They have repudiated Foundationalism with a large "F"—the grand theory; however, I do think they are moving to another foundation (small "f"): Toulmin in philosophy and ethics, and Wartofsky in epistemology. Both illustrate what philosophy *qua* philosophy brings to medicine, namely, critical reflection, a perspective on ultimacy, and an attempt to grasp its general as well as its particular meanings. Toulmin and Wartofsky are perhaps too reticent about what they, as philosophers, have to contribute to praxis, and perhaps too generous about what praxis can do for philosophy. But they have called our attention to praxis as a keystone; they seem to neglect the arch necessary to support the keystone. Keystones need arches too.

Georgetown University Medical Center
Washington, D.C., U.S.A.

REFERENCES

1. Aristotle: 1941, 'Metaphysics', in R. McKeon (ed.), *The Basic Works of Aristotle,* Random House, New York.

2. Bernstein, R. J.: 1966, *John Dewey,* Ridgeview Publishing, Atasca Doro, CA.
3. Bernstein, R. J.: 1971, *Praxis and Action: Contemporary Philosophies of Human Activity,* University of Pennsylvania Press, Philadelphia, PA.
4. Burleigh, M.: 1994, *Death and Deliverance: "Euthanasia" in Germany c.1900-1945,* Cambridge University Press, Cambridge, U.K., and New York, NY.
5. Dinan, S. A.: 1983, 'The particularity of moral knowledge, practical reasoning', in D.O. Dahlstrom (ed.), *Proceedings of the American Catholic Philosophical Association* LVIII, 65-72.
6. Frey, R. G.: 1988, 'Moral standing, the value of lives, and speciesism', *Between the Species* 4, 191-201.
7. Gadamer, H. G.: 1981, *Reason in the Age of Science* F. Lawrence (trans.), Massachusetts Institute of Technology Press, Cambridge, MA.
8. Gadamer, H. G.: 1986, *The Idea of Good in Platonic and Aristotelian Philosophy,* Yale University Press, New Haven, CT.
9. Galen: 1991, *Galen: On the Therapeutic Method, Books I and II,* R. J. Harkinson (trans.), Clarendon Press, Oxford, U. K.
10. Hippocrates: 1972, 'Ancient medicine', in W. H. S. Jones (trans.), *Hippocrates I,* Loeb Classical Library Volume 147, Harvard University Press, Cambridge, MA.
11. Jaeger, W.: 1957, 'Aristotle 's use of medicine as a model of method in ethics', *Journal of Hellenic Studies* LXXVII, 54.
12. James, W.: 1956, 'The moral philosopher and moral life', in *The Will to Believe and Other Essays in Popular Philosophy,* Dover Publications, New York, NY.
13. Jonsen A.R. and S. Toulmin: 1988, *The Abuse of Casuistry: A History of Moral Reasoning,* University of California Press, Berkeley and Los Angeles, CA and London, U. K.
14. King, L. S.: 1978, *The Philosophy of Medicine,* Harvard University Press, Cambridge, MA.
15. Leavitt, F.: 1992, 'Let 's keep metaphysics out of medical ethics: Critique of Poplawski and Gillett', *The Journal of Medical Ethics* 18, 206-209.
16. Lobkowicz, N.: 1967, *Theory and Practice*, University of Notre Dame Press, Notre Dame, IN.
17. Maritain, J.: 1944, *The Dream of Descartes, Together with Some Other Essays,* Philosophical Library, New York, NY.
18. Marx, W.: 1954, *The Meaning of Aristotle's 'Ontology',* Martinus Nijhoff, The Netherlands.
19. McDonald, C. J.: 1996, 'Heuristics: The silent adjudicators of medical practice', *Annals of Internal Medicine* 124, 56-66.
20. Parkin, C.: 1995, 'Metaphysics and medical ethics', *The Journal of Medical Ethics* 21, 106-111.
21. Pellegrino, E. D.: 1966, 'The generalist function in medicine', *Journal of the American Medical Association* 198, 541-545.
22. Pellegrino, E. D.: 1976, 'Philosophy of medicine: Problematic and potential', *The Journal of Medicine and Philosophy* 1, 5-31.

23. Pellegrino, E. D.: 1979, 'Toward a reconstruction of medical morality: The primacy of the act of profession and the fact of illness', *The Journal of Medicine and Philosophy* 4, 32-56.
24. Pellegrino, E. D.: 1983, 'The healing relationship: The architectonics of clinical medicine', in E. Shelp (ed.), *The Clinical Encounter: The Moral Fabric of the Patient-Physician Relationship,* Philosophy and Medicine Series, Volume 4, D. Reidel, Dordrecht, The Netherlands, pp. 153-172.
25. Pellegrino, E. D.: 1988, 'Medical ethics: Entering the post-Hippocratic era', *The Journal of the American Board of Family Practice* 1, 230-237.
26. Pellegrino, E. D.: 1996, 'What the philosophy *of* medicine *is*', forthcoming in a volume edited by Giovanni Russo, Rome, Italy.
27. Pellegrino, E. D. and D. C. Thomasma: 1981, *A Philosophical Basis of Medical Practice: Toward a Philosophy and Ethic of the Healing Professions,* Oxford University Press, Oxford, U. K., and New York, NY.
28. Pellegrino, E.D. and D.C. Thomasma: 1987, *For the Patient's Good: The Restoration of Beneficence in Health Care,* Oxford University Press, Oxford, U. K., and New York, NY.
29. Pellegrino, E. D. and D. C. Thomasma: 1993, *The Virtues in Medical Practice,* Oxford University Press, Oxford, U. K. and New York, NY.
30. Rorty, A. O.: 1989, 'Socrates and Sophia perform the philosophic turn', in A. Cohen and M. Dascal (eds.), *The Institution of Philosophy: A Discipline in Crisis,* Open Court, LaSalle, IL.
31. Scarborough, J.: 1991, 'Review of *Galen: On the Therapeutic Method, Books I and II,* R. J. Harkinson (trans.)', *The Bulletin of the History of Medicine* 69, 628-629.
32. Smith, J. E.: 1983, *The Spirit of American Philosophy* (Revised Edition), State University of New York Press, Albany, NY.
33. Toulmin, S.: 1982, 'How medicine saved the life of ethics', *Perspectives in Biology and Medicine* 25, 736-750.
34. Toulmin, S.: 1996, 'The primacy of practice', in this volume, pp. 41-53.
35. Wartofsky, M.: 1978, 'Editorial', *The Journal of Medicine and Philosophy* 3, 265-272.
36. Wartofsky, M.: 1996, 'What can the epistemologists learn from the endocrinologists? Or is the philosophy of medicine based on a mistake?', in this volume, pp. 55-68.
37. Wildes, K.: 1993, 'The priesthood of bioethics and the return to casuistry', *The Journal of Medicine and Philosophy* 18.

H. TRISTRAM ENGELHARDT, JR.

BIOETHICS AND THE PHILOSOPHY OF MEDICINE RECONSIDERED

I. CONTEMPORARY BIOETHICS AND THE PHILOSOPHY OF MEDICINE: THE ATTEMPT TO GROUND SECULAR HEALTH CARE POLICY

In 1975, the first volume of the Philosophy and Medicine series appeared with a concluding section: 'The Role of Philosophy in the Biomedical Sciences: Contribution or Intrusion' ([12], pp. 211-234). Those reflections had as their ancestor a roundtable discussion in the first Trans-disciplinary Symposium on Philosophy and Medicine, held May 9-11, 1974, at the University of Texas Medical Branch at Galveston. This roundtable was held just as bioethics and the philosophy of medicine were about to claim an accepted position in the academy and public policy. The material speaks with a confidence in philosophy, a concern about how to understand bioethics[1] and the philosophy of medicine, and a sense for the contributions they can make, even if it is unclear as to the nature and conceptual independence of the philosophy of medicine.

The six participants included Chester R. Burns, through whose labors the Institute for the Medical Humanities came into existence out of a history of medicine division, of which he had been the director. He correctly understood the import and interplay between the history and philosophy of medicine. Another member of the panel was Edmund D. Pellegrino, the founding editor of the *Journal of Medicine and Philosophy,* who brought the medical humanities to be an accepted cluster of academic undertakings. As a theologian, Kenneth Vaux made a bridge between the new philosophical interests in medicine and the theological roots out of which much of the debates concerning bioethics had just emerged. Two of the other participants were philosophers, the last was a physician/philosopher. Of the six participants, three held doctorates in medicine.

The turn to philosophy in the mid-seventies was significant. In fact, the reflections in bioethics and the philosophy of medicine began as if there were no prior history of sustained analysis and reflection. Scholars turned to the issues at hand as if there were no need to take account of the conceptual distinctions that

R. A. Carson and C. R. Burns (eds.), Philosophy of Medicine and Bioethics, 85-103.

had been framed in the theological reflections reaching back centuries [8]. There was as well a significant history of philosophical reflections concerning medicine [1], [3], [4], [51]. There was even a secondary literature on the history of philosophies of medicine [64]. Despite this considerable historical depth of reflections in the philosophy of medicine, it was not significant for the debates of the early seventies. So, too, the reflections in theology, which had begun to take shape in the late forties and developed into a sustained discussion in the fifties, sixties, and early seventies [21], [26], [29], [31], [38], [56], [57], [63], no longer commanded a central role in most academic bioethical debates in the late seventies and eighties. Those who continued to work in religious bioethics tended to develop arguments in ways heavily influenced by ongoing secular bioethical reflections. There has only recently been a renewed attempt to articulate religious bioethics in religious terms [17].

This initial flowering of theological interest in bioethics was driven by a complex set of forces. There was the straightforward need to apply old principles to new situations. Much of the Roman Catholic reflection in bioethics is of this sort. One finds numerous handbooks to direct the physician, priest, and the nurse in a changing environment of health-care delivery [5], [6], [19], [20], [34], [39]. Many of the studies are innovative and not merely wooden applications of past reflections. This body of medical ethical reflection, what we would now largely term bioethics, is characterized by its constituting a single coherent community of research. The members share a common set of assumptions and procedures. They are working on common problems under one paradigm. They constitute one community of thought. There is progress and development of insight into new areas; new challenges are explored and addressed. There is no sense of crisis, but continuity, indeed, confidence in the ability to meet the new technological developments and to provide guidance. As such, this literature contrasts with the moral and intellectual chaos that followed after Vatican II even in Roman Catholic bioethics [30].

All of this changed. Bioethics, as we know it, emerged during the cultural upheavals that marked the sixties. There was a questioning of authority and taken-for-granted moral traditions [7], [22], [35]. Since traditions of moral reflection presuppose concrete moral understandings, particular ways of characterizing problems, and particular content-full notions of what is appropriate, periods when traditions are brought into question change the way in which moral problems are articulated. They undermine the coherence required, to recast a metaphor from Thomas Kuhn, for normal ethical reflections [32]. A period of crisis reflection is engendered during which the community no longer possesses a common paradigm of what is at stake or of how problems are to be solved. In the theological bioethics of the time, Joseph Fletcher was the herald of many of the changes that marked bioethics. He played the role of the revolutionary, indicating the unavoidability of the crisis that was underway and to which he contributed, while Paul Ramsey and Stanley Hauerwas [24], [25] attempted to protect the paradigm of the traditional Protestant approach to bioethical issues.

This crisis or rupture within Protestant bioethical reflections was soon mirrored within Roman Catholicism as it passed through Vatican II and its aftermath. The Council, which was heralded as a source of renewal and as a sort of Second Pentecost, was followed by the large-scale departure of priests and nuns, as well as a fundamental reexamination of church traditions and modes of doing theology. Roman Catholic theology and bioethics entered a period of crisis reflection. Alexander Schmemann, for example, in reflecting on Roman Catholic liturgics, noted that the Latins became characterized by experimentation and anarchy "inspired by an . . . indeed deeply anti-traditional set of aspirations" [61]. The effects on the broader culture of the chaos in Roman Catholicism have generally been underestimated. An institution that had resisted modernism [42] now embraced it, calling for an *aggiornamento* [33]. These events were read by many as an invitation to radically rethink theology and ethics in order to embrace modernity. It invited fundamental reexplorations of basic issues. It promised an opportunity to come to terms with the new advances in medicine and changing social circumstances.

The development of medical ethics or bioethics in the sixties and seventies had deep roots in the modern philosophical and Enlightenment projects of discovering how human behavior ought to be conducted without an appeal to a particular religious faith (e.g., Christianity) or a particular cultural perspective (e.g., Greece or the West). This goal of discovering a universally valid moral perspective had been undertaken through examinations of reason, through studies of human moral sensibility, through explorations of human sympathies, through studies of human nature, etc. This project had drawn its force from reactions against the religious turmoil that marked the West from the sixteenth century onward. Against this turmoil, philosophy held out the promise of discovering a universal secular basis for human community that could be articulated without appealing to particular religious, cultural, or traditional moral commitments or insights.

In the process, the culture of the West was recast. The Enlightenment shaped the French Revolution, which led to secularization in central Europe in 1803 [10] and the fall of papal secular power in 1870. These changes marked, if not the end, then at least the serious weakening of established Christendom in Europe and the significant fracturing of moral traditions [40]. Western Europe was departing from the monotheistic, monocultural, homogeneous vision of values and reality it had begun to fashion as its dominant self-understanding at the time of Charles the Great and that remained intact for nearly three quarters of a millennium. Even if Europe was not in danger of becoming non-Eurocentric, there was ever less clarity as to where to find the moral center of focus.

Against these changes and uncertainties, it is only too understandable that formal reflections on medical ethics would be undertaken (Gregory, 1770; Percival, 1803). As established traditions weakened and fragmented, there was the need reflectively and formally to determine the nature of proper behavior. Given the vast implications of the changes, it is only too understandable that many would deny their full sweep. Changes were noted, but their significance was usually discounted. For example, the author of a British text in secular medical ethics published in 1902 acknowledges, "It is not sufficient to say, as some people do, that medical ethics may be summed up in the Golden Rule, or that a man has only to behave like a gentleman" ([60], p. 2). He recognizes that mores are changing so that "what was regarded as customary and even proper some years ago, has often come to be universally condemned" ([60], pp. 2-3). Since the author is primarily addressing issues of medical customs and etiquette, he is unwilling to entertain the view that "our conception of Christianity and chivalry has undergone a complete revolution within the same period" ([60], p. 3). Even so, significant changes were underway, which culminated in the fracturing of paradigms for medical ethics, both within the profession of medicine and within particular religions (e.g., among the Roman Catholics and the Protestants).

Philosophy promised for health care in the twentieth century what it had offered in the seventeenth and eighteenth centuries in response to the divisions that fueled the Thirty Years' War and the Civil War in England: a standard of moral conduct that could bind humans as such. If philosophy were able to establish a content-full account of human rights, duties, proper character, virtue, sentiments of care, etc., these would then be available for all reasonable persons to apply in directing choices in the face of new temptations and new technologies. The interest in the medical humanities in the sixties and seventies can also be seen as trading on an appeal that had supported the first, second, and third humanisms.[2] In the sixteenth century, the first humanism claimed a basis for human unity over against the emerging religious dissatisfactions and divisions of the time. The second humanism at the end of the eighteenth and beginning of the nineteenth centuries [50] promised a cultivation proper to humans. The third humanism and so-called New Humanism at the end of the nineteenth and beginning of the twentieth centuries [28], [59] were invoked to place the new sciences and technologies within the context of human values. The medical humanities, including bioethics and the philosophy of medicine, were invoked in the sixties and seventies toward similar ends and with similar hopes: disclosing the values and goals that are truly human.

Against this background, one can charitably construe the seemingly pleonastic invocation of "*human* values" in various programs in medical education (as, for example, in the Society for Health and *Human* Values) as an appeal to special moral and aesthetic excellences of humans. After all, of what other values would one be speaking (e.g., canine values, porcine values?). The intent does not seem to be to contrast with divine values. The underscoring of human can instead be understood as an attempt to answer Heidegger's question, "In what does the humanity of humans consist?" ([27], p. 319), so as to provide a general framework for moral deportment and the good life. What is added is the heuristic character of the medical context. For the medical humanities, the answer to Heidegger's question was to be found in the encounter between the liberal arts and medicine, because medicine addresses the whole of the human drama.

Such assumptions appear to frame Edmund Pellegrino's provocative rallying cry, "Medicine is the most humane of sciences, the most empiric of arts, and the most scientific of humanities" ([52], p. 17). Moreover, the humanities were not just to be an academic undertaking but a moral calling.

> The humanist must also be "authentic." The medical setting requires that the humanist incorporate the values he or she professes and the character traits that are embodiments of the liberal arts teachings, to be human if not humane . . . truly, the humanist must be "holier than thou" ([53], p. 26).

The humanities were not simply to convey intellectual perspective or a greater command of the cultural context within which health care is delivered. The humanities would improve the conduct of physicians who came under their influence.

> So far as ethics and the humanities go, they undoubtedly raise the sensitivity of students and faculty to ethical and values questions . . . . Almost everywhere, as a result, patients are better apprised of their part in clinical decisions, and of the value and moral issues woven into their relationships with the physician. This is a result to be desired in a society that is democratic, educated and pluralistic in its value systems. Whatever personalizes and particularizes healing will make it more humane ([53], p. 51).

The humanities would humanize medicine and its professionals.

Bioethics and the philosophy of medicine were regarded as of one fabric with the endeavor to disclose and sustain a core set of commitments proper to humans. In particular, they claimed that a universal human community could be articulated and justified around a universal set of moral commitments. Given the considerable variety of Roman Catholic and Protestant bioethics, given the disarray into which these religions were falling, and given the pressing need to frame policy regarding the proper approach to organ transplantation, third-party-assisted reproduction, and critical care, secular philosophical bioethics appeared to offer what was very much needed. A common moral theory was demanded; although traditional sources appeared unable to supply that ethic, philosophy promised to unite all around a set of fundamental moral commitments.

The development of a bureaucratic or governmental bioethics in the seventies appeared to substantiate the secular bioethical claim of being able to discover a common moral vision binding all. The National Commission for the Protection of Human Subjects fashioned a set of recommendations and a set of guiding principles that purported to give canonical, content-full moral direction to health-care policy. It developed an ethic to govern research ranging from issues of fetal research to psychosurgery [43], [44], [45], [46], [47], [48]. It even provided a trinity of ethical principles to guide research involving human subjects [49]. Public bioethical reflection could claim to have created the secular equivalent of a moral theology, along with guidelines for a pastoral theology, all of which could then issue in secular canon law (i.e., secularly morally justified regulations for research involving human subjects). This accomplishment seemed to vindicate the project of discovering a content-full, canonical morality that could bind all in one moral community.

The claim of being able to disclose an apparent foundational moral communality was buttressed further by the appearance of Beauchamp and Childress's *Principles of Biomedical Ethics* [2], and the availability of their four moral principles of autonomy, beneficence, non-maleficence, and justice. The authors of *The Principles* and many of those who apply them are able to reach across theoretical divisions and realize common conclusions about moral controversies. The principles are regarded as able to help health professionals, health-policy makers, and bioethicists come to common decisions regarding diverse cases, despite divergent theoretical commitments. While religious bioethics remained plural in their numerous approaches and moral understandings, secular bioethics appeared united around a common understanding of the field of bioethics as an academic endeavor that could provide reliable moral guidance. The shift of interest in bioethics from a focus on religious bioethics to secular bioethics seemed vindicated. Secular bioethics appeared able to provide what religious bioethics could not: a neutral, but still content-full, vision of proper moral deportment that all could endorse and that would constitute the bioethics of the human community. On the basis of this success, bioethics could take its place not only in the academy but in the public policy arena.

II. WHY THERE IS NO CANONICAL, CONTENT-FULL SECULAR BIOETHICS

The difficulty is that this apparent triumph of secular bioethics is hollow, indeed false. There are as many secular accounts of morality, justice, and fairness, as there are major religions. Despite the apparent moral unanimity found in the recommendations of the National Commission, actual public-policy debates and political campaigns, not to mention bioethical disputes, reveal a significant range of conflicting moral visions and moral rules. There is, in fact, no one secularly justified, canonical sense or account of justice for the allocation of health-care resources. There is, in fact, no settled general secular view of the morality of abortion and euthanasia. More significantly, there is no content-full, generally accepted, or even dominant account of the good life and of proper conduct. It is not simply that Western culture has fractured into different religious groups. Religious groups have themselves often fractured into conservative and reformed factions. Over against those who would wish to be true to millennia-old moral commitments and traditions are those who smell the opportunity of fashioning all anew as the next millennium begins. In addition, the culture itself has seen the general disestablishment of Christianity, so that one can, with some justification, speak of the West having entered a post-Christian era.[3]

This cacophony, this polytheism of moral visions and contentions, characterizes post-modernity with its loss of a dominant moral account or of a generally accepted moral narrative.

> In contemporary society and culture—postindustrial society, postmodern culture—the question of the legitimation of knowledge is formulated in different terms. The grand narrative has lost its credibility, regardless of what mode of unification it uses, regardless of whether it is a speculative narrative or a narrative of emancipation ([36], p. 37).

There is no longer a single, generally accepted moral narrative within which to embed society, so that it can plausibly be said to locate its members within one moral community.

This is not to say that one cannot communicate across moral narratives and across moral communities. The moral strangers one encounters may differ only in the fundamental circumstance that they rank the values cardinal to their lives in ways different from one's own. Even though one lives fully committed to a particular content-full moral vision, one can still understand those whom one acknowledges as moral strangers. One can communicate and collaborate with them in undertakings in the market and as citizens within limited democracies. All this is possible, though one is, in fact, separated in moral vision and commitment. Even Hassidic Jews and the Amish can meet and trade in the same market, though they remain separated in moral understandings and religious convictions.

Other than in such special cases in which communities stand out because of their dress and deportment, many may not notice the substantive differences that divide, since members of substantially different moral communities may still be willing to communicate within what Alasdair MacIntyre has termed the "internationalized languages of modernity" ([41], p. 384). Others may actually live their lives within a "rootless cosmopolitanism, [which is] the condition of those who aspir[e] to be at home anywhere—except that is, of course, in what they regard as the backward, outmoded, undeveloped cultures of traditions" ([41], p. 388). Such "citizens of nowhere" may in fact dismiss and discount traditions of real moral substance. Yet, real differences remain.

To appreciate the real moral alternatives available in the contemporary world, one need only consider the reflections of Chinese policymakers as they contemplate the differences between refashioning China in the image and likeness of Cambridge, Massachusetts, versus the image and likeness of Singapore. Which choice appears more inviting will, in part, be a function of which background thin theory of the good is embraced and by whom. If, in approaching a new understanding of civic friendship, one gives priority to prosperity and security over liberty and equality, a society structured according to the general lineaments of a dictatorial capitalism will be morally preferable to one made in the image of Rawls's account of the liberal polity. The choice is between different moral worlds. The protagonists of each will be moral strangers in the sense of not sharing sufficient common basic premises in order to be able to resolve moral controversies together.

Such differences in moral perspective reveal not just a heterogeneity of moral perspectives as a social fact of the matter. The differences indicate, in addition, the impossibility of resolving substantive moral controversies by appeal to sound rational argument. One cannot calculate which approach will have the better consequences without a prior agreement regarding the proper ways to compare consequences. In the previous example, calculations of the consequences resulting from the choices of different political systems depended on how one ranked liberty, equality, prosperity, and security. These ambiguities cannot be resolved by an appeal to a hypothetical chooser or a group of hypothetical contractors to resolve fundamental differences in approach unless one already knows what moral sense, thin theory of the good, or moral account one should impute to such choosers or contractors. However, the choice of correct moral sense, thin theory of the good, or moral account is what was at stake to begin with. Nor will an appeal to maximizing preferences resolve such difficulties. In order to make such calculations, one must be able to compare rational versus impassioned preferences, corrected versus uncorrected preferences (in addition, one will need to know how appropriately to correct preferences). One will need, as well, to know how to compare present versus future preferences. One must know God's discount rate for time.

Nor will it be possible to resolve matters by an appeal to a notion of moral rationality without knowing what content that rationality should incorporate. So, too, appeals to human nature and human sentiments will reveal moral diversity, not agreement. In a post-Darwinian world, one finds humans with diverse and varied constellations of inclinations. One is despoiled of a morally canonical understanding of the typically human that can be taken self-evidently to be morally normative for humans.

There is no way to resolve fundamental moral controversies without a view of what is morally relevant in any appeal to facts, to nature, to human sensibilities, to human sympathies, or to the character of rationality. In order to discover a content-full vision of moral probity that others will share, one must already share basic moral premises. One must already have in hand a guiding moral sense, a guiding understanding of how to discern between moral noise and information. To bring a moral argument to any particular conclusion, one must already have some value content in hand. The result is that one must either engage in an infinite regress or beg the question. The babble of post-modernity besets us not simply as a de facto catastrophe, but as an epistemological condition from which reason cannot liberate us. For bioethics, as for all applied ethics, this means that one cannot simply apply ethics. One must determine which ethics one should apply, recognizing that there is no way to discover outside of a particular moral perspective or tradition which ethic should be canonical.[4]

It is for this reason that arguments regarding whether health care should be provided equally to all or as a basic minimum with the opportunity to purchase more, regarding whether organs should be sold on the open market or whether such transactions should be forbidden, regarding whether there should be commercial surrogacy or whether such contracts should be forbidden, regarding whether physician-assisted suicide and euthanasia should be allowed or whether such undertakings should be forbidden, etc., go on forever without resolution, if one tries to approach them as content-full moral issues that should be resolved by sound rational argument. In order to resolve such controversies by sound rational argument, one must already share very important common moral premises and understandings of reality. Absent such understandings, controversies regarding what should count as the correct choices continue without resolution not only as a matter of fact, but in principle the participants fail to share sufficient common moral premises and common rules of moral evidence to deliver a principled closure [14].

How, then, could it have seemed that the National Commission and other deliberative bodies succeeded in discovering appropriate practical resolutions of moral disputes? How could they have seemed to have come to such convincing agreement regarding fetal research or the use of prisoners, if matters are as difficult as just alleged? The answer lies in the political construction of agreement, and the ways in which various interest groups shape the character of such deliberative commissions. In fashioning such commissions, it is prudent not to appoint individuals of radically differing viewpoints. One can only imagine how different discussions concerning the conduct of fetal research would have been, had the National Commission been composed of representatives from Operation Rescue, a delegate from the Pope of Rome, a libertarian, a pro-choice advocate, a committed atheist, a Maoist Communist, etc.[5] If one appoints individuals from significantly different communities of moral understanding, then such commissions will not, in fact, agree either in theory or in practical recommendations. To avoid endless debate, if not chaos and disorder, indeed, in order to endorse the policies of a particular dominant group, individuals are selected who are able to collaborate. The agenda is focused on areas where agreement of a particular character is likely. Consensus and agreement are constructed not just through the careful direction of debate and reflection by whoever chairs the meetings, but by the prior selection of participants. Consensus and agreement are manufactured by shaping the structure of the debates that will take place and by determining who will participate in them.

The seeming success of the Beauchamp and Childress approach sheds further light on the phenomenon of apparent agreement in bioethics. Part of the plausibility of their middle-level principles lies in the fact that Beauchamp advocates a teleological account of morality, while Childress endorses a deontological account. Yet the two can endorse similar judgments about how one should act in particular cases. That individuals from such different theoretical perspectives can agree regarding the application of the four principles suggests that these middle-level principles can indeed bridge theoretical and moral differences. Yet, Beauchamp and Childress began by sharing a very similar ideological, political, and moral vision of the world, although they then proceeded to reconstruct that vision or understanding, one in a teleological, the other in a deontological theoretical framework. It is not at all unanticipated that they can use their four principles to reach across their theoretical differences to come to similar practical conclusions regarding real cases, given the similarity of the points of departure. The source of their success lies in their original similarity in ideological and moral commitments. Had they embraced quite different moral visions as points of departure, then the middle-level principles would have disclosed not agreement but disagreement.

Imagine a committed Rawlsian and a committed Nozickian invoking the principle of justice to determine how to allocate resources in particular problematic cases. In such circumstances, the invocation of the principle of justice will disclose differences rather than similarities. So, too, if an individual who regards abortion as morally unproblematic discusses the issues of beneficence involved in providing free abortions to indigent women with a convinced prolifer who recognizes abortion as tantamount to murder, their conclusions will be different because the frameworks within which they interpret benefits and harms will be substantively different.

Commissions and the use of middle-level principles appear to succeed in bridging moral disagreements and in conveying the sense of a common underlying morality. In reality, they only show that individuals with antecedently similar moral, ideological, and/or political views can come to agreement regarding particular judgments about how to proceed with respect to particular cases and policies. One can also understand why this success is not substantial and camouflages the larger moral disagreements that separate moral communities and continue to fuel ongoing bioethical disputes. One can understand as well why this failure will not be readily acknowledged. To confront these failures frankly would require reconsidering the project of bioethics as it has been understood by many for the last twenty years. One would need to look for a foundation for the authority of bioethical agreements in some ground other than sound rational argument, existing moral consensus, overlapping moral consensus, or agreement derived from practical judgment and middle-level principles.

These conclusions are not grounds for abandoning the project of secular morality. Rather, they require its reconception. If individuals do not already share a common ideology, moral view, or religion, such that they can appeal to sound rational argument (or recognize certain individuals as in authority), so as to resolve moral controversies and to authorize communal endeavors, and since sound rational argument will not disclose a canonical moral vision that will endorse particular conclusions and authorize the realization of a particular public policy, authority can only be drawn from the consent of those who participate. The foundational principle under such circumstances is permission. The authority to which persons can appeal when they do not share a common ideology, moral vision, religion, or content-full political understanding is the authority of the consent of collaborators.

An element of the modern philosophical project had been the hope to show that particular moral conclusions are rational, such that one could dismiss competing conclusions as irrational and then with the warrant of rationality impose policy in accordance with those conclusions. Under such circumstances, the force employed would not be alienating but would restore humans to their true rational and moral nature and conduct. Since the project of discovering through rational examination a content-full morality has failed, and in its own terms is in principle doomed to fail, authority for the moral endeavors that bind moral strangers must be sought elsewhere. The authority that moral strangers can always recognize is the authority that they convey by their permission to particular undertakings and those forbearance considerations that are integral to this practice. For this reason, one can understand the much greater success of bioethics in achieving agreement regarding issues of consent and forbearance rights than regarding positive entitlements or the nature of virtuous conduct in health care. Indeed, if anywhere there is a semblance of consensus, it is where direct violations of forbearance rights are involved.[6]

Much would need to be said about the implications of these conclusions. One would need to warn that this understanding does not convey any particular value to autonomy, permission, or freedom. One would need to underscore that, in addition to showing that one can in general secular moral terms justify the moral standing of robust forbearance rights and rights to privacy, as well as the moral standing of free market contracts and limited democracies, one can also show how welfare rights can be created. But just as there will not be a general secular value to assign to liberty, autonomy, or permission, neither will one be able, outside of a particular moral tradition, ideology, religion, or content-full political perspective, to assign any general secular value to the particular choices or structures (e.g., particular health-care or welfare systems) created through the peaceable consent of those involved. Such choices and structures will be beyond secular good and evil, since good and evil cannot be specified in general secular terms. Institutions will simply either possess or not possess the authority that moral strangers can understand that they have conveyed to their common endeavors.

Many puzzles will remain concerning how one ought to judge whether individuals are acting freely and competently. Generally, the presumption will be in favor of the authority of individuals to do with themselves and consenting others as they choose, no matter how wrong, evil, or improper such undertakings might be. Individuals will in general be at liberty peaceably to do with themselves and consenting others that which is wrong.[7] Problems will arise with regard to children. It will be very difficult to show the general secular immorality of infanticide (though, rest assured, the author of this paper, within his own moral perspective, recognizes the immorality of infanticide). In many cases it will be possible to prohibit acts against children, which those children are likely to recognize as abuse and mutilation once they are competent. That is, it will be possible to establish grounds to recognize the forbearance rights of future individuals with respect to actions undertaken against them before they are moral agents.[8]

Despite these puzzles, at least one knows where one cannot succeed. In particular, the project of discovering a normative consensus in general secular terms will always fail. One is left instead with the challenge of drawing moral authority from the agreement of diverse individuals who often live their lives in divergent moral communities. Twenty years after its first flush of Enlightenment optimism, bioethics will need to reassess its abilities and its prospects.

III. CARE, NARRATIVE, AND MORAL DIFFERENCE

In the face of reason's failure, if one is disinclined to turn to permission as the source of moral authority when moral strangers meet, one may decide instead to turn to emotions, affections, and moral sensibilities. For example, one may appeal to an ethics of care and caring. Here the old problems return, for there is not one understanding of care, not one way to be caring. Does one show care by providing welfare, or is it better to have people be schooled in the pedagogy of necessity and the market? The answers will not turn simply on the facts of the matter, but on how one compares the moral significance of particular experiences, skills, and outcomes. Is it caring to provide physician-assisted suicide and euthanasia? Is it caring to respect freedom or to seek equality?

Accounts of caring are embedded in particular moral narratives, supported by particular moral communities. It is only within such narratives and communities that one can, in fact, learn in a content-full fashion what it is to care or to have a good moral character. Within such narratives and the communities that sustain them, one learns the substantial differences between vice and virtue, between vicious and virtuous people. One can learn contentfully about the relationships among rights and duties, virtues and vices. Depending on the narrative and the community, one may even learn about holiness and evil. Because there are numerous narratives, there will be numerous possibilities for initiation into an understanding of character and virtue. Such communities and their understanding will compete with each other. Coming to understand a narrative may cause one to change narratives, to change moral communities, to convert.

The human other is rarely inscrutable and opaque. Very often, the differences between oneself and the other turn on different orderings, different compositions of the values that shape and support human existence. Such differences are significant. They support different understandings of justice and fairness and of the priority of freedom and equality. Such different orderings and compositions separate and distinguish moral visions, moral narratives, and moral communities. They convey with them different understandings of caring and failing to care. The differences can be very deep, as when they turn on concerns with values and morality that step beyond secular interests in virtue and vice and involve themselves instead with holiness and evil. Despite all these gulfs and differences, which separate moral communities, if the secular authority of the general society is understood to be drawn from peaceable agreement, there will be the possibility of living together in peace, and with commonly understood secular moral authority. This will require recognizing that the larger society is not one's own particular moral community, nor one's own moral community the larger society ([16], pp. 211-224). If these conclusions hold, both the structure of society and the character of bioethics will in the end need to be reconsidered.

IV. SOME CONCLUSIONS: WHY THIS IS ALL SO HARD

There are many plausible sources of the reluctance to recognize the actual limits and capacities of secular philosophy, as well as secular bioethics. Though many have lost religious faith, many still seek something like the doctrinal certainty and authority of Western Europe's religious past. They would hope to be able to establish both for themselves and for others a content-full vision of the moral life. Once established, they would not need to live in a controverted context in which one must often recognize the right of others, individually and communally, to do what one knows to be deeply wrong. There is often an invocation of a secular consensus as once Western Christians had invoked a consensus of the faithful.

The claim of such a moral consensus as the secular ability to discover a content-full moral vision can have political force. As Karl Marx understood some one hundred fifty years ago, ideas are important in controlling a society and achieving particular interests and goals. Ideas function as "the ideal expression of the dominant material relationships, the dominant material relationships grasped as ideas; hence of the relationships which make the one class the ruling one, therefore the ideas of its dominance" ([37], p. 39). Even if one were to qualify Marx's stress on economic relations, his account can still help illuminate major elements of bioethics' short past. Government has turned to bioethicists not just to determine what health-care policies can be justified, but also in order to provide a rationale for the policies governments wanted to pursue. In these terms, one can perhaps better understand the Clintons' hasty attempt to create an ideological justification for their health-care reform proposal ([65], pp. 3-10; [9], pp. 421-431; [62]). Bioethics, even if it cannot, in fact, deliver a canonical, secular, content-full moral understanding, can still be recruited to aid those who wish to have power, and as a consequence it can provide the opportunity for bioethicists to share in power. Here Karl Marx again is useful in his characterization of "conceptive ideologists, who make the perfecting of the illusion of the class about itself their chief source of livelihood" ([37], p. 40).

Over the last twenty years, bioethics has become part of the everyday secular culture of most industrialized countries and integral to their reflections on health-care policy. This has not only given a place for bioethics as a cultural phenomenon and a place for bioethicists in the academy, it has in addition given bioethicists an opportunity to wield political influence and direct day-to-day health-care decision-making. In religious-based hospitals where particular moral commitments may be expected of clinical ethicists, this is less problematic. One can more readily recognize the content-full commitments that the bioethicists bring with them. In secular hospitals, bioethicists, if they are not subtly to import their own ideological and content-full philosophical commitments, will need to provide something equivalent to informed consent concerning their moral commitments, if they are not to be mere ideologues. When they enter public policy-making, this concern must take on special weight. In these new niches far from the Academy, bioethics and the philosophy of medicine will need to reexamine their foundations and make clear their implications for the ways in which they touch on the lives of individual patients and public policy generally. Over the intervening years, the philosophy of medicine and bioethics have justified the confidence that marked the roundtable discussion in 1974 ([12], pp. 211-234). Now there is a need to explore the character and significance of this success.

Center for Medical Ethics and Health Policy
Baylor College of Medicine
Houston, Texas, U.S.A.

NOTES

1. In this essay, I will without any critical distinction move among the terms bioethics, medical ethics, health-care ethics, and biomedical ethics, recognizing that these terms intertwine as one moves from the seventies into the nineties. Medical ethics came to be replaced by biomedical ethics and then by Van Rensselaer Potter's term bioethics [55], which achieved a new meaning and salience through the *Encyclopedia of Bioethics* [58].

[2] The reader should note the strategic ambiguity of the term humanism, which encompasses at least five major and quite different undertakings: (1) humanism as humane concern for others, (2) humanism as a body of learning, (3) humanism as a body of philosophical and moral theory, (4) humanism as refinement, and (5) humanism as a cluster of cultural and philosophical acts ([15], pp. 43-52).

[3] Compare, for example, Gilbert's study of post-Christian Britain with T. S. Eliot's plea for a Christian society [22], [11]. Eliot remarks "We must remember also that the choice between Christianity and secularism is not simply presented to the innocent mind, *anima semplicetta*, as to an impartial judge capable of choosing the best when the causes have both been fully pleaded. The whole tendency of education (in the widest sense—the influences playing on the common mind in the forms of 'enlightenment') has been for a very long time to form minds more and more adapted to secularism, less and less equipped to apprehend the doctrine of revelation and its consequences. Even in works of Christian apologetic, the assumption is sometimes that of the secular mind" ([11], p. 190).

[4] I have explored these issues in greater detail in [13]. They are further developed in [18].

[5] I am here in deep debt to my discussions with Kevin Wm. Wildes, S. J., and his forthcoming volume, *A View from Somewhere.*

[6] My contention is not that adopting a permission-based account of moral authority will usher in the millennium and the concurrence of all persons. The contention is rather that, if there is any way to proceed towards a principled account of moral authority that can be justified to moral strangers in their common actions, it will be found in terms of deriving general secular moral authority from permission.

[7] The wrongness of the actions of others will always be understood within the perspective of one's own content-full moral vision. Embedded within the constraints of general secular morality, one will often be constrained to say "X has a right (in that X has secularly justifiable forbearance rights) to do A, but A is wrong (given the special insight I possess as a member of my moral community)."

[8] This is a complex and difficult area, where the author would like to conclude to much more substantive moral condemnations than he sees himself able to do in terms of general secular morality. For an attempt to make some headway in these matters, see the "Principle for Intervention on Behalf of a Ward" in [13], pp. 286-7).

REFERENCES

1. Bartlett, E.: 1844, *Essay on the Philosophy of Medical Science,* Lee and Blanchard, Philadelphia, PA.
2. Beauchamp, T. L. and Childress, J. F.: 1979, *Principles of Biomedical Ethics,* Oxford University Press, New York, NY.
3. Bieganski, W.: 1894, *Logika Medyzyny,* Kowalewski, Warsaw, Poland.
4. Blane, G.: 1819, *Elements of Medical Logic,* T. and G. Underwood, London, U. K.
5. Bonnar, A.: 1944, *The Catholic Doctor,* Burns Oates & Washbourne, London, U. K.

6. Coppens, C.: 1897, *Moral Principles and Medical Practice,* 3rd ed., Benziger Brothers, New York, NY.
7. Cox, H.: 1966, *The Secular City,* Macmillan, New York, NY.
8. Cronin, D. A.: 1958, *The Moral Law in Regard to the Ordinary and Extraordinary Means of Conserving Life,* Typis Pontificiae Universitatis Gregorianiae, Rome, Italy.
9. Dubler, N. N.: 1993, 'Working on the Clinton Administration's Health Care Reform Task Force', *Kennedy Institute of Ethics Journal* 3 (December), 421-431.
10. Eichendorff, J.: 1958, 'Über die Folgen von der Aufhebung der Landeshoheit der Bischöfe und der Klöster in Deutschland', in *Werke und Schriften,* Cotta-sche, Stuttgart, Germany, vol. 4, 1133-1184.
11. Eliot, T. S.: 1982, *The Idea of a Christian Society,* Faber and Faber, London, U. K.
12. Engelhardt, H. T. Jr. and Spicker, S. F. (eds): 1975, *Evaluation and Explanation in the Biomedical Sciences,* D. Reidel, Dordrecht, The Netherlands.
13. Engelhardt, H. T. Jr.: 1986, *The Foundations of Bioethics,* Oxford University Press, New York, NY.
14. Engelhardt H. T. Jr. and Caplan, A. L. (eds.): 1987, *Scientific Controversies,* Cambridge University Press, New York, NY.
15. Engelhardt, H. T. Jr.: 1991, *Bioethics and Secular Humanism: The Search for a Common Morality,* Trinity Press International, Philadelphia, PA.
16. Engelhardt, H. T. Jr.: 1994, 'Sittlichkeit and Post-Modernity: An Hegelian Reconsideration of the State', in *Hegel Reconsidered: Beyond Metaphysics and the Authoritarian State,* H. T. Engelhardt, Jr. and T. Pinkard (eds.), Kluwer, Dordrecht, The Netherlands.
17. Engelhardt, H. T. Jr.: 1995, 'Moral Content, Tradition, and Grace: Rethinking the Possibility of a Christian Bioethics', *Christian Bioethics* 1 (March), 29-47.
18. Engelhardt, H. T. Jr.: 1996, *The Foundations of Bioethics,* 2nd ed., Oxford University Press, New York, NY.
19. Ficarra, B. J.: 1951, *Newer Ethical Problems in Medicine and Surgery,* Newman Press, Westminster, MD.
20. Finney, P. A.: 1922, *Moral Problems in Hospital Practice,* 2nd ed., Herder, St. Louis, MO.
21. Fletcher, J.: 1954, *Morals and Medicine,* Princeton University Press, Princeton, NJ.
22. Gilbert, A. D.: 1980, *The Making of Post-Christian Britain,* Longman, London, U. K.
23. Gregory, Dr.: 1770, *Observations on the Duties and Offices of a Physician,* Strahan, London, U. K.
24. Hauerwas, S.: 1974, *Vision and Virtue,* Fides, Notre Dame, IN.
25. Hauerwas, S.: 1977, *Truthfulness and Tragedy*, University of Notre Dame Press, Notre Dame, IN.
26. Healy, E. F.: 1956, *Medical Ethics,* Loyola University Press, Chicago, IL.
27. Heidegger, M.: 1976, 'Brief über den "Humanismus"' in *Wegmarken,* Vittorio Klosterman, Frankfurt/Main, Germany.

28. Hoeveler, J. D., Jr: 1977, *The New Humanism,* University Press of Virginia, Charlottesville, VA.
29. Jakobowits, I.: 1959, *Jewish Medical Ethics,* Bloch, New York, NY.
30. Kelly, D. F.: 1979, *The Emergence of Roman Catholic Medical Ethics in North America,* Edwin Mellen, New York, NY.
31. Kelly, G.: 1958, *Medico-Moral Problems,* Catholic Hospital Association, St. Louis, MO.
32. Kuhn, T.: 1962, *The Structure of Scientific Revolutions,* University of Chicago Press, Chicago, IL.
33. Küng, H.: 1961, *The Council, Reform and Reunion,* Sheed and Ward, New York, NY.
34. La Rochelle, S. A. and Fink, C. T.: 1944, *Handbook of Medical Ethics,* M. E. Poupore (trans.), Newman Book Shop, Westminster, MD.
35. Lerner, D.: 1962, *The Passing of Traditional Society,* Free Press, Glencoe, Illinois.
36. Lyotard, J. - F.: 1984, *The Postmodern Condition,* G. Bennington and B. Massumi (trans.), Manchester University Press, Manchester, U. K.
37. Marx, K. and Engels, F.: 1960, *The German Ideology,* International Publishers, New York, NY.
38. McFadden, C. J.: 1946a , *Medical Ethics,* Davis, Philadelphia, PA.
39. McFadden, C. J.: 1946b, *Medical Ethics for Nurses,* Davis, Philadelphia, PA.
40. MacIntyre, A.: 1981, *After Virtue,* University of Notre Dame Press, Notre Dame, IN.
41. MacIntyre, A.: 1988, *Whose Justice? Which Rationality?* University of Notre Dame Press, Notre Dame, IN.
42. Motzkin, G.: 1992, *Time and Transcendence: Secular History, the Catholic Reaction, and the Rediscovery of the Future,* Kluwer, Dordrecht, The Netherlands.
43. National Commission for the Protection of Human Subjects of Biomedical and Behavioral Research: 1975, *Research on the Fetus,* HEW, Washington, DC.
44. National Commission for the Protection of Human Subjects of Biomedical and Behavioral Research: 1976, *Research Involving Prisoners,* HEW, Washington, DC.
45. National Commission for the Protection of Human Subjects of Biomedical and Behavioral Research: 1977, *Report and Recommendations on Psychosurgery*, HEW, Washington, DC.
46. National Commission for the Protection of Human Subjects of Biomedical and Behavioral Research: 1977, *Psychosurgery* (Appendix), HEW, Washington, DC.
47. National Commission for the Protection of Human Subjects of Biomedical and Behavioral Research: 1977, *Research Involving Children,* HEW, Washington, DC.
48. National Commission for the Protection of Human Subjects of Biomedical and Behavioral Research: 1978a, *Research Involving Those Institutionalized as Mentally Infirm,* HEW, Washington, DC.
49. National Commission for the Protection of Human Subjects of Biomedical and Behavioral Research: 1978b, *The Belmont Report: Ethical Principles and Guidelines for the Protection of Human Subjects of Research,* HEW, Washington, DC.
50. Niethammer, F. I.: 1808, *Der Streit des Humanismus und Philanthropismus in der Theorie des Erziehungsunterrichts unserer Zeit,* Jena, Germany.

51. Osterlen, F.: 1855, *Medical Logic,* G. Whitney (ed. and trans.), Sydenham Society, London, U. K.
52. Pellegrino, E. D.: 1979, *Humanism and the Physician,* University of Tennessee Press, Knoxville, TN.
53. Pellegrino, E. D. and McElhinney, T.: 1982, *Teaching Ethics, the Humanities, and Human Values in Medical Schools,* Society for Health and Human Values, Washington, D.C.
54. Percival, T.: 1803, *Medical Ethics,* Russell, Manchester, U. K.
55. Potter, V. R.: 1971, *Bioethics, Bridge to the Future,* Prentice-Hall, Englewood Cliffs, NJ.
56. Ramsey, P.: 1970a, *Fabricated Man,* Yale University Press, New Haven, CT.
57. Ramsey, P.: 1970b, *The Patient as Person,* Yale University Press, New Haven, CT.
58. Reich, W. T. (ed.): 1978, *Encyclopedia of Bioethics,* Free Press, New York, NY.
59. Rüdiger, H.: 1937, *Wesen und Wandlung des Humanismus,* Hoffmann & Campe, Hamburg, Germany.
60. Saundby, R.: 1902, *Medical Ethics: A Guide to Professional Conduct,* Wright, Bristol, U. K.
61. Schmemann, A.: 1990, *Liturgy and Tradition,* Thomas Fisch (ed.), St. Vladimir's Seminary Press, New York, NY.
62. Secundy, M. G.: 1994, 'Strategic Compromise: Real World Ethics', *The Journal of Medicine and Philosophy* 19 (October), 407-17.
63. Smith, H.: 1970, *Ethics and the New Medicine,* Abingdon Press, Nashville, TN.
64. Szumowski, W.: 1949, 'La Philosophie de la médecine, son histoire, son essence, sa dénomination et sa définition', *Archives Internationales d'Histoire des Sciences* 9, 1138.
65. White House Domestic Policy Council: 1993, *The President's Health Security Plan,* Times Books, New York, NY.

HENK TEN HAVE

FROM SYNTHESIS AND SYSTEM TO MORALS AND PROCEDURE: THE DEVELOPMENT OF PHILOSOPHY OF MEDICINE

I. INTRODUCTION

Anybody interested in the philosophy of medicine nowadays will have a difficult time identifying up-to-date literature. In the internationally renown journals in this area, the focus is almost exclusively on bioethics. What was welcomed as a revival of the philosophy of medicine twenty years ago, has more and more restricted itself to the domain of values and moral issues. The work of philosophers and critics, such as Foucault, Illich and Van den Berg, who at that time had a reputation for scrutinizing a broad range of issues in medicine and health care, and for developing an encompassing, albeit not generally endorsed, theoretical perspective, has not been followed by an oeuvre of similar stature.

Reflecting on the last two decades, it seems that at least three interrelated developments have contributed to the virtual invisibility of philosophy of medicine as a theoretical and practical endeavor.

The first of these is the 'ethicalization' of the philosophy of medicine. Instead of covering all the branches of philosophy in general, philosophers of medicine are increasingly engaged in bioethical studies. Many of them have renamed themselves "bioethicists." Philosophical anthropology and social philosophy are almost non-existent in relation to medicine and health, while the philosophy of medical science, medical epistemology, and medical ontology are very rare products of scientific labor.

The second phenomenon is the technicalization of ethics. If "bioethics" is the appropriate label for most of the studies undertaken in the philosophy of medicine, then the majority of bioethical studies can no longer be characterized as "moral philosophy". Bioethics is considered an autonomous discipline; its aim is to contribute to the solution of difficult dilemmas in health care. Theoretical and elemental studies criticizing the foundations of medical thinking and acting are not useful from a perspective in which bioethics is a "linguistic" technology for a specific set of practical problems.

R. A. Carson and C. R. Burns (eds.), Philosophy of Medicine and Bioethics, 105-123.

The third phenomenon is difficult to characterize. In postmodern culture moral views are usually regarded as private; our moral convictions are the result of personal choices. Since people make different choices, moral convictions also differ. Given this pluralistic situation, what is most important is respect for the moral judgments of other human beings. This feature of privatization is often associated with relativism: if different moral convictions exist, an absolute judgment concerning these convictions is impossible. The best we can do in this situation is to develop neutral rules and formal procedures that give the variety of convictions equal time. Privatization, relativism, and proceduralism as features of postmodernism all seem to imply anti-realism. Objective values and substantial moral convictions are simply impossible, in much the same way as realistic ideas about the world or nature are completely out of fashion in the philosophy of science. Within a constructivist perspective, moral values as well as disease, diagnosis, therapy, and technology are all human constructs, ideas of the ingenious human mind, instruments to create order in a man-created world. Ethicists should examine how moral values are created, maintained, challenged, and recreated. Bioethicists in particular should concentrate on analyzing scientific discoveries and technological innovations in health care, and study how these developments transform medicine's moral order. Being focused on the social construction of scientific knowledge and its ontological and moral order, ethics is part of science and technology studies. It is no longer possible to develop and apply a normative point of view, other than the one revealed through a sociological approach. Whether we can judge actions as morally good does not depend on some intrinsic quality or extrinsic norm. Programs of action that go unchallenged have themselves to be regarded as a standard for judging past and present actions [32].

If the past two decades of philosophy of medicine can be plausibly described in terms of processes of "ethicalization", technicalization and anti-realism, the question then is what this implies for philosophy of medicine's prognosis. My thesis is that the current situation, as diagnosed above, is a *specific* articulation of the philosophy of medicine that presents only a *partially correct* response to the motivations and inspirations leading to philosophizing about health care and medicine. That is, the present-day domination of philosophy of medicine by bioethics increasingly brings us to acknowledge that pragmatism is not sufficient, and that more fundamental questions must be addressed.

II. THE DISAPPEARANCE OF PHILOSOPHY OF MEDICINE

Glancing at *The Journal of Medicine and Philosophy, Theoretical Medicine,* and the *Philosophy and Medicine* book series, one is struck by the fact that the majority of contributions address issues in bioethics. It seems that philosophy of medicine has come to end, or that it has been transformed into bioethics. The argument that scientific endeavors are finite is not an uncommon one. A discipline (e.g., chemistry) can subdivide and generate various other disciplines; some disciplines (e.g., medical semiotics) may completely disappear; a discipline may merge with another creating a new one (e.g., medical informatics). However, the case of philosophy is different. Philosophy is considered to be the origin of all other sciences; once these sciences developed, what need is there for philosophy? It is also questioned whether there is real progress in philosophy; progress is generally accepted in the other sciences, but for many, Whitehead's adagium is valid, that philosophy is nothing but "footnotes to Plato". Therefore, can it be argued that the philosophy of medicine has come to an end, at a time when we also proclaim the "end of history"? Can it be argued that bioethics will also come to end as soon as major moral issues are codified and regulated by law?

The End of Philosophy

> The development of philosophy into the interdependent sciences . . . is the legitimate completion of philosophy. Philosophy is ending in the present stage. It has found its place in the scientific attitude of socially active humanity. But the fundamental characteristic of this scientific attitude is its cybernetic, that is, technological character. . . . The end of philosophy proves to be the triumph of the manipulable arrangement of a scientific-technological world and of the social order proper to this world. The end of philosophy means the beginning of the world civilization based upon Western European thinking ([12], pp. 376-77).

With these words, Martin Heidegger announced his famous thesis that we have arrived at the end of philosophy. In the course of history, philosophy has brought sciences into being but now that they have been established, they go their way independent of philosophy. There is no longer any use for philosophical thinking; the sciences are taking over what philosophy has tried to do. With the end of philosophy, Heidegger does not mean the interruption or termination, but the completion of philosophy (*Vollendung*). Philosophy is evolving into the sciences, into a form of knowledge that surpasses philosophy in its final stage.

However, Heidegger does not mean the perfection of philosophy. We have no criterion to evaluate and compare different philosophies. We cannot maintain that Hegel's philosophy is more perfect than Kant's: "Each epoch of philosophy has its own necessity" ([12], p. 375).

The thesis regarding the end of philosophy is not uncommon in the history of philosophy. In most cases philosophical activity has continued after the proclamation that philosophy has ended. Some thinkers argued that philosophy must be superseded by a new discipline such as sociology, according to Comte, or archaeology of knowledge in the view of Foucault. Indeed, these attempts give evidence of at least two things. They confirm the general pattern, described by Heidegger, of philosophy as the queen of the sciences. And they show the viability and versatility of philosophical reflection—finding new problems, inventing new methods, continuously transforming its self-interpretation.

Only rarely is it argued that philosophy comes to an end, since all the definite answers have been given. Philosophizing is no longer necessary, because we now have a philosophical system solving all the perennial riddles. Instead of seeking wisdom, we now are wise or can be wise if we want to be. This conviction is, for example, stated by Wittgenstein in the Preface to his *Tractatus*: "... the *truth* of the thoughts communicated here seems to me unassailable and definitive. I am, therefore, of the opinion that the problems have in essentials been finally solved" ([33], p. 29). He indeed gave up philosophy to work as a gardener and country schoolteacher in Austria—but only temporarily.

Growth of Knowledge

The recurrent theme of the end of philosophy throws a peculiar light on philosophy itself. In what sense is there any advance of philosophical knowledge? The question can be answered at two levels. First, it is obvious that as a matter of fact we have more philosophical knowledge available now than in the past. In the course of its history, philosophy has indeed produced ever more concepts, ideas, systems, and books. This expansion of knowledge is a fundamental characteristic of such an intellectual activity as philosophy. The situation is not different in philosophy of medicine. Another matter is how we should evaluate this accumulation of philosophical knowledge. Can we say that the state of philosophy nowadays is better, higher, or more developed than in the past? This question refers to the question of progress in philosophy.

Expansion of Philosophy

In philosophy of medicine in particular, and in philosophy in general, the number of books, journals, institutes, centers, and professionals has steadily increased. What is most striking is the growth of the literature. This product of philosophical reflection, however, shows particular aspects of the process itself. A philosophical analysis of a specific problem usually starts with a discussion of theories of other philosophers, preferably from antiquity until the present. Textbooks in philosophy of medicine, such as those by Pellegrino and Thomasma [21], Wulff, Pedersen and Rosenberg [34], von Uexküll ([30], pp. 133-162) draw largely on historical expositions. This tendency to historicism points to the fact that, with respect to philosophy, history plays an entirely different role than it does in other disciplines. Historical research reveals that the same types of problems as well as solutions seem to return in ever-changing forms and formulations. Philosophies of the past are never simply history in the sense that they can be disregarded as irrelevant or as relics of a more primitive level of thinking. This special relation between philosophy and its history brings out again the question of progress, but it also leads to an ongoing controversy over the process of philosophizing itself: what exactly is philosophy? Notwithstanding the publication of fifty volumes in the *Philosophy and Medicine* book series, and the availability of hundreds of articles in specialized journals, authors continue to question the existence and role of the philosophy of medicine ([4], pp. 67-77; [35], pp. 79-85; [19], pp. 201-205).

Such continuous, self-referential activity is paradoxical. Philosophers argue that they start with curiosity; they wonder and want to find out about the world and themselves, about suffering, disease, healing. They examine the assumptions that form the basis of other disciplines. They do not accept dogmatism and intellectual authority, not even in their own discipline and tradition. Philosophers, as Kant [14] told his pupils, do not teach philosophy but how to philosophize, not ideas to reiterate but how to think. At the same time, philosophy seems oriented towards reflecting on its own products. Students of philosophy develop, through learning names, dates, concepts, theories, and systems, into experts in philosophy, and not necessarily into philosophers. Lévi-Strauss [18], describing his philosophical education at the Sorbonne, calls it a rally through doctrines, theories, and hypotheses. In less than two years, he resigned his post as a philosophy teacher because he did not wish to repeat his lessons year after year. He left for Brazil, preferring cultural anthropology to philosophy, paradoxically to become the proponent of a new philosophical theory: structuralism. Lévi-Strauss's decision is indicative of what has always been considered distinctive of philosophy: it is an activity that cannot be delegated to another. The reflective activity is an individual process; it requires thinking for yourself, enquiring for yourself. Questions and answers from the tradition of philosophy must be formulated anew, made problematic in view of the present situation, and mastered in a personal way. Thinking for yourself implies authenticity; authentic activities can lead to new products, new philosophies. The relationship of product and process is actually quite complicated. New schools of philosophy have originated from a new method of philosophizing: Kant's transcendental method, Hegel's dialectics, and Husserl's phenomenological method. Every authentic activity of philosophizing creates its own subjects. Growth of knowledge in philosophy is less important than the activity of philosophizing itself; but new knowledge is at least a sign of an expanding activity.

Progress in Philosophy

Though there is an accumulation of philosophical knowledge, it is doubtful whether wisdom has increased in the last two thousand years. What kind of progress can we expect to make in philosophy, if there is progress at all? When there is growth in philosophical knowledge, it is not a continuous development towards a mature state, a final goal, or a conclusive system. Every great philosopher seems to start anew, building on the ruins of his predecessors' work. The main problems and their solutions recur without definitive answers that would make those questions obsolete. But this observation should not bring us to the conclusion that we only witness eternal repetition, and that there is no progress at all in philosophy [22]. This conclusion is not warranted by the activities of philosophers themselves.

First, philosophizing is not equivalent to touring an intellectual museum; in a sense it is philosophy only if it results in the production of new knowledge. The development of the sciences creates new problems; reflecting upon such problems, as is obvious in bioethics, means rethinking traditional philosophical positions, amending and refining them, or even uncovering and identifying new ones.

Second, activities to enhance the coherency and consistency of philosophical systems often do lead to philosophical progress; for example, the evolution of utilitarianism from Bentham to Mill and Sidgwick to Brandt, Hare, and Parfit.

Third, many philosophical problems are essentially insoluble, because progressive insight rather than resolution is the intended aim. In contrast with science, where progress is possible through increasing consensus, philosophy appears to thrive on diversity. Instead of describing this fundamentally incompatible series of systems and schools of thought as the "scandal of philosophy", we should take it as the hallmark of philosophy. It is not so much the fact that so many different theories and philosophical schools coexist, but the fact that other people's ideas and theories are often disregarded; this is the real scandal of philosophy ([11], pp. 99-103). Diversity and plurality testify that philosophy is perspectival. In retrospect, each time and culture confronts the same basic problems, and each answer is tentative. Progress lies in the multiplication of visions and perspectives of mankind and world. This makes us realize the complexity of the world as well as the limitations of individual reflection. If anything, this awareness can be viewed as an advance in wisdom, i.e., a better knowledge of ourselves and a better understanding of what we claim to know.

III. THE APPEARANCE OF THE PHILOSOPHY OF MEDICINE

From today's perspective it might appear that the philosophy of medicine is a very recent affair. The new interest in philosophy of medicine in the 1970s began with the idea that the two disciplines have coexisted without significant interchanges for so long. The editor of the first issue of the *Journal of Medicine and Philosophy* (March, 1976) mentioned the adventurous exploration of largely uncharted regions ([36], pp. 3-4). In the enthusiastic mood of those days it was easily overlooked that there existed in fact long and heterogeneous traditions of philosophizing about medicine. What should interest us is not so much the recent rebirth of the philosophy of medicine but the processes leading to its development in the second half of the last century.

The actual birth of the philosophy of medicine was the consequence of a reinterpretation of medicine as well as philosophy, producing a new orientation of philosophy with respect to medicine.

During the nineteenth century, socio-political processes, more than the internal evolution of scientific knowledge, led to the creation of a homogeneous and organized medical profession. But professional cohesion and uniformity of medical practice, once brought about by external forces, are in need of internal justification; therefore, these factors motivated a search for the identity of the now unified discipline. Can modern medicine claim special legal and financial protection because it is founded on a superior scientific base? What are the specific characteristics of medicine? Formulating answers to this basic question becomes the *modus operandi* of the new discipline of the philosophy of medicine.

However, a second change was required in philosophy. As will be argued later, in the same period, philosophy moderated its encompassing claims. Instead of interpreting itself as queen of the sciences, it began to look to the sciences for methodological guidance and theoretical models. Moreover, science itself came to be one of the most interesting objects of philosophical study. We should realize that a major shift in the level of reflection and analysis was involved. The traditional focus on an objective reality or natural order was replaced by a new concern with the scientific construction of reality. Instead of operating at the same conceptual level as science and examining its objects, philosophy moved to a meta-level, analyzing scientific conceptualization as its proper activity. A similar repositioning took place with regard to medicine. In the history of medicine, speculative philosophical systems have existed along with empirical and experimental approaches. Well-known examples are the medical systems propounded by idealist philosophers in nineteenth-century German medicine ([23], pp. 72-92). So-called "philosophical medicine" was felt to be competitive with the new ethos of medicine as a science. The long-standing relationship between medicine and philosophy was therefore increasingly regarded as an antagonistic one. From this period dates the negative attitude towards speculative thinking and, sometimes, towards philosophy in general—an attitude which was dominant during medicine's development into a natural science. Famous examples of such negativism are provided by Bernard and Bleuler. The best philosophic system in the opinion of Claude Bernard [1] consists in not having any. And, according to Eugen Bleuler, philosophy has produced nothing but a graveyard of theoretical systems; it is useless, having at best historical value. Medicine must not be contaminated by philosophy; it would be "a mixture of garlic and chocolate" ([2], p. 8). Contamination is dangerous because philosophy prevents the formation of realistic ideas. Bleuler even went so far as to insinuate that philosophizing itself may be a symptom of a morbid mind ([2], p. 9).

The conflict between medicine and philosophy was resolved at the end of the nineteenth century by redefining the role and object of philosophy vis-à-vis medicine. Philosophy was relocated as a science of science at another level of abstraction, not concerned any longer with medicine's object of study and intervention, but with the concepts and methods employed by medicine in studying and manipulating its objects. The outcome of this relocation was the emergence of the philosophy *of* medicine as an important intellectual discipline. Conceiving philosophy as a meta-activity, a competitor was turned into a critical ally.

Philosophy was no longer in competition with medicine; on the contrary, it was the inevitable manifestation of the need to interpret and explain the nature of medicine itself. The price paid for this pacification was that the philosophy of medicine no longer is focused on interpreting and explaining illness, suffering, and dying, but primarily concerned with medicine's conceptualization and management of disease, pain, and death.

IV. CONTINUITY AND DISCONTINUITY

The repositioning of philosophy in regard to medicine was therefore an important determinant for the recent development of philosophy of medicine. However, this does not imply that interaction between philosophy of medicine and contemporary philosophical schools was lacking. Only the aspirations and goals of philosophical reflection on medicine have changed considerably. Developing comprehensive systems was no longer a viable goal; understanding what was going on in medical praxis itself was difficult enough. Given these redefined aspirations, philosophy of medicine seems to have more continuity then we generally acknowledge.

Interaction with Philosophy in General

"Let us give up the construction of great systems; let us cultivate small philosophy which will lead to a reform of all intellectual pursuits". This programmatic remark from his inaugural lecture at the University of Warsaw was spoken by the Polish philosopher, Tadeusz Kotarbinski, more than three-quarters of a century ago. Kotarbinski (1886-1981) was one of the most prominent representatives of twentieth-century Polish philosophy. What is important here is that Kotarbinski was one of the leaders of the so-called Lwow-Warsaw school of Polish philosophy during the period between the world wars [24]. The credit for the emergence of this school is generally given to Kazimierz Twardowski (1866-1938), who, as a gifted teacher in Lwow University from 1895 to 1930, inspired many young scholars, not only in philosophy but also in other areas, such as medicine. Before World War II, his disciples held most of the chairs of philosophy in Polish universities. Some of them gained an international reputation in the 1930s. For example, the contributions of Lukasiewicz, Ajdukiewicz, and Tarski became familiar to students of mathematical logic throughout the world.

This philosophical movement itself is identified as Polish *analytical* philosophy. Many representatives of this school made original contributions to philosophy and developed their own theoretical systems, but what united them was not so much a common doctrine as a particular methodology and a specific style of philosophizing. Their activity is called 'analytical' because of the following characteristics:

> a) they aim at the clarification of language as a container and transmitter of ideas. The recognition of the special role of language in philosophy goes together with high standards of exactitude and stringent demands for conceptual clarity and the precise formulation of problems,
>
> b) they were preoccupied with meticulous and often piecemeal analyses of philosophical problems. There is a predilection for small philosophy: anatomizing particular and detailed sub-problems rather than constructing grand philosophical syntheses, and
>
> c) they were seriously concerned with epistemology. They were investigating the conditions and possibilities of acquiring knowledge of the world, and the truth-value of such knowledge. The relationship between language, logic, and reality was their central problem.

From these characteristics it can be seen that this school, although it has had little influence outside Poland, had much in common with British analytical philosophy. The founder of this school, Twardowski, is also connected with others, thereby illustrating the interdependency as well as the antagonism of the major philosophical schools of the twentieth century: analytical philosophy, phenomenology, structuralism, marxism, and critical theory. Twardowski had studied philosophy in Vienna under Franz Brentano (1838-1917) ([5], pp. 117-129).

Brentano, an impressive personality, inspired many talented pupils—notably Husserl, Meinong, and Marty—to make original contributions to philosophy. Brentano was primarily interested in the basic disciplines of philosophy: metaphysics, ethics, and epistemology. What we have said about Polish philosophy is also true for Brentano's disciples: they developed quite different doctrines: phenomenology, theory of objects, and semantics, to name a few ([9], pp. 15-28). Yet what they have in common is their philosophical attitude, their rational style of philosophizing. They hoped to develop philosophy by using rational methods, the instruments of logic, and a clear and concise language. This was their inheritance from Brentano. Disappointed by the failures of nineteenth-century speculative philosophy, Brentano was impressed by the success story of science as a way to overcome intellectual controversy, stagnation, and impotence. The true method of philosophy must be no other than that of natural science ([3]; [16], pp. 11-27; [29]). This point of view implies a concern with painstaking analysis of specific problems, but it is also concerned with two epistemological views: anti-idealism and language criticism (*Sprachkritik*). Brentano rejected idealism—that things in the world and facts about things are dependent for their existence upon the operations of a mind. He maintained the objectivity of objects of thought. Objects are recognized by the mind, not made by it. This anti-idealism he joined with criticism of natural languages. There is a problematic discrepancy between our words and *the* world. Language is misleading since it suggests that because a word exists a corresponding entity or thing exists as well. The task of philosophy is therefore therapeutic; it has to purify ordinary language and to systematically reformulate misleading expressions. In doing so, it tries to construct an ontologically ideal language.

Brentano is no doubt an important landmark in the landscape of modern philosophy. Dissatisfied with the philosophical work of the nineteenth century, he opened up new avenues for twentieth-century philosophy.

A similar landmark in the history of philosophy of medicine is Wladyslaw Bieganski (1857-1917) [20]. He can be considered as a *pars pro toto* for the development of philosophy of medicine during precisely the same period as Brentano's activities. That does not mean that one should assume the existence of a direct relationship between Brentano's and Bieganski's work; both are more or less points of orientation within the panorama of philosophy and medicine. Brentano is a symbol for a reorientation of philosophy toward the sciences, whereas Bieganski is a symbol for a similar reorientation of medicine toward philosophy.

For most of his professional life, Bieganski worked as chief physician at the local hospital of a provincial town, as factory physician, and as a general practitioner in private practice. He managed, however, to write 116 publications in philosophy and medicine. At least ten of them can be classified as major works. They explored internal medicine, infectious diseases, but also medical ethics and the logic of medicine, as well as the theory of logic, general ethics, and epistemology. But what is most remarkable is that Bieganski was not an exceptional case. In a series of historical surveys, Szumowski ([26]; [27]; [28], pp. 1097-1139) pointed out that philosophy of medicine in Poland was developed by three generations of physicians. Having completed his medical education, Szumowski himself studied philosophy in Lwow under Twardowski. Along with Bieganski and Twardowski, he was a member of the editorial board of the *Medical Weekly,* published in Lwow ([8], pp. 179-215). The Polish tradition, though very strong and institutionalized, was not exceptional either. The phenomenon of philosopher-physicians could be observed in many other countries, France and Germany in particular. Only a few of these traditions have been studied so far, for example, the Polish tradition by Löwy [20] and the anthropological movement that had some influence in German and Dutch medicine from the 1930s until the 1960s [6]; [15].

In sum, the development of the philosophy of medicine from the turn of the nineteenth century was obviously related to the vitality and the inspirational energy of contemporary philosophers. Although it is sometimes difficult to expose and to reconstruct the basic pattern of such relations, philosophy in general seems to have contributed especially to the maturation of the philosophy of medicine.

Traditions in the Philosophy of Medicine

If philosophy in general is concerned with understanding and explaining the world and mankind's place in it, then the driving force of philosophy of medicine is the need for self-interpretation in medicine. The development of philosophy of medicine over the last hundred years can be described in many ways: for example, one can focus on national and cultural traditions. But it is also possible to concentrate on conceptual structures. Doing so, we can distinguish three different traditions.

a) The epistemological tradition

Initially, from the second half of the last century to the first decades of the present one, the identity and intrinsic coherence of modern medicine is specified in *epistemological* terms: medicine is characterized as a natural science. Crucial from this perspective is the method of acquiring and applying knowledge of *disease.* Medicine can only progress through unbiased observation in the laboratory, the dissection room, and at the bedside—through collecting facts and deducing conclusions. The artistic element, the art of medicine, is set aside in this mode of interpretation. To be a healer or an expert in the art of healing, one does not need a scientific education.

The scientific conception of medicine was a very productive one. But precisely its success created problems. With more and more data accumulating and new facts multiplying, it became difficult to detect any cohesion in the rapidly growing body of medical knowledge. It became hard to believe in a comprehensive identity of medicine given the proliferation of independent specialties. It was felt that the unity and coherence of medicine were endangered by its own development. When medicine became a *natural* science, it seems to have lost its identity precisely because of its *scientific* progress.

This problem is well paraphrased in contemporary philosophy of medicine by the slogan "medical synthesis". According to this view, the problems of medicine are twofold: first, medical knowledge is fragmented and medical practice is one-sided because of specialization; second, the patient as the object of medicine is no longer adequately addressed since the conceptual tools of medicine are insufficient and too simple ([37], pp. 453-456). The solution to these problems is *synthesis* in the realm of medical practice as well as in the field of medical thinking. Many agree that philosophy should provide a solution. Philosophy can distinguish between relevant and irrelevant knowledge, trivial and crucial findings, so that it uncovers the general framework and the core concepts of medicine. This philosophical activity of sorting out and weighing data is particularly important for medical education. Medicine should be presented within a broader perspective during philosophical propaedeutics for every medical student, as Szumowski [27] proposed.

An example of the philosophical issues with which philosophers of medicine in this phase have been concerned can be found in the work of Edmund Biernacki ([20], p. 49 ff). As for most of his colleagues in this period, the central problem for Biernacki was the inadequate and problematic relationship between medical science and the art of healing. However, his outline of synthetic medicine proceeds from a radical separation between the theory and practice of medicine. The problem originates from the fact that medicine is composed of two different disciplines. Instead of attempting to reconcile the impossible, we should clearly separate the science of disease on the one hand and the art of healing on the other. Such separation is fruitful for practice, since diagnosis is based on the first and treatment on the second. Through such separation, we will no longer confuse medical science and medical practice; it will also enable us to reverse the usual relationship between the two disciplines. Priority is commonly given to diagnosis, postulating the idea that there are disease units in reality; only when the diagnosis is clear, is therapy thought to be rationally possible. Biernacki wants to eliminate this order of priority: therapy should not be based on diagnosis. The core of medicine is the art of healing; therapy, therefore, is the essence of medical work. The science of disease can be helpful, but therapy involves more than scientific knowledge. Curing diseases is possible, and has always been possible, even if knowledge about diseases is incomplete.

At a more fundamental level, with the unity of medicine now in jeopardy, two different epistemological strategies were deployed to respond to the need for synthesis.

The first was an attempt to strengthen the conceptualizing grip of the knowing subject on his world. Proliferating knowledge can be organized and understood as long as the epistemological subject is applying a rigorous methodology.

Synthesis, therefore, is not a goal but a method. Such a project requires a commitment to analytic study and to the search for interconnections among the results, cultivating synthesis not by speculating but by reducing the multiplicity of phenomena to basic laws and fundamental processes.

The second strategy was questioning the status of the knowing subject itself. Instead of encouraging the subject to advance methodically towards objectivity and preciseness, it called for appreciating its own subjectivity. Central to this strategy was the self-interpretation of medicine as an art. The German physician, Richard Koch, for example, defined medicine as *Heilkunst*, "a healing art" ([17], p. 30). If medicine is characterized as an art, special attention must be given to the character of the doctor. For it is the personal qualities of the artist that determine his achievements. The conclusion to this philosophical position is summarized by the German physician-philosopher Georg Honigmann: "Being a physician means being a whole human being" ([13], p. 303). It was recognized, however, that medicine was more concerned with *acting* than *knowing*. If the personal qualities of the physician are so important for the accomplishments of medicine, serious doubts could easily arise concerning the central role of epistemological issues in medicine.

b) The anthropological tradition

The interpretation of medicine as an art evolved into a new conception of medicine: anthropological medicine. It was popular, particularly in Germany and the Netherlands, from approximately 1930 until 1960, the principal figures being Buytendijk, Von Weizsäcker, and Von Gebsattel ([10], pp. 1-42; [31], pp. 133-162). The majority of its protagonists were physicians working in clinical practice or laboratory research. What is important here is the strong tendency to *idealize* the doctor-patient relationship. The programmatical demand of Von Weizsäcker, for example, to introduce the *subject* into the life sciences and medicine implies not only acknowledging the subjectivity of the knowing and acting physician, but also that of the patient. Doctor as well as patient experience themselves as psychosomatic organisms. This means, for example, that medicine should examine the relation between personal biography and illness. In short, the identity of medicine is constituted through internal determinants, e.g., the personal qualities of its practitioners, but, more significantly, also by external determinants, e.g., the individual qualities of patients. Medicine is a unique profession in that it systematically and methodically attends to the patient as person.

c) The ethical tradition

It is surprising that this anthropological orientation has been so rapidly superseded by the ascending interest in bioethics since the 1960s. After all, there is a clear historical continuity between these two orientations. By concentrating on the subjectivity of the patient, anthropological medicine made visible, so to speak, the moral dimension of medicine. It did so by criticizing the presuppositions underlying the dominant conception of medicine as a natural science, and by incorporating analytical methods and the prevailing mechanistic image of mankind in a broader framework, an authentic science of humanity *(Geisteswissenschaften).*

The point to stress is that anthropological medicine demonstrated that the internal characteristics of medicine were not identifiable unless one involved the external context. In fact it is clear that anthropological medicine is itself a *normative* program, an extended argument for a normative science of personal life. Medicine's understanding of mankind is limited so long as it concentrates on the *physiological* aspects of life. But it was through the hermeneutical *deconstruction* of the intrinsically normative dimension of medicine (the latter truly essential to its identity), that the anthropological tradition has led to more recent interest in the moral issues in medicine.

V. PHILOSOPHY OF MEDICINE RECONSIDERED

It is characteristic of the philosophy of medicine today that it explicitly encompasses all these perspectives, as Spicker ([25], pp. 163-175) has illustrated. Spicker also points out that something has radically changed in the meta-position of philosophy with respect to medicine. Analyzing the recent evolution of the philosophy of medicine leads to the conclusion that the identity of medicine is no longer solely based upon internal characteristics (the special kind of knowledge, the art-like application of knowledge, or the personal qualities of its practitioners), but also upon external determinants (such as the personal qualities of patients). But that insight translates into the view that medicine itself is part of society and culture; indeed, one of the most powerful expressions of culture. Dominating cultural values (such as our preoccupation with immortality and our existential invulnerability) are clearly articulated in contemporary medicine. The question about what medicine is essentially, can no longer be answered by medicine itself. The response requires attention to the impact of ideas and values that prevail in contemporary societies and cultures. Any distinction between medical and other issues seems to be the result of complicated social negotiations, cultural preferences, and, especially today, economic factors.

The increasingly *external* determination of the identity of medicine has at least two fascinating effects:

First, it has changed the relationship of medicine and philosophy. When medicine is primarily directed by cultural values and social influences, it is no longer an activity of a select group that has its own philosophical meta-domain. The recurrent debate over the last twenty-five years (whether or not there exists a discipline called 'Philosophy of medicine') is symptomatic of these changes. The claim that philosophy of medicine should be regarded as a *subdiscipline* of philosophy is particularly significant. Not too long ago, simply explicating the self-understanding of medicine would have been regarded as part of medicine itself.

Second, since medicine is fundamentally embedded in culture, it cannot be understood without attention to cultural values. If medicine is formally characterized as an applied, practical science, then what it materially implies is determined by the ends of *knowing* and *acting*. But these ends are intrinsically fragile and constituted in a complex process of social consensus. Under the influence of economic constraints, the process of establishing medicine's ends is now receiving even greater attention. This surely involves the setting of limits to medicine. These issues, however, require a preliminary philosophical examination of the meaning of our actions. As Engelhardt argues in the decennial issue of the *Journal of Medicine and Philosophy*: "With a science and technology as powerful as medicine, one needs to decide what among all the things medicine can do, should be done and under what circumstances" ([7], p. 5).

It would appear that from our standpoint epistemological, anthropological, and ethical perspectives are all indispensable to address the new problems. Instead of stressing the *objectivity of medical knowledge,* today the *value of knowledge* is of paramount importance. Instead of attending to the existential meaning of illness, the social meaning of medical action is now seen as in need of clarification. Today's agenda for philosophy of medicine is in transition. The substantial question is not "What do we know?" but "What do we want to do with our knowledge?"

At present, medicine is identified as a complex of meaningful knowledge and actual practices. But this is a fluid identity. To determine that identity more precisely, we must take a closer look at the societies and cultures that influence this identity. Such a task requires a critical comparison of national traditions and their cultural values if we are to understand how medicine will be transformed by a variety of philosophical perspectives.

Department of Ethics, Philosophy and History of Medicine,
Catholic University of Nijmegen,
Nijmegen, The Netherlands

REFERENCES

1. Bernard, C.: 1865, *Introduction à l'étude de la médecine expérimentale,* Baillière, Paris, France.
2. Bleuler, E.: 1921, *Naturgeschichte der Seele und ihres Bewusztwerdens. Eine Elementarpsychologie,* Springer Verlag, Berlin, Germany.
3. de Boer, Th.: 1989, *Van Brentano tot Levinas. Studies over de fenomenologie,* Boom, Meppel/Amsterdam, The Netherlands.
4. Caplan, A.L.: 1992, 'Does philosophy of medicine exist?' *Theoretical Medicine* 13, 67-77.
5. Dambska, I.: 1978, 'François Brentano et la pensée philosophique en Pologne: Casimir Twardowski et son école', in R. M. Chisholm and R. Haller (eds.): *Die Philosophie Franz Brentanos,* Rodopi, Amsterdam, The Netherlands.
6. Dekkers, W. J. M.: 1985, *Het bezielde lichaam.Het ontwerp van een antropologische fysiologie en geneeskunde volgens F.J.J.Buytendijk,* Kerckebosch, Zeist, The Netherlands.
7. Engelhardt, H. T.: 1986, 'From philosophy *and* medicine to philosophy *of* medicine', *The Journal of Medicine and Philosophy* 11, 3-8.
8. Giedymin, J.: 1986, 'Polish philosophy in the inter-war period and Ludwik Fleck's theory of thought-styles and thought-collectives', in R. S. Cohen and T. Schnelle (eds.): *Cognition and Fact—Materials on Ludwik Fleck,* D. Reidel, Dordrecht, The Netherlands.
9. Haller, R.: 1985, 'The philosophy of Hugo Bergman and the Brentano School', *Grazer Philosophische Studien* 24, 15-28.
10. ten Have, H. and van der Arend, A.: 1985, 'Philosophy of medicine in the Netherlands', *Theoretical Medicine* 6, 1-42.
11. ten Have, H., Bergsma, J. and Broekman, J.: 1987, 'Preface', Thematic Issue 'The philosophical foundations of medical practice', *Theoretical Medicine* 8, 99-103.
12. Heidegger, M.: 1978, *Basic Writings,* D. F. Krell (ed.), Routledge and Kegan Paul, London and Henley, U. K.
13. Honigmann, G.: 1924, *Das Wesen der Heilkunde. Historisch-genetische Einführung in die Medizin für Studierende und Aerzte,* Felix Meiner Verlag, Leipzig, Germany.
14. Kant, I.: 1964, 'Nachricht von der Einrichtung seiner Vorlesungen in dem Winterhalbjahre von 1765-1766', *Werke* II, Suhrkamp, Frankfurt am Main, Germany.
15. Kasanmoentalib, S.: 1989, *De Dans van Dood en Leven. De Gestaltkreis van Viktor von Weizsäcker in zijn wetenschaps historische en filosofische context,* Kerckebosch, Zeist, The Netherlands.

16. Katkov, G.: 1978, 'The world in which Franz Brentano believed he lived', in R. M. Chisholm and R. Haller (eds.): *Die Philosophie Franz Brentanos,* Rodopi, Amsterdam, The Netherlands.
17. Koch, R.: 1930, 'Philosophische Grenzfragen der Medizin. Der Begriff der Medizin', *Vorträge des Instituts für Geschichte der Medizin an der Universität Leipzig,* G.Thieme Verlag, Leipzig, Germany, vol. 3, pp. 9-31.
18. Lévi-Strauss, C.: 1955, *Tristes Tropiques,* Librairie Plon, Paris, France.
19. Loewy, E. H.: 1994, 'Philosophy and its role in medicine: Inaugurating a new section', *Theoretical Medicine* 15, 201-205.
20. Löwy, I.: 1990, *The Polish School of Philosophy of Medicine: From Tytus Chalubinski (1820-1889) to Ludwik Fleck (1896-1961),* Kluwer, Dordrecht, The Netherlands.
21. Pellegrino, E.D. and Thomasma, D.C.: 1981, *A Philosophical Basis of Medical Practice. Toward a Philosophy and Ethic of the Healing Professions,* Oxford University Press, New York and Oxford, U. K.
22. Plessner, H.: 1946, *Is er vooruitgang in de wijsbegeerte?* Wolters-Noordhoff, Groningen-Batavia, The Netherlands.
23. Risse, G. B.: 1976, '"Philosophical" medicine in nineteenth-century Germany: An episode in the relations between philosophy and medicine', *Journal of Medicine and Philosophy* 1, 72-92.
24. Skolimowski, H.: 1967, *Polish Analytical Philosophy: A Survey and a Comparison with British Analytical Philosophy,* Routledge & Kegan Paul, London, U. K.
25. Spicker, S.F.: 1990, 'Invulnerability and medicine's "promise" of immortality: Changing images of the human body during the growth of medical knowledge', in H. A. M. J. ten Have, G.K. Kimsma and S.F. Spicker (eds.), *The Growth of Medical Knowledge,* Kluwer Academic Publishers, Dordrecht, The Netherlands.
26. Szumowski, W.: 1933, *Coup d'oeil sur l'évolution de la médecine en Pologne,* Cracovie, Drukarnia Narodowa, Poland.
27. Szumowski, W.: 1937, *L'Histoire de la médecine et la réforme des études médicales,* Cracovie, Librairie Gebethner & Wolff, Poland.
28. Szumowski, W.: 1949, 'La philosophie de la médecine, son histoire, son essence, sa dénomination et sa définition, *Archives Internationales d'Histoire des Sciences* 2, 1097-1139.
29. Taljaard, J.A.L.: 1955, *Franz Brentano as wysgeer. 'N Bydrae tot die kennis van die Neo-positiwisme,* T.Wever, Franeker, The Netherlands.
30. von Uexküll, T.: 1988, *Theorie der Humanmedizin. Grundlagen ärztlichen Denkens und Handelns,* Urban & Schwarzenberg, München/Wien/Baltimore, MD.
31. Verwey, G.: 1990, 'Medicine, Anthropology and the Human Body, in H.A.M.J. ten Have, G.K. Kimsma and S.F. Spicker (eds.), *The Growth of Medical Knowledge,* Kluwer, Dordrecht, The Netherlands.
32. Wackers, G.L.: 1994, *Constructivist Medicine,* Universitaire Pers Maastricht, Maastricht, The Netherlands.
33. Wittgenstein, L.; 1922, *Tractatus Logico-Philosophicus,* Routledge and Kegan Paul, London, U. K.

34. Wulff, H.R., Pedersen, S.A. and Rosenberg, R.: 1986, *Medical Philosophy*, Blackwell, Oxford, U. K.
35. Wulff, H.R.: 1992, 'Philosophy of medicine—from a medical perspective', *Theoretical Medicine* 13, 79-85.
36. Zaner, R.M.: 1976, 'Toward a philosophy of medicine' (editorial), *Journal of Medicine and Philosophy* 1, 3-4.
37. Zimmerman, H.: 1934, 'Ueber Aerztliche Synthese', *Klinische Wochenschrift* 13, 453-456.

STUART F. SPICKER

THE PHILOSOPHY OF MEDICINE AND BIOETHICS: COMMENTARY ON TEN HAVE AND ENGELHARDT

I

Since the early '70s, an area of intellectual reflection and publication has slowly re-emerged that has focused not only on bioethical issues in health care but more generally on other conceptual, i.e., philosophical issues in medicine—although Henk ten Have is surely correct when he points out that most of these issues have never quite made it out of the shadows as rapidly as the bioethical problems have come into the ascendancy; in his words, contemporary "bioethics is a biased articulation of the philosophy of medicine." Indeed, if one turns to the literature of the late-nineteenth century, one discovers a plethora of books and articles (even a journal titled *Medical Critique* [*Krytyka Lekarska*], published between 1897 and 1907) that contain the writings of a school of Polish physician-philosophers. However, the generic title for this turn-of-the-century Polish school is more likely filed under the rubric "philosophical anthropology" rather than the philosophy of medicine. Nevertheless, ten Have calls our attention to the writings of this Polish school of physician-philosophers, who were, as he puts it, "preoccupied with meticulous and often piecemeal analyses of philosophical problems" rather than dedicated to "constructing huge philosophical syntheses."

Here is not the place to recapitulate ten Have's salient points, but simply to attend to a claim and a question that in my view are worth underscoring if his reflections are to prove useful for further research:

Claim:

Specific philosophical (i.e., nonbioethical) problems were frequently confronted by members of the Polish school of the philosophy of medicine. Indeed, ten Have points out that these problems are the "same types of problems as well as solutions" with which we today are concerned. This is especially important since he claims that there is progress in philosophy, and philosophical "plural-

R. A. Carson and C. R. Burns (eds.), Philosophy of Medicine and Bioethics, 125-133.

ism"—evidenced by multiple theories and schools—ought to be properly celebrated rather than lamented or deplored.

Furthermore, ten Have's tripartite classification of the historical moments of the philosophy of medicine—what he calls the epistemological, anthropological, and ethical traditions, from the 1850s to the present day—is surely useful; but more importantly, he seems to be on the right track when he calls our attention to the "central problem" of Edmund Biernacki: the confusion among Biernacki's contemporaries concerning the relationship between medical science and medical practice. Even today, the tendency is to stress the distinction between theory and practice, between the "science of disease" (and its relevance to diagnosis) and the "art of healing" (the selection of treatments). Strikingly, ten Have reminds us, Biernacki gave priority to therapy, not diagnosis—emphasizing doing over knowing—since doing or treating "involves more than scientific knowledge." The implications of this turnabout are surely worth pursuing, since Biernacki is not merely suggesting that in many cases physicians act properly when they initiate treatment modalities *before* they fully understand the disease processes at work in the patient.

Since ten Have complains about the excessive or inappropriate attention paid today to the bioethical, it is appropriate that he emphasize the role of cultural values, social influences, the attention to narrative and story, and the rather recent emphasis on the "acting subject" and medical *practice* in contrast to cognition, differential diagnostic reasoning, and *theorizing* on the part of physicians.

Question:

To what end is it useful to proffer a set of options concerning possible "prognoses" with respect to the future of the philosophy of medicine? Perhaps ten Have's question is rhetorical: Does he believe that the philosophy of medicine will

(1) come (indeed, quite soon) to an end?
(2) become transformed into bioethics?
(3) become subdivided into the philosophy of medicine and bioethics?
(4) become codified as positive law?
(5) be replaced by a further preoccupation with medical technology and the biomedical sciences (*viz.*, Heidegger)?
(6) become identified with the projects of medical sociology and anthropology (undertaken by those concerned with the social meaning of medical actions)?
(7) reflect a combination of 1-6 noticed above?

The answer to this question clearly requires further reflection.

II

As we noted, the past two decades have witnessed a striking cultural phenomenon: as H. Tristram Engelhardt, Jr., has observed (reminding us of Edmund Pellegrino's earlier observations), there has been a renewed social interest in medicine and the health professions that has occasioned a revitalization of and added new intellectual vigor to the activity of philosophizing itself. Or, as ten Have has remarked, recent interest in the philosophy of medicine reflects an "historical continuity." Furthermore, contemporary medicine helped stimulate the creative energies of philosophical reflection by furthering the scholarly analysis of concepts drawn from medicine and health care more generally (often unexplored in the literature of theoretical and clinical medicine) that showed themselves to be significant to cultures like our own, cultures in which healing (if not curing) was and remains an on-going, humane activity. It is also well known that medicine's recently renewed interest in philosophy has led to a disciplined, reflective appreciation of philosophical speculation on the part of physicians and other health care professionals. Indeed, courses on bioethical issues that include other conceptual issues in medicine are now integrated into United States and European medical school curricula; medical students and their physician mentors now have a significantly greater appreciation of bioethical problems than was the case over twenty-three years ago when I was appointed to the Faculty of Medicine of the University of Connecticut. In short, a continuous though somewhat "invisible" development has culminated in a renewed interest in the philosophy *of* medicine, that branch of philosophical inquiry which, ten Have correctly observes, now requires a careful assessment if we wish to offer a reasonably sound prognosis of its future as we transition to the twenty-first century.

III

Rather than focus our attention on the details of Engelhardt's version of the "story of the West," it may prove useful to mention his general strategy: once he lifts the Veil of Maya and reveals the unhappy truth of our present postmodern circumstance, we can come to appreciate truly the fact that (although he does not take joy in or celebrate his principal conclusion), unlike our predecessors, who could far more easily arrive at consensus on serious theological and moral issues—like the morality and meaning of the intended interruption of pregnancy or ideal ways of dying—our present predicament is precisely that—a predicament! As he says, we who live in the postmodern period are condemned to mere "babble." Could any worse conclusion be drawn before a readership that includes philosophers and physicians? Indeed, a second one comes to mind: we, who live in postmodernity, are compelled not only to "embrace modernity" but inescapably to confront the failure of reason to guide our *conduct* (to be distinguished from the psychologist's broader concern with *behavior*). In Engelhardt's words, the Enlightenment project has failed. Even worse, there is a third conclusion: we are compelled to confront the fact that we have been duped; moreover, many of us continue in this foolishness: we continue to believe that we can rely on "the grand narrative of the past" to guide our moral conduct, and we fail to see that we no longer possess a "common paradigm." Indeed, all such paradigms, including those paradigms proffered for the analyses of issues in medical ethics, are "fractured," since this follows logically from the fact that we cannot at all rely on "a universally valid moral perspective." Finally, we are told that we really do not even share, as we continue to delude ourselves in assuming, "a coherent community of research."

Although his strategy as well as his line of argument is original and impressive, I am *not yet* convinced of one of his principal conclusions: that we cannot any longer rely on "a universal secular" basis for resolving, for example, a large number of bioethical dilemmas; moreover, I believe his distinction between "moral friends and "moral strangers" is overdrawn and misleading—that is, following his strategy, he divides us by suggesting that we are impotent due to "substantive differences"—iterating MacIntyre's appeal to the "international languages of modernity" that separate us even more sharply than my fence comfortably separates me from my Bahai neighbor. Why can't we think (and act) our way through to the view that we are all citizens of *everywhere*, rather than nowhere, or a particular somewhere? In short, I ask the following:

Question 1: Has the Enlightenment project truly failed? To claim that it has, of course, requires further argument. Does this mean that any appeal to what we take to be the structure of sound, rational argument is simply based on another confusion? I do not believe that we are easily convinced of any "failure," given this sense of reason, though we surely should appreciate the *limits* of reason, especially when we appeal to the "political construction of agreement" as we often do, and as is typically reflected in the process of bureaucratic bioethics.

Question 2: Is there nothing *sacred* in the secular—in law-governed, democratic societies? Can't a secular Jew have a more than adequate, practical, consensual (perhaps sacred) understanding with his neighbor who is devoted to living by the canonical views of Greek Orthodoxy? Can't we begin with mutual "communication and collaboration" in the marketplace? Are secular Jew and Greek Orthodox Catholic, physician and patient, really "moral strangers" as the former stands at the bedside of the latter, the sick and dying patient? Finally, do we *all* need to be bound together in a single moral community, after all? Put another away, why give up so easily on the moral importance of a "bureaucratic governmental bioethics" that might serve, under positive law and regulation, to provide a sufficient but not perfect moral vision "binding all"? To be sure, this may not prove to be "the secular equivalent to a moral theology," but it may indeed "provide reliable moral guidance," as does the law and even equitable regulation. Why, then, is this assumed to be morally trivial? [NB: Here, I should point out, I am neither defending the so-called principles-based bioethical theory nor other moral theories, for that matter, theories especially familiar to this readership; rather, I am simply calling for caution in concluding, as Engelhardt does, that there are devastating implications from the fact that people make "divergent theoretical commitments" when they pay allegiance to the canonical views of a wide variety of nonsecular institutions.] In short, there is perhaps more to learn from the recent "shift of interest in bioethics from a focus on religious bioethics to secular bioethics." And if so, need this be sold as the "solution for modernity"? Have any of the world's religions provided such a solution during premodernity? Does Engelhardt not simply lay another trap that yawns for those of us who are naïve, but not displaying quite the same naïveté he described earlier? Since pluralism is, as he suggests, an "essential ingredient" of the secular, can we not reply that it is also an essential ingredient of the nonsecular? What's philosophically to be gained by an appeal to a *poly-nonsecularism* in contrast to a *poly-secularism*?

IV

I asked: Is there nothing sacred in the secular? Do various types of interventions (with consent) or invasions (without consent) into the human body, for example, suggest that our lived bodies are sacred, that there are morally licit and illicit choices of actions, beyond forbearance actions? Does our autopoetic, organic life (the "lived-body" of the phenomenologists) suggest a universally common (if not quite shared) moral starting point? Can we find *the* moral locus in permission or consent with respect to interventions bearing on our lived bodies, especially when harmed? Since Engelhardt calls for a "reconceptualization" does *this* not merit pursuit? If not, why not? [Here I note that I am not sounding a retreat to other mid-level principles as proffered, for example, by Tom L. Beauchamp and James F. Childress in their *Principles of Bioethics* [1], but urging the exploration of other approaches to ground *shared* moral commitments that could conceivably unite us. [Again, note here that I am not suggesting an approach similar in any way to those who appeal to some version of the principle or virtue of *solidarity*, a term principally current in Europe to connote an effective social bonding or morality of all humanity.] Borrowing Wittgenstein's notion of "family resemblance," however, might we not consider the lived-body of *anthropos* (with its various properties, features, and vulnerabilities) as the "metaphysical locus," so to speak, of what we have in common, of what could provide the ground for an "overlapping moral consensus"? Here I call attention to the practice of female genital "circumcision" ("mutilation" to those of us in the West) that is endorsed and permitted in a number of cultures, a concern that Loretta Kopelman discusses in this volume. The critics of this practice are not, in my view, simply appealing to our "moral sentiments" or "inclinations." The French expresses it best, but still in slang: *"Le centre du monde est le con de la femme"* (this is by no means simply a cryptic allusion to ontic pleasure). [Note that neither English nor French has a word to denote the inviolability of the woman's body; only slang prevails: *'le con'* (masculine gender!), as well as words equivalent to 'vagina,' and 'uterus'. Even 'womb' (*'la matrice'*) has a medical denotation, and *'le sein'*, usually translated 'breast', sometimes denotes the genital interior of the woman's body.] In short, can one locate good and evil in harmful acts of invasion and/or omissions or refrainings with respect to the lived-body? Indeed, who better to address this concern than Engelhardt? After all, he has dedicated himself to the philosophical exploration of morbidity, infirmity, disability, pain, suffering, and dying, though he stresses the point that persons in different cultures report their experiences somewhat differently.

To summarize this concern: Engelhardt has not yet convinced me that "the project of discovering a normative consensus in general secular terms will *always* fail" (my emphasis), though he has, I believe, shown us that it has failed thus far. However, I have suggested that perhaps moral authority can be located elsewhere: in the integrity of the lived-body. Again, what I have in mind is not simply an ethics of caring, nor is there any obligation for those who dwell in the secular world to pay allegiance to the canons of a particular nonsecular community. Moreover, there is no need at this time to abandon the search (the rational search, by the way) for what we essentially have in common, if not truly share. If *this* approach proves successful, then it *would* follow (in contrast to Engelhardt's conclusion) that not only would the "larger society" be identical to our moral community, but the converse would also be true.

Having raised these questions, and having offered another approach to their resolution, let me be clear that I do not anticipate that we shall ever quite achieve either a "secular consensus" or a "consensus of the faithful," but we might reasonably hope for a "secular experientialism" (to borrow a term from Shadworth H. Hodgson). Whatever new directions we might venture to explore, one thing is clear: we shall, thanks to Engelhardt, be better equipped to avoid the abyss of futile hope of "perfecting an illusion" so long as we step carefully during our search for the full meaning of the private, existential, embodied (and perhaps inviolable) self.

It is therefore no mere accident that modern medicine has turned out to have provided a new impetus to philosophy, indeed, a rebirth of philosophizing in our time. This, of course, is not a new idea; for as Edmund Pellegrino remarked in his Annual Oration before the members of the Society for Health and Human Values in 1974, "Medicine could provide the powerful stimulus to philosophy which Christian theology provided in the middle ages" ([2], p. 20). Over two decades later, it is clear that his earlier prognosis was indeed prescient.

AFTERWORD

During the past twenty-five years, attention to the moral issues in medicine and health care (bioethics) has more than demonstrated a capacity to attract the interest of scholars in the health professions as well as in the humanities. Numerous scholarly publications have appeared, along with the publication of hundreds of monographs, as well as the second edition of the *Encyclopedia of Bioethics* and this fiftieth volume of the *Philosophy and Medicine* series.

Future scholarly and critical appraisals of the philosophy of medicine, then, must of necessity include reflections on the extent to which this field is to remain an independent area of scholarly investigation and endeavor. Indeed, the Philosophy and Medicine Symposium from which this volume developed, obliged those present to consider the promise of the field not only as an interdisciplinary movement, but as the initial step further to develop the philosophy *of* medicine—analogous to the philosophy of science, or the philosophy of biology.

In terms of scholars and teachers, the field of bioethics and the philosophy of medicine continues to command the energies of hundreds of individuals throughout the world. Interest in this field has in part stemmed from concerns similar to those that gave rise, as Engelhardt points out, to the New Humanism (and the Third Humanism on the Continent) at the end of the nineteenth and beginning of the twentieth centuries: a recognition, however, that the medical, scientific, and technological endeavors of society have slowly become disengaged from the humanities.

Twenty-five years ago, the philosophy of medicine and bioethics appeared to offer medicine and the biomedical sciences an opportunity for critical self-consciousness and appraisal of the concepts, values, and images that serve to structure an ever-more significant element of contemporary life—health and well being. Many within medicine and the biomedical sciences greeted this development as bearing an additional cultural task as well; on the other hand, the philosophy of medicine and bioethics was thought to be capable of providing philosophers and others engaged in the *Geisteswissenschaften* with the occasion to direct their critical powers in matters of moment to both society and, in particular, public policy. In short, during the past few decades philosophy and the medical humanities have brought an important scholarly focus not only on issues in the biomedical sciences and bio-technologies, but on a vast array of societal issues in health care as well. For their part, the biomedical sciences and bio-technologies have provided philosophers and other humanists with an opportunity critically to assess major societal endeavors: the economics of health care policy and reform as well as the future implications of genetic engineering.

In closing, it should be noted that those who convened in Galveston, Texas (February 15-17, 1995) and those who contributed to this volume have paid attention to the relationship between the conceptual foundations of ethics and the philosophy of medicine and to the ways in which the philosophy of medicine and bioethics have developed. We have at least tacitly agreed on how the philosophy of medicine originated and developed and why its existence is no longer a merely putative claim, as was suggested by Jerome Shaffer in his contribution to the Round-Table Discussion that convened in Galveston in May, 1974.

The shared hope is that we have articulated particular themes and issues unique to the philosophy of medicine and bioethics, and have underscored the stages that have led to the clarification of (1) particular ethical and nonethical concepts: e.g., health, disease, pain, suffering, dying, action and cognition, the tacit logic of medicine, and medicine's decision rules, and (2) the relation that obtains between public policy interests and the ways in which the philosophy of medicine and bioethics have developed in the context of various ideologies.

Finally, bioethics and the philosophy of medicine stand now at a unique juncture: their problems are becoming recognized as requiring an international field of scholarly research. Here it should be mentioned that this period of the development of the field has implications not only for scholars who reside and work in the United States because it offers them a rare opportunity to better appreciate the assumptions and presumptions of their work through a critical reassessment of their field, but also for international scholars who conduct research in the medical humanities and participate in the sustained dialogue with those of us in the United States.

Since we must surely conclude that the project for the field twenty-five years ago has not yet been fully realized, it remains for us and others to identify and address those areas that should receive the serious attention of physicians and philosophers.

Center for Medical Ethics and Health Policy
Baylor College of Medicine
Houston, Texas, U.S.A.

REFERENCES

1. Beauchamp, T. L. and Childress, J. F.: 1979, *Principles of Biomedical Ethics,* Oxford University Press, New York.
2. Pellegrino, E. D.: 1974, 'Medicine and philosophy: Some notes on the flirtations of Minerva and Aesculapius', Society for Health and Human Values, Annual Oration (Nov. 8th).

SECTION II

PRACTICE AND THEORY

LARRY R. CHURCHILL

BIOETHICS IN SOCIAL CONTEXT

I. MAPPING THE MORAL TERRAIN

What is it like to be in a moral quandary? How do people describe this kind of experience? They often say they feel anxious, as well as intellectually perplexed. They describe themselves as emotionally vulnerable and conflicted. Typically, there is a sense that a great deal is at stake in terms of how others will view them, or how they view themselves, as well as what their future life prospects may hold. The affective dimensions center on being ambivalent, anxious, perhaps guilty, and sometimes overwhelmed. The cognitive aspects are uncertainty, ambiguity, or lack of clarity about values—a sense of facing problems that resist simple resolution or exceed one's capacities. Overall, the experience is not high on the hedonic scale. Anyone who has wrestled with a serious moral dilemma, or who works with persons who do, will recognize this list and could easily extend it.

I begin with this abbreviated description of the human experience of ethical conflict in order to emphasize that ethics is about a concrete human phenomenon, or set of phenomena. Ethics is about what happens to people in the process of living a life, not a matter of what happens in the heads of academics in their studies. Ethics is first and foremost a matter of bumps and bruises encountered as people seek moral direction, as they strive to live lives they can respect, or at least tolerate upon reflection. And it is indelibly part of human nature to so reflect, to make judgments about ourselves and others that can only be called *value* judgments—judgments not reducible to facts and of much greater importance than matters of taste or decorum. I begin in this way as a reminder that we are the species who says "ought," and to refresh our memories about what the experience of saying "ought" is like. To be helpful in illuminating the choices and guiding the actions of patients and health professionals, the academic study

R. A. Carson and C. R. Burns (eds.), Philosophy of Medicine and Bioethics, 137-151.

of bioethics will have to be responsive to these experiences, which are the basic ingredients of moral life.

A related question is equally important. How do people resolve moral quandaries? What sorts of strategies, maneuvers, processes, and steps do they take to escape the experience of value conflict? These strategies and maneuvers, like the experience of a moral quandary, are constantly in evidence, and are of an endless variety. There are, no doubt, common features among persons stemming from cultural or social similarities. Yet upon careful examination, we find the movements in each ethical problem-solving process display unique configurations, just as each person has a unique signature, and no person ever quite duplicates exactly even his or her own signature.

How do people resolve moral quandaries? Often they tell stories about themselves, about the sort of persons they are or hope to become. They tell such stories for many reasons—as a means of reassurance, as imaginative rehearsals, as trial balloons, or in the hope of determining a course of action that will fit what they take to be the ongoing narrative thread of their lives. They ask what others would think of them were they to pursue one course of action or another. They consult the living and the dead, and in certain biomedical situations, even the unborn, trying to configure their relationships appropriately. They seek advice from authorities—kinship networks, friends, people they respect, people on whom they mirror themselves, from communities to which they belong, or hope to belong. These communities may be specifically identifiable as one's neurosurgical colleagues, or as utilitarians, or orthodox Jews, or one's family, and so on. However designated, a reference community is composed of important others who might approve or disapprove a choice, and accept or reject the person involved. These sorts of strategies are typical of those encouraged by the ethical philosophies that focus on character and virtues. But people consult not only reference communities; they also consult rules and principles, and occasionally oaths sworn in professional initiations. Ethical problem-solving usually involves critical, applied analysis of such rules and principles, as well as reflective talking, remembering and imagining.

The style, and not just the content, of individual problem-solving varies also. Some persons seek, primarily, to think themselves out of moral difficulty, looking for consistency between situations. They may employ formal tests, such as asking what choices they would approve in others as a measure of what they should choose for themselves. Those who identify strongly with formal codes of ethics may undertake strenuous intellectual work in interpreting and applying these codes. For example, physicians often work to parse the meaning of "First, do no harm," and Christians often seek actions that embody "the love of God."

For others, the essence of moral problem-solving is getting "in touch" with one's true feelings, with the idea that deep emotions are morally instructive, but not readily accessible.

This list of strategies and styles of moral problem-solving is, of course, far from complete. I intend it only as a reminder of the "stuff" of ethics—the real world thoughts and actions that bioethical theories and analyses are intended to address and refine.

My point here is that human moral activity is complex and multifaceted. It includes decision and action, reflection and contemplation, sustaining habits over time, nurturing character, and many other kinds of activity. It is part logic, part acts of will, part turning and tuning emotional sensibilities, part feats of imagination. Because moral activity is complex and multifaceted, ethical theory must be multifaceted as well. Since there is no single, uniform subject matter in view when we say 'morality,' since it is always necessary to probe to discover just what aspect of morality is being considered, the need for diverse kinds of ethical theory should be evident. Yet often this need for theoretical diversity is not recognized, and a mismatch results. For example, moral theories designed to clarify the thinking of the muddle-headed are sometimes brought to bear on the emotional facets of a situation, or on what are better described as failures of character or will. Or a rule or principle that is clarifying and insightfully used in one context is misapplied in another. Such mismatches and misapplications are an inevitable part of every person's struggle for morally mature judgments, for ethical wisdom. Yet often bioethics exhibits just the opposite of careful attention to the complexity of moral experience, and fails to engage persons as they seek resolution to their moral quandaries. This failure to engage is not just unfortunate. It is damaging. Bioethicists too often do not simply fail to read complex moral situations, they also oversimplify them. They tend to substitute their own idealized version of experience for the moral phenomena described above.

Using the metaphor of cartography, let me put this in another way. Because moral experience is complex, it requires careful mapping. Yet often in the study of bioethical problems, academics bring their own maps and ignore the sketches provided by the natives. Instead of constructing a map after walking across the terrain and talking with those who live there, academics often arrive with their own requirements for what the lay of the land must be. Such preconceived moral topography is of interest to other academics, but of little use to persons who are trying to find their way. When academics or professionals have power to impose their maps on moral travelers, the problems become acute, as I shall illustrate in the two scenarios that follow.

II. NOBODY SAID HELLO

In a regular feature of the *Cambridge Quarterly of Healthcare Ethics* called "Bedside Story," an ethics committee meeting is recounted from the perspective of the patient ([1], pp. 185-186).

The patient was a young HIV-positive woman. After she underwent exploratory surgery for pelvic pain, it was discovered that her fallopian tubes were blocked. She requested surgery to remove the obstruction, explaining to her gynecologist that she and her husband wanted the opportunity to have a child in the future, in case a cure for AIDS is found. Because this patient was HIV positive, any baby would have a thirty percent risk of being born with HIV, and the patient herself might not live to raise the child. [This case preceded 1993 studies indicating that zidovudine can reduce HIV transmission to newborns by two-thirds.] The patient's request for surgery became an ethics "case" when her physician referred her request to an ethics committee.

The patient's perceptions of her interaction with the ethics committee are given in some detail, and make distressing reading. According to the patient, "Nobody said hello . . . " and the demeanor of the committee embodied many of the standard complaints directed toward physicians in tertiary-care settings—impersonal, businesslike exchanges, directed by anonymous professionals, and questions that presumed the priority of the professionals' agenda. But the agenda, at least as the patient perceived it, was overtly moralistic rather than professional, as the committee probed the cause of the patient's HIV infection and her reasons for wanting to bear a child. Some committee members "seemed angry at me and asked how I could be making plans so far into the future . . . ." Finally, the committee made not a recommendation to the physician, but a decision delivered to the patient by the ethics-committee chair.

In sum, in this patient's view the committee treated her like "a child," put her "on trial," found her "guilty" and denied her medical services. She summarized her feelings: "I didn't need them practicing morality on me."

The patient's story was followed in the next issue of the *Cambridge Quarterly* by a response from the ethics committee chair ([2], pp. 285-286). While acknowledging that the patient may have felt as if the committee "ganged up" on her, the chair portrayed the patient as ambivalent in her desire for surgery. She maintained that "the consultation process did facilitate this couple's ability to resolve their dilemma." The chair also listed ways the committee could improve the consultation process, which she portrayed as too dominated by the "didactic manner" of experts.

What is distressing about the response of the ethics-committee chair is the subtle discrediting of the patient's perspective and the ease with which this chair dismissed legitimate moral concerns as matters of process and protocol. The chair repeatedly said of the patient's views that they were "interesting." But most telling is the summary statement of the chair following her list of proposed improvements: "We feel that instituting some or all of these measures will improve the consultation process and help to make it a positive and rewarding experience for all concerned." The chair seemed oblivious to the connection between the consultation process and the conclusion the committee reached, as if an encounter that left the patient feeling infantilized and morally judged can be separated from whether the committee rendered good advice. It is unclear whether this ethics committee did more harm than good.

Can an ethics consultation be good when nobody says hello, that is, when the discussion about what is good or right lacks human qualities of respect and sympathy? A decision on the part of the patient to do what most persons might agree is "best" is morally bankrupt if the price of getting the patient to this decision is questioning the patient's character. Not just the decision, but the decisional process, must be patient-centered. The only person likely to have been well-served by this consultation was the physician, who was given the ethics committee's approval for not doing what he apparently did not want to do.

These brief accounts from a patient and an ethics-committee chair raise profound questions about whether ethics committees can serve both patients and caregivers, to say nothing of institutions. But my concern here is with whose notion of "ethics" is at work in this committee. We are not told exactly what the ethics committee's understanding of "ethics" is, except that, in the chair's words, it involved "principles" and a "process." Yet whatever principles or process were in use, it is clear that the committee's concept of ethics was dominant over the patient's. The committee effectively substituted their own version of "the case" for the patient's moral experiences. It should not be surprising that the patient felt disenfranchised and angry. If this is how ethics committees work, patients will correctly view them as representing only the interests of doctors and hospitals, and as indifferent or at worst hostile to the patients' perspectives.[1]

III. SACRIFICIAL ATONEMENT

Fifteen years ago, a patient-physician interaction was pivotal in demonstrating to me the inadequacy of much that goes under the name "bioethics."[2] This interaction involved a sixty-year-old woman being seen in the clinic for pain in the groin area and a palpable lump. Some years earlier, she had undergone amputation of her left leg but remained ambulant through the aid of a prosthesis and crutches. The physician strongly suspected that the mass represented a recurrence of the malignancy that had required taking her leg, but the patient, after hearing a careful explanation of the need for surgery, refused. A deeply religious person, she believed that God afflicted her with cancer in order to save souls. She saw the amputation of her leg as a "giving up to the Lord," an offering, or sacrifice, in order that through her example others could be reconciled to God. The loss of her leg was an atonement for the transgressions of a member of her local congregation—an interpretation which she announced and which was affirmed within her community. As an example of her witness, she told of a dream in which she was again whole, walking to visit a neighbor whom she knew to have end-stage carcinoma of the head and neck. Interpreting this dream as a sign to her, she and her husband went the following day to visit this neighbor. She later reported that this visit had a conversional and healing quality.

The physician interpreted this patient's refusal of treatment as a problem of noncompliance. He saw a clear need for a biopsy, and very probably surgery and/or radiation. The principle-oriented bioethicist in attendance saw the situation as a classic conflict of respect for autonomy with paternalism. The "issue" was how aggressively to test the authenticity of the patient's refusal, and whether beneficent paternalism (pressing for treatment) might be justified in this situation.

What is impressive here is just how wide of the mark both the physician *and* the bioethicist were. Neither "noncompliance" nor "autonomy" were meaningful terms for this patient. To suggest that either of them captured the true significance of her choice would be like treating a myth as a propositional statement; for example, treating the creation stories in *Genesis* as literal accounts, or Aesop's fable of the hare and the tortoise as an historical occurrence. As Maurice Merleau-Ponty put it, it would be "like applying our own grammar and vocabulary to a foreign language" ([15], p. 121).

The first and most basic task in ethics is paying attention to the moral phenomena. Paying attention requires not only suspending initial judgments of right and wrong, but also suspending our usual categories of interpretation. If ethics is not one thing but many—not one kind of human activity, but a complex variety of activities held together for each person in unique ways—then it is essential to let people speak for themselves, and in their own terms, about the moral dimensions of their lives. Interpretation, of course, always follows, but it should not precede and preempt the voices of the morally perplexed , nor should it presume finality of interpretive power over these voices.

The patient requesting surgery for tubal blockage may have been heard by the ethics committee, but there is no evidence for it in either her comments or those of the committee chair. The chair analyzed the difficulties as administrative in nature, a problem of how to handle a demanding and distraught patient. Judging from the written record, the ethical problem was defined and then managed by the committee, not *with* the patient, but *for* the patient. The patient never appeared: "Nobody said hello."

The patient who refused surgery in order to continue her evangelical mission simply went unheard. The key interpretive concepts of her life (sin, and salvation through sacrificial atonement) had priority for neither physician nor philosopher. Neither shared her universe of discourse. To the physician she appeared as noncompliant and irrational. To the philosopher her decision took the form of an ill-advised exercise of autonomy. Neither physician nor philosopher was prepared to see (much less affirm) her refusal of treatment as part of a larger healing ritual for her and her community.

IV. BIOETHICS IN SOCIAL CONTEXT

The two scenes described above display presumptions of professional privilege, both medical and bioethical. These presumptions have practical implications; they not only distort, they disable as well. They distort patient (and professional) experiences, and they disable persons from using their own moral resources at critical life junctures when they need these resources most.

Bioethics conceived as part of a larger social inquiry can help to disabuse us of this sort of parochialism. Social research places physicians and patients in a cultural and historical, and not just a medical and moral, interaction. It attends to power differentials and to the nuanced exchanges that characterize physician-patient interactions. Socially informed perspectives see illnesses not just as standard pathological entities but as dynamic concepts, dependent for their meaning on personal and institutional settings. Social research recognizes the powerful ways that relationships can either enhance or defeat a therapeutic course. Bioethics conceived as part of this larger social location of disease and illness, patients and doctors, brings into focus the local and distinctive meanings of moral quandaries, meanings forged in social exchanges rather than in theoretical systems. The typical mode of doing bioethics neglects these dimensions in favor of the search for general meanings (rather than particular ones) and sees these as the products of intellectual acumen and logical rigor (rather than social meanings and institutional forces).[3]

Historically, ethics in medicine has been a construct of physicians. As such it has emphasized the development of character, typically in terms of the given period's orthodox requirements for custom and decorum. The influx of philosophically trained ethicists over the past two decades has substantially changed this focus, away from character, custom and decorum, and toward the application of rules and principles. At the same time, a reaction against principled approaches is evident in the emphasis on narrative ethics. Yet narrative approaches still frequently carry an assumption typical of principled approaches about application, i.e., that biomedical ethics is fundamentally "applied ethics"—if not the application of antecedently known principles, then the application of narrative methods and concepts to moral problems. The idea that the experiences of patients and doctors might themselves be key formative items for ethics, altering expectations for methods and requiring multiple approaches, has been only a minor note in the bioethics movement to date.

The philosophical, principle-oriented version of "applied ethics" is the most familiar. Here bioethics means the uses of Kant, Mill, or Rawls (or pick your favorite theorist!) in the context of medicine and the health sciences. Tom L. Beauchamp and James F. Childress have done the most to popularize this view. Their book, *Principles of Biomedical Ethics,* is now in its fourth edition. The title bespeaks the methodology. Principles, they say in the first edition preface, bring "order and coherence" to the frequently "disjointed approach" that relies on discussion of cases ([5], p. vii). Principles provide a way to justify choices, to say why one action is morally preferable to another. Principles save us from ad hoc decisional processes and moral inconsistency. Biomedical ethics is, then, largely defined as the skillful use of the principles of autonomy, nonmaleficence, beneficence and justice in situations of moral complexity in health care. The fourth edition of *Principles of Biomedical Ethics* is less revisionist in its tone, and more eclectic in its approach, with greater attention to nonprincipled methods. Moreover, Beauchamp and Childress never claim that their approach describes the whole of the moral life, but only the justification of choices when faced with dilemmas. Those who employ their methods, however, are seldom as careful, or as cognizant of the limits of this approach.

Narrative ethics, with its emphasis on stories, substitutes what Stephen Crites calls the "narrative quality of experience" for principled reason ([10], pp. 291-311). Narrative ethics broadens what will count as morally significant, but it often retains a key prejudice characteristic of the heavy-handed use of principles. It retains the universalizing assumption of principled reasoning—the conviction that the job of ethics is to describe, if not the universal values, the universally correct metaethical position. It is this assumption that often gives the theoretical debates between Kantian deontologists, Benthamite utilitarians, and narrative theologians such passion. Each is contending for dominance of the entire field, not just the validity of their perspective *among* others, but *over* all others. Crites, for example, working from the Augustinian notion that human experience is always time-laden or "tensed," extends this observation into a requirement for ethics, claiming that "ethical authority . . . is always a function of a common narrative coherence of life" ([10], p. 310). Stanley Hauerwas, with a more direct critical focus on bioethics, says that principled reason always functions in a story context, explicitly or implicitly, and that "character and moral notions only take on meaning in a narrative" ([12], p. 15). Alasdair MacIntyre asserts in *After Virtue,* "I can only answer the question 'What am I to do?' if I can answer the prior question 'Of what story or stories do I find myself a part?' " ([14], p. 201).

So while narrative bioethicists give us new concepts with which to work, they tend to overextend the range of these concepts, making them the new methodological requirements for ethics generally, and bioethics in particular. It may be true, as J. Hillis Miller claims, that "without storytelling there is no theory of ethics" ([16], p. 3), but ethical theorizing is not reducible to storytelling. Again, it is the overextension of a valid moral insight that is the source of the problem. As much as the principle-oriented bioethics which it eschews, narrative bioethics is too often in the service of an urge to provide not only *A* model *of* moral experience, but *THE* definitive model *for* moral experience.[4]

I criticize principled and narrative approaches because of their prominence, but the same criticism could be directed at the superordinate claims that occasionally come from those who espouse casuistry, hermeneutical approaches, feminist theories, and a variety of others.[5] Each of these approaches can add a valuable dimension to bioethics, but none alone has adequate range or depth to encompass the field or fold the others under its interpretive grid.

The universalizing urge of bioethical theorists blocks recognition of a central, empirical datum, i.e., that the moral experiences of human life are more varied and complex than the theories we employ to grasp them conceptually. This is true generally, but especially apparent in the moral experiences that are part of health care. Practice continues to outrun and elude theoretical encapsulation. Ethical theories are best seen, then, not as competitors for the whole of moral life, but as potentially complementary modes of approach, each with a useful, but limited, range of application. In bioethics this means that no one theory can do full justice to the experiences of patients who are ill and those who care for them. We should be suspicious of theories that claim universal purchase, and cautious about the drive for all-encompassing formulations. The theories will prove to be procrustean beds, and the drive will distract us from attending carefully to the variety of human moral dynamics.

The appropriate methodological stance for dealing with this experiential complexity is not an anti-theoretical posture, but one open to the multiplication of theories. What our experiences require is not *the* theory (or no theories), but several theories. Differing theories are best seen as complements rather than competitors, with each describing a facet of moral life, each supplying something other theories lack. Bioethicists do us a disservice if they make us choose between Kant and Mill, or Rawls and MacIntyre, or suggest that the resolution of competing theories (deontological vs. utilitarian, etc.) should be sought in a supertheory that transcends the others. If human moral experiences were simple and shallow, a single theoretical framework might capture them all. Since they are not, we are free to welcome many theories, and perhaps periodically revive old ones and concoct new ones to keep the variety and depth of moral experience displayed before us. We must resist the philosopher's platonic hankerings for moral ideals or decisional processes that will both transcend human experiences and somehow become the measure of them all.

The urge for theoretical hegemony can take several forms. Some bioethicists have suggested that the inadequacy of any single type of theory to cover the wide diversity of moral experiences should lead us to devise a supertheory that will teach us how to use the right theory in the right way at the right time. For example, in the introduction to the second edition of their widely used anthology, *Ethical Issues in Modern Medicine,* John Arras and Robert Hunt say, "It is our contention that, if answers to ethical issues are to be found, they are to be found through the development of a cogent and comprehensive ethical theory—a theory that will both explain the principles of morality and give us a guide to their application" ([3], p. vii). While recognizing the need for theoretical pluralism, these authors seem to be holding out the hope that the problems of skillful use of the tools and methods of ethics can still be resolved at the theoretical level, in some sort of "comprehensive" theory. While this is an improvement over the view that some single principle or theory will have universal validity, it still seeks resolution in the conceptual realm, rather than through a closer and more perceptive grasp of the social and cultural factors of the situations in which theories are used. Interestingly, the most recent edition of *Ethical Issues in Modern Medicine* drops the aspiration for comprehensiveness. Here Arras and his new collaborator, Bonnie Steinbock, make more modest claims and affirm the dynamic flux of experience and judgments, saying the best we can hope for is greater, but not complete, systematic coherence ([4], p. 39). This recognition is the beginning step to the more full contextualization of ethics I am arguing for in this essay.

How does ethics in the context of a larger social inquiry help to avoid this tendency toward universalist assumptions and superordinate theoretical processes in ethics? By insisting that ethics be done in a cross-disciplinary fashion, using the tools and perspectives of disciplines like literary criticism, medical sociology and cultural anthropology. Let me explain what I mean through an example, an example of moral exchange central to health care, but one which is elusive to the usual ways of doing ethics.

John Berger and Jean Mohr's account of a British G.P., entitled *A Fortunate Man* describes the following scene:

> Once he was putting a syringe deep into a man's chest: there was little question of pain but it made the man feel bad: the man tried to explain his revulsion: "That's where I live, where you're putting the needle in." "I know," Sassell said, "I know what it feels like. I can't bear anything done near my eyes, I can't bear to be touched there. I think that's where I live, just under and behind my eyes" ([6], pp. 47, 50).

What is interesting about this slice of conversation is how it conveys experiences of suffering and empathy—items central to the moral life of medicine. But notice, there are no decisions portrayed here, no problems delineated, no priorities sorted. Rather what is displayed is what sociologist Arthur Frank calls "recognition" ([11], pp. 49-55).[6]

Describing his own experiences first with a heart attack, then with cancer, Frank says that a critical illness is like a journey that takes its travelers to the margins of life. He believes that the central task in giving care to patients is recognizing the journey they are on. He reserves the name "caregivers" to people who listen to the ill, witness to the particulars of patients' illnesses and "recognize" them—a feat Frank believes most hospital staff have little time for ([11], pp. 48-49). Frank concludes that the power to recognize others on their journey of suffering is "fundamental to our humanity" ([11], p. 104).

My point in this example is that bioethical inquiry that is methodologically conceived and practiced in the context of the other social sciences and humanities stands a better chance of noticing and accrediting these non-decisional aspects, like "empathy," "journey" and "recognition"—aspects which are not peripheral but central to medicine and the health professions, and which figure heavily in the shape and meaning of moral quandaries. Ethics has as much to do with how well we attend as with how well we decide, and this is a skill characteristic of good fieldwork in sociology and anthropology. Conceptions of ethics which pay no attention to attending, which have no category called "recognition," are weakened by this absence.

Principle-oriented, applied-ethics bioethicists do, of course, use terms like "empathy" and "recognition" in their analyses. Yet they are usually pressing to get beyond these items to a formulation of "the problem" or "the issue" that will allow them to apply their principles. They tend to see empathy and recognition as preliminaries to the real work of ethics. In the gadarene rush to proper formulation, they sometimes miss the items of major importance.[7]

Ludwig Wittgenstein put it well in *Zettel*:

> Here we come up against a remarkable and characteristic phenomenon in philosophical investigation: the difficulty—I might say—is not that of finding the solution but rather that of recognizing as the solution something that looks as if it were only a preliminary to it. "We have already said everything.—Not anything that follows from this, no, *this* itself is the solution!"
>
> This is connected, I believe, with our wrongly expecting an explanation, whereas the solution of the difficulty is a description, if we give it the right place in our considerations. If we dwell upon it, and do not try to get beyond it.
>
> The difficulty here is: to stop ([19], para. 314).

The difficulty Wittgenstein attributed to philosophical investigation is also characteristic of bioethical inquiry. The difficulty in "recognizing" persons, in stopping and accrediting the great variety of moral experiences and valid problem-solving strategies, is part of American bioethics. I am not, to be sure, arguing that bioethics need *only* describe and not explain or guide. I am arguing that privileging theory in the way described above gives description too little place in ethics. And giving description a more robust place will enhance our skills in recognizing and accrediting what we are prone to rush past.

V. CONCLUSION

In summary, I have argued that this difficulty in attending and recognizing is tied to restrictive methodological assumptions about the field of bioethics. These restrictive assumptions are not just unfortunate blindspots, but potentially disabling forces to those who are most vulnerable and powerless in medical interactions. I have also argued that placing bioethical inquiry in the larger context of other humanities and social science disciplines will help to counter this methodological parochialism. The distinct advantage of placing bioethics in social context is that an interdisciplinary setting makes it less likely that we will be seduced by the intellectual glamour, or the intuitive emotional appeal, of any single approach to moral problems. Resisting a hyper-theoretical approach to the methods of bioethics will make for greater agility in problem-solving, and more resilience in facing those problems that cannot be solved. It will, in the end, make for better health professionals and for better patient care.

The University of North Carolina
Chapel Hill, North Carolina

NOTES

[1] An interesting move in the right direction is Susan M. Wolf's "Toward A Theory of Process" [20]. Wolf argues that the lack of attention to process poses a serious threat to patients.

[2] I have discussed this experience previously in "Bioethical Reductionism and Our Sense of the Human" [9].

[3] Thomas H. Murray, writing in a spirit congenial to this essay, analyzes the Dax Cowart films with attention to the social, psychological, and political detail (and not just the conflicting principles) [17].

[4] In contrast to the sweeping claims of some narrative bioethicists, Rita Charon articulates a more balanced view, signaled in the title of her essay "Narrative Contributions to Medical Ethics." Here she argues for a necessary place for narrative among other methods of ethics, and that bioethicists need as much facility in listening and interpreting as they have in applying principles; "narrative competence is among the required capacities of any biomedical ethicist" ([8], p. 275). Charon's point of view is entirely congenial with my perspective.

[5] See, for example, the excellent work by Albert Jonsen and Stephen Toulmin, *The Abuse of Casuistry* . Toulmin, however, occasionally carries his zeal for practical reasoning to an anti-theoretical extreme. For example, he has recently claimed that the casuistical analysis of actual cases in clinical medicine is "pretheoretical," by which he means that the moral significance of these cases can be stated in a way that is neutral as between theoretical constructs. See his essay "Casuistry and Clinical Ethics" ([18], p. 310ff). Yet casuistry is not pretheoretical, or neutral, but theory-laden, insofar as it is committed to a tradition of naturalistic, Aristotelian interpretations about how ethical problems arise and how they are properly solved. Casuistry is, after all, one ethical method of problem-formulation and problem-solving among others, not the theory-free description Toulmin claims. Rather than trying to get behind, or beyond, ethical theory, the more productive effort lies in spotting the *misuses* of theory. Theories, of course, can obscure and frustrate, as well as illuminate and facilitate, problem-solving. The trick lies in using theories judiciously. In brief, Toulmin is on target with his criticisms, but wrong in his view that somehow casuistry is exempt from these criticisms. An alternative, and perhaps more accurate, interpretation of Toulmin is that the target of his criticism is only the kind of

theorizing that assumes hegemony over the terms of moral debate and insight, and that he would allow for the kind of low-altitude theorizing I want to endorse here, and which I find evident in his own embrace of casuistry. If this is the correct interpretation of Toulmin's polemic against theory, then I share his critical agenda, but remain skeptical about the preeminence of casuistical analysis.

[6] For an excellent development of the concept "recognition" that also draws from Arthur Frank's work, see Ronald A. Carson's "Beyond Respect to Recognition and Due Regard" [7].

[7] The original "gadarene rush" is recorded in *Matthew* 8:27 ff., and refers to the manner in which a demon-possessed herd of swine found their way down a steep bank and into the Sea of Galilee.

REFERENCES

1. Anonymous: 1992a, 'Bedside story', *Cambridge Quarterly of Healthcare Ethics* 2, 185-186.
2. Anonymous: 1992b, 'Bedside story', *Cambridge Quarterly of Healthcare Ethics* 3, 285-286.
3. Arras, J. and Hunt, R. (eds.): 1983 (2nd ed.), *Ethical Issues in Modern Medicine,* Mayfield, Palo Alto, CA.
4. Arras, J. and Steinbock, B. (eds.): 1995 (4th ed.), *Ethical Issues in Modern Medicine,* Mayfield, Palo Alto, CA.
5. Beauchamp, T. and Childress, J.: 1979 (1st ed.), 1995 (4th ed.), *Principles of Biomedical Ethics,* Oxford University Press, New York, NY.
6. Berger, J. and Mohr, J.: 1976, *A Fortunate Man,* Writers and Readers Publishing Cooperative, London, U. K.
7. Carson, R.: 1995, 'Beyond respect to recognition and due regard', in *Chronic Illness: From Experience to Policy,* S. K. Toombs, D. Barnard and R. Carson (eds.), Indiana University Press, Bloomington, IN.
8. Charon, R.: 1994, 'Narrative contributions to medical ethics: Recognition, formulation, interpretation, and validation in the practice of the ethicist', in *A Matter of Principles? Ferment in U. S. Bioethics,* E. DuBose, R. Hamel and L. O'Connell (eds.), Trinity Press International, Valley Forge, PA.
9. Churchill, L.: 1980, 'Bioethical reductionism and our sense of the human', *Man and Medicine* 5, 229-242.
10. Crites, S.: 1971, 'The narrative quality of experience', *Journal of the American Academy of Religion* XXXIX, 291-311.
11. Frank, A.: 1991, *At the Will of the Body,* Houghton Mifflin, Boston, MA.
12. Hauerwas, S.: 1977, *Truthfulness and Tragedy,* University of Notre Dame Press, Notre Dame, IN.
13. Jonsen, A. and Toulmin, S.: 1988, *The Abuse of Casuistry,* University of California Press, Berkeley, CA.

14. MacIntyre, A.: 1981, *After Virtue,* University of Notre Dame Press, Notre Dame, IN.
15. Merleau-Ponty, M.: 1964, *Signs,* R. McLeary (trans.), Northwestern University Press, Evanston, IL.
16. Miller, J.: 1987, *The Ethics of Reading: Kant, deMan, Eliot, Trollope, James and Benjamin,* Columbia University Press, New York, NY.
17. Murray, T.: 1993, 'Moral reasoning in social context', *Journal of Social Issues* 49, 185-200.
18. Toulmin, S.: 1994, 'Casuistry and clinical ethics', in *A Matter of Principles,* E. DuBose, R. Hamel and L. O'Connell (eds.), Trinity Press International, Valley Forge, PA.
19. Wittgenstein, L.: 1967, *Zettel,* G. Anscombe (trans.), Basil Blackwell, Oxford, U. K.
20. Wolf, S.: 1992, 'Toward a theory of process', *Law, Medicine and Health Care* 20, 278-289.

JUDITH ANDRE

THE WEEK OF NOVEMBER SEVENTH: BIOETHICS AS A PRACTICE

I. THE WEEK OF NOVEMBER SEVENTH

I traveled many city miles that week and found myself part of many different institutions. I learned in many different ways: from reading, from listening, from watching, from helping out. And there were many ways to share what I came to understand.

Monday

Mrs. Bactri[1] was eighty-eight-years old, and stubborn. Also, possibly, not competent to make decisions about her own care: but whenever she *could* speak, what she said was, "No. No gastrostomy. No stomach tube." Her physician asked for an ethics consultation, and I was one of the group who met with Mrs. Bactri's daughter, Catherine. Everything about the situation was familiar: the issue (refusal of treatment); the procedure (we listened, we talked; we asked how Mrs. Bactri had lived her life and what she had said she wanted); and the resolution (a patient or her surrogate has a right to refuse treatment, even when the refusal risks shortening her life). Much of it was satisfying: Catherine Bactri was in tears when we started, and weary: "I'm only trying to do what I always promised my mother I would do." We listened with attention and respect and, when she burst out that the attending physician "made her feel terrible," reminded her quietly that she was free to change doctors.

As I recorded our consultation, I realized that twenty-five years ago, in graduate school, I never dreamed that I would one day be writing something in a hospital chart. I also wondered about what we had accomplished, and about the cost. Four professional people—doctor, nurse, philosopher, and biochemist—had each spent several hours on the case. Consultations are a slow and expensive process. Mrs. Bactri's life was probably shortened slightly: a few months, perhaps, one percent or less of her nearly ninety years. Her daughter

R. A. Carson and C. R. Burns (eds.), Philosophy of Medicine and Bioethics, 153-172.

felt far better: perhaps that was the major positive result. The physician probably felt more comfortable, too. I don't know if he had asked for the consultation hoping that it would protect him legally, or because he was genuinely uncertain about the right thing to do. In either case I felt uneasy; the issues were not ethically complex; the physician should have been able to sort things out, and should not be using an ethics consultation as a form of legal protection. But, of course, people do that all the time. I worried, too, that we were making it easier for the doctor not to take the time to really talk with Catherine Bactri. I was somewhat consoled by the young family physician on our ethics team, who was a model of unperturbed attention. She reminded me of the fictional doctor in the first chapter of my colleague Howard Brody's *The Healer's Power,* a young physician who also had learned to talk and to listen. A new generation. Perhaps.

Back in my office I turned to my own writing. Ten years ago I was writing about the moral status of self-regarding acts, using as analytic tools universalizability and action theory; I submitted it to a number of philosophy journals, appreciated or resented the reviewers' comments, and eventually published it. I'm not sure that anyone ever read it. Today my topic is very-low-birth-weight babies; I was asked to write it, no referee will pass judgment on it, it will be printed in a newsletter published here at the Center—and people will read it. It is much less careful, much less technical, not even original—but may well make a difference in what some people do.

Tuesday

Election day. For me it begins in an 8:00 A.M. undergraduate bioethics course. I've taught philosophy to undergraduates for almost twenty years, although I do so relatively infrequently now. The years of twelve classroom hours a week are over for me, and I am grateful. Classroom teaching has given me great pain and great joy: by now I appreciate the chance to buffer it with other activities.

Today I need all the buffering I can get. Our topic is justice and health care, and at the end of class I ask each student to write down an interesting or important unanswered question. Probably a fourth of them write something like, "Why should people who make good money pay for health care for those who don't?" Two weeks from now they will hand in papers on a microallocation issue: if there's only one kidney available but two people need it, how should we choose? Virtually everyone will argue that the single kidney should go to "Mrs. Benson" rather than to the unidentified accident victim: if it goes to Mrs. Benson, they argue, we know it's going to someone worthy. The danger of rewarding someone undeserving looms large; they would not take that risk. One writes with stunning, if unconscious, cruelty that "the poor have worse outcomes, so treating the affluent is a better investment."

The ease with which they write such things, their unawareness of competing arguments and of a need to defend their own, exposes my failure. But their positions also reveal a lot about general public opinion, particularly that of white suburban Michigan. I know already what today's election is likely to bring, and these student papers underscore that the Democrats are likely to do very badly.

I welcome the chance to get away, to drive across town for an IRB meeting—an IRB is an Institutional Review Board, responsible for protecting human subjects of scientific research. This particular IRB is part of the Michigan Department of Public Health. New to the committee, I have to get a sense of who the other members are, what the acronyms and jargon mean, what procedures everyone is used to. Handed a voluminous sheaf of government regulations, I appreciate, grudgingly, some of the resentment voters will take into the booths today. After the meeting someone from the substance abuse program lingers. Especially aware today of hostility toward the government, I ask her whether she minds being called a "bureaucrat." But she's used to it, and believes that mid-level civil servants are advocates for the people in a way that no one else in the country is. My experience with family-planning nurses inclines me to believe her. But how unknown their dedication is, and what rigid, unthinking anger is being cultivated in America. One hundred and twenty miles from where we sit talking, the Michigan Militia meets. In a few months almost 200 civil servants will die in Oklahoma City, expendable simply because they are government employees.

Next, back into town to meet with a hospital ethics committee. We look at a draft of a policy on "Do Not Resuscitate" (DNR) orders. The policy encourages doctors to talk with their patients, and tries to protect the wishes of incompetent patients even when their families have contrary wishes. This committee understands many things they didn't a few years ago: that hospital lawyers try to protect the institution even at the expense of patients and that doctors sometimes yield to the wishes of the family even though the patient would have chosen otherwise. Families, after all, are conscious and able to sue. I've learned a lot, too; most recently, why families of dying infants sometimes press desperately for every intervention, no matter how useless and painful, and why it is hard to resist them: difficult not only psychologically but ethically. I've seen "ethics consults" where the major dynamic had nothing to do with moral reasoning, but rather with supporting a lone, desperate mother learning to abandon hope for her baby.

Today, however, the committee wrestles with the concept of futility. They listen with affectionate noncomprehension as I repeat that the concept is muddy, and that adding the adjective "medical" does not help. We struggle with language and promise to continue the conversation.

Finally home, I refuse to watch the election results; tomorrow will be soon enough. Instead I pick up an issue of *The Journal of Philosophy,* one of my few remaining efforts to keep up with mainstream philosophy. Confused in mid-page, I start again, and find that I had read the word "epistemology" as "epidemiology." Finally anchored in a world more abstract than I've been in all day, I read a discussion of contextualism that is a pleasure and a revelation. "Contextualize" is used widely, self-righteously, and vaguely in much of medical ethics; here I find it distinguished from coherentism and defended with vigor and precision ([7], p. 627ff).

Wednesday

This morning I finally hear the election results. From my point of view it is worse than I could have imagined. Trying to remember whether I felt this bad in 1972, or in 1980, I can't really recall. What bothers me most in what I hear is its angry and punitive tone. Political conservatism, as such, need have nothing to do with hatred of anyone, let alone of the poor; a fierce belief in individual liberty could be combined with an equally intense compassion. In Grand Rapids, for instance, a city where conservative religions are powerful, a newspaper covered the death of an influential corporate executive by describing first—and at length—his civic spirit. His position in the business world was described three paragraphs down. Principled attacks on taxation could be joined with vigorous, effective calls to private charity. In Grand Rapids, too, a single mother I know was urged by her employer to keep in touch with her mother in India, and to call her from work so the company could pay for it. "When your children need you," he went on, "go home; we'll work around it."

But the public voice of conservatism this election day is different. I recall a Marxist saying that a welfare state isolates and stigmatizes those who do not succeed, in a way that a socialist state does not. The language of this campaign bears him out.

Thursday

The implications of the election begin to unfold. For some (certainly not all) of what has happened I blame the press, who seem to lack much ability to reflect upon themselves. How welcome a Journalism Ethics movement would be. But the forces that gave birth to bioethics twenty-five years ago—technology forcing dramatically new choices, and a vigorous civil rights movement—have no analog in journalism. Its new technologies (computers, satellite dishes, the Internet, and so on) do not force hard questions of the kind that kidney transplantation did, nor do they add weeks or years of misery to people's lives as ventilators and gastrostomies have. I cannot imagine, offhand, what social forces would lead the media to the sustained, public self-examination in which health-care professionals engage.

For that matter, I don't know what would lead universities to do so, either. My first project this morning is developing a panel on academic ethics,[2] to be called "Devouring our Young: The Mistreatment of Job Applicants, Part-Timers, and Junior Faculty." The panel addresses the way academics treat one another, rather than how we treat students, and this gives me pause. Bioethics, in contrast, has always focused primarily on how patients are treated, and rightly so. As newly elected legislators promise to support higher education, I listen with mixed feelings. Like journalists, we professors need to be considerably more self-critical. The most fundamental questions in higher education, I would argue, are about how many people actually go to college, and why; perhaps we should close a significant number of colleges and universities, and put the money into primary and secondary education. Perhaps not. But there is no one positioned to raise the questions, and many positioned to defend the status quo. Here, as everywhere, the questions that are asked in a public forum are only a fraction of those that could be asked, and their appearance is largely a result of the distribution of social power [10].

This thought reminds me of persistent criticisms of bioethics for its attention to individual cases and practitioners rather than to the social structures that distribute power. The criticism has never seemed fair to me. One need only recall the evolution of questions about kidney allocation, an issue with which the field began. A truly individualistic discussion would have remained at the level of individual choices: whether Mrs. Benson or the unidentified trauma victim deserved the single kidney. In fact, however, the discussion quickly moved on; the appropriateness of deciding on the grounds of social worth was soon questioned, and public policy almost everywhere framed to forbid doing so (in theory, and usually in practice). Questions of whether organs should be for sale, and of whether one's body should be considered property, soon developed. These are not questions about how one individual should solve one problem. All of us in the field help develop policy, from DNR policies in a hospital through fetal tissue guidelines in federal research.

In fact one of my three colleagues, Leonard Fleck, has devoted his professional life to issues of justice, to the consideration of what kind of national health-care policy would be fair. More than that, he has developed a technique for working with audiences that he hopes will launch a more sophisticated public discourse on the topic. (I recall with dismay that health care played no part in anyone's campaign this fall.) This afternoon two of my colleagues plan a workshop with a major managed-care health system in Detroit. Within a few weeks the *Hastings Center Report* will feature an article on "The Ethical Life of Health Care Organizations" [12]. I conclude that the atomism of bioethics is vastly overestimated.

A colleague drops by with brochures for the London course we teach each summer. An interdisciplinary class, its core is observation of English health-care workers on the job. An early assigned reading, however, in the sociology of health ([5], pp. 19-35) argues that health care doesn't make much difference; that it is poverty and affluence which most affect morbidity and mortality. In spite of much lower per capita spending on health care, and facilities that are often worn and shabby, health statistics in the U. K. are not very different from those in the U. S. And in spite of quite an egalitarian health-care system, the health of the rich is far better than the health of the poor.

With last summer vividly in mind, I begin to write a review of David Hilfiker's *Not All of Us Are Saints* [8]. Again I am writing for a local newsletter, and I keep in mind Detroit physicians who might appreciate Dr. Hilfiker's "journey with the poor." In London's East End last summer, we saw the poorest neighborhoods in the United Kingdom; I saw nothing like the devastation so common in Detroit. London poverty is nonetheless real, and I saw its connection with sickness. One diabetic woman, for instance, wore tattered, pointed-toe shoes, relics of a fashionable youth, which now restricted her circulation dangerously. In contrast, Hilfiker describes a diabetic who sleeps on a Washington street and gets frostbitten; a man who cannot wash his wounds, cannot refrigerate his insulin. The English woman, the American man, are both in danger of gangrene and amputation, but his danger is more severe. Even if we had national health care, he would still be sleeping on the streets.

Suddenly all these considerations fit together, like pieces of a jigsaw puzzle. Yes, it matters whether we are asking the right questions; on the other hand, no, bioethics does not confine itself to the separate choices made by individuals. What bioethicists do, however, is help divert attention from the most serious health problem in the country, and one of the most serious moral problems: the grinding, killing poverty in which our underclass lives.

Friday

My first appointment this morning is with Lauren, a graduate student in our interdisciplinary health and humanities program. A childbirth educator, she believes in home birth in most cases; she has data showing it to be ordinarily safer. She wants to understand why birth in the United States becomes ever more high-tech, more interventionist, more—in her terms—brutal. Once again I remember England, and a talk toward the end of our summer course from Marjorie Tew. Now close to eighty, Ms. Tew is a statistician and, as she points out, about as neutrally positioned as one could be on the issue of medicalized birth. She is not a doctor, not a midwife, and many years past her own childbearing. When she set out decades ago to find the data connecting improved perinatal rates with health care, she found none; she has finally concluded that in childbirth, as in so much else, most of the credit for better outcomes should go not to medicine but to improved standards of living. (Nutrition is especially important. Rickets, for instance, deforms pelvises and makes normal delivery difficult.)

Thanks to Lauren, my standard public talk on what most people call "maternal-fetal conflict" now begins with the story of a laboring woman who locked herself into her hospital bathroom. She wanted a non-medicalized birth, had negotiated for this with her obstetrician, and thought everyone concerned had agreed. In local labor and delivery suites, though, an intravenous line is routine, and this barricaded woman could find no other way to fend it off. "Maternal-fetal conflict" is a bad label for a complex issue. What is ordinarily at issue is a disagreement *between a pregnant woman and a health-care professional,* both of whom care very much about the baby's welfare. The too-common description, however, assumes a conflict *between a woman and her baby,* between her rights to her body and the fetus's safety. In the textbook I'm using this semester only one article mentions, and that in passing, a crucial case in which a woman whose baby was "certain to die" unless she had a Caesarean section nevertheless delivered, vaginally, a perfectly healthy child. The author does not realize that this is, in fact, the usual outcome (in the few cases where the woman was able to avoid the intervention and where we know what happened) [9]. My own awareness comes from reading work in medical anthropology—and from listening to Lauren, and to Marjorie Tew, and to the British G. P. with whom we worked in London.

Again I realize that bioethicists have been asking the wrong question—or, more precisely, not asking an essential one. It does matter whether a woman's rights over her body outweigh the claims of her fetus, the standard form of the "maternal-fetal conflict" discussions. But it also matters that the predictions of doctors about harm to the fetus are unreliable. It matters in part because the moral conflict is less severe when the danger to the fetus is less certain. But it matters more deeply because the standard account, omitting iatrogenic harm, not only exaggerates the role of medicine in saving lives, but glosses over unnecessary surgeries, episiotomies, nosocomial infections, and so on.

My point is not that medicine can do damage. Whether a woman gives birth at home or in the hospital, she and her baby are likely to be fine. My point is rather a broadening of my earlier thesis. Just as bioethics can divert attention from issues like poverty—death-dealing poverty—it can inflate the role of medicine in good health. We don't usually *say* that access to doctors is the most important factor in health and happiness (although the rhetoric of prenatal care does just that). But bioethics is newsworthy; we in the field get calls from the media and invitations to speak, and attention to us is perforce attention to health care, which comes to seem more important to well-being than it is.

But all that attention, that ease of access to the media, means that we have a bully pulpit. If I continue to give a voice to those who oppose medicalized childbirth, I will reach more people than can Lauren and her network. If bioethicists decide to pay more attention to poverty, we have some chance of being heard.

Saturday

The week ends with a hellishly difficult and unsatisfying ethics consultation. Many people are involved, most of them unpleasant, hostile to me and to one another. The parents of a badly impaired three-year-old child are resisting further treatment; their family physician agrees, but the pediatrician involved wants to report them to Child Protective Services. Sitting in are the family lawyer and a representative of the insurance company. I wear a red blazer, trying to remember whether that counts as a power suit; almost immediately I realize that we should not have met at all without much more preparation. I can do nothing to facilitate respect, communication, or emotional processing. I act like a judge and write an opinion like one. "Parents may decide their child's treatment unless it is clear they are acting against its best interests; given medical disagreement we cannot conclude that this is the case. . . ." Afterward I'm anxious for days, worried about whether the little girl will live—knowing that what I wrote makes her death more likely. Doctors live with this responsibility all the time, and I understand their burden better now.

Philosophically, however, I'm stimulated by the case, for which I think standard terms like best interests, risks and benefits, are not completely adequate. We need a vocabulary about possible futures and steps that commit us to one rather than another. The language of risk, strictly speaking, will do the job, but "risk" is a reifying and static word. We need something which suggests more explicitly a sequence of events, each stage having burdens and benefits, each choice making the following ones more or less likely. The pediatrician at this meeting told us the parents were willing to risk their child's death rather than have minor surgery. But the girl will have nothing like a normal life span in any case, and the parents did not trust the only doctor available. They preferred to chance an earlier death for their daughter rather than setting out on what might in retrospect seem a futile course of botched interventions.

The human elements, however, were more striking than the philosophical ones. One of my team members mused a few days later, "No one felt better when that consult was over." And she was right; no one did. Is there any place in the literature of ethics consults that uses the criterion "Participants felt much relieved?" Probably not; but there should be. I've come to think of it as a necessary condition—not a sufficient one—of success.

It was at that point that I realized how different my professional life had become. This is not just different in degree from my life ten years ago; it is different in kind.

II. BIOETHICS AS A PRACTICE

From 1973 to 1991 I taught philosophy in traditional undergraduate programs. My move to Michigan State in 1991 was more dramatic than I expected, a move not just to new subjects but to new activities, interactions, and purposes: to a new form of professional life. Alasdair MacIntyre's ([11], pp. 169ff) notion of a practice is valuable here: a practice is a coherent and complex set of activities, socially constructed, changing and developing through time. Each practice is lived out in a specific way: the lives of a physicist, a neurosurgeon, a portrait painter, are all different. As MacIntyre uses the term, chess is a practice, checkers is not; architecture is, and bricklaying is not.

Bioethics may not be sufficiently complex and coherent to count as a practice; perhaps it resembles bricklaying more than architecture. But that is not my concern. As a practice, or a near-practice, bioethics should be evaluated not only for its scholarship but more broadly, for its practical impact. Recognizing that, Ron Carson and Chester Burns asked contributors to this volume to address, not a body of scholarship, but "this work." "Has it made an appreciable difference? . . . Are the sick better off?" I would like to hold up an even higher standard, and ask whether the world is better off. First, however, let me bolster my contention that the work bioethicists do constitutes something like a practice.

Bioethics is a Practice

My professional life involves not just reading, thinking, and putting words on paper, but many different kinds of learning and sharing. The week of November 7 was not unusual. Teaching, writing, consulting, contributing to hospital and public policy, offering comfort and support, organizing conferences, speaking to the public, and so on and so on: this is a complex life. Probably no two bioethicists lead lives of exactly the same shape; the various profiles may share only a family resemblance. For that reason I hesitate to call bioethics a practice in the strictest sense; yet this does not really matter. What we do have, I will argue, is a new form of professional life, sufficiently coherent and complex to make MacIntyre's questions illuminating.

To begin with, however, we must remember that *every* academic field is a practice, at least in this loose sense. Not one of them consists simply of disembodied texts. These texts are written by individuals chosen according to explicit and to implicit criteria, for a variety of motives and with assorted results. Postmodernists have been pointing this out rather colorfully for the past decade, as have sociologists more quietly for the past century. But many fields remain unself-conscious about their fuller selves and therefore unself-critical about, for instance, the treatment of adjunct and junior faculty, and about the question of who should attend college and why. The newness of bioethics is startling to me, and clear to anyone who received traditional academic training. This newness may help us see what the field really is, in all its practical as well as theoretical implications.

It is Not a Subset of Philosophy

I would argue that bioethics is not a subfield of philosophy. In the 1970s many thought of bioethics as the application to health care of the insights of philosophy, in the way that biology (we thought) serves medicine. But we know from the history and philosophy of science that practical fields do not just benefit from theoretical ones, but feed them as well. Lens-makers do not just learn from physicists but teach and equip them, give them crucial data and make new kinds of inquiry possible. Biology, too, learns from medicine: learns that some procedures and medications work, long before we know why. Aspirin is one well-known example.[3]

The same is true of health care and philosophy, as Stephen Toulmin was one of the first to note, and as Henry Richardson [13] has lately demonstrated. Richardson describes the relationship between abstract norms and concrete decisions not as an application but as a specification: he is working toward a new theoretical model of the relationship between general principles and the ethical life.

As I thought back over my week of November 7, for instance, I remembered thinking that relatively little was at stake in whether Mrs. Bactri was given a gastrostomy. I now believe differently. Little is at stake if we think only of a few months added, or not added, to an already long life. A description in terms of respect or autonomy captures more. But the most philosophically interesting description concerns the way a death can change the meaning of a life. (This is most obvious in deathbed conversions.) Catherine Bactri's stubbornness on her mother's behalf, and our respect for it, vindicated her mother's determination and proved the strength of their relationship. Any other outcome would have been a critical defeat for both women.

Life and health care, in other words, supply philosophers with important issues and insights. The relationship is not one-way but reciprocal.

Bioethics is a New Practice

Doing bioethics, then, demands both more and less than graduate training in philosophy. Ethicists can and do come from other disciplines (religion, philosophy, social science, health care, law); those from philosophy must learn a great deal of empirical material, those from other fields must learn some ethical theory and acquire some analytic skills. Whatever their original discipline, bioethicists need to master a growing literature specific to bioethics. They will need to learn a language, to know paradigmatic cases, and to acquire some interpersonal and institutional skills. They will have to meet standards quite different from those they are used to.

To some extent every academic field is different from the others. Our vocabulary, and institutional structures within higher education, can obscure this. The word "research," for instance, means quite different things to a laboratory scientist, an anthropologist in the field, and a philosopher in the library. The first may supervise a team of twenty people and a budget of a hundred thousand dollars, the second live with stone-age people in the Kalahari Desert, the third immerse herself in the language of Aristotle. Each needs different abilities and has different experiences. Bioethics is equally distinct.

One mark of this new field is that the traditional professorial duties, teaching, research, and service, take such new forms; in fact they become difficult to distinguish. Certainly we teach medical students and residents. But in some sense we also teach many others, in grand rounds, in ethics consultations, in developing hospital policy, and so on. One could say these are efforts to transform the institutions which shape our paying and official students, but that, I think, would be a rationalization. For the most part we are trying to help medicine become more humane and more ethical, in whatever ways we can. Recognizing this sheds some light on the traditional category of "teaching," which is more commercially defined than is obvious. What separates students from others who learn is that students have paid for the privilege, and will receive a credential for their money and effort. Our special obligations to them arise neither from their seriousness nor their vulnerability but from the conditions upon which we receive our salaries. In bioethics, I suppose, there is a special obligation to medical students and residents. Yet medical school policy may require only thirty-five hours a year teaching these "official" students, where an undergraduate professor will spend from 180 to 360 hours a year in the classroom.

Just as bioethics teaching and service intermingle, so do research and service, particularly because "research" has come to mean "publications." We write for many different audiences and purposes. When the same point is made for several different professional journals—say, radiologists, oncology nurses, and physical therapists—it can feel as if one is writing with a computer's Search and Replace function.[4] Yet the point—about futility, say—may be new to each audience, including the bioethicists in that audience. The article will be at once a service to practitioners in the field and a part of the advancement of bioethical understanding.

Perhaps what distinguishes bioethics most is its relationship to the lived world. Ethicists aim at changing it, and that becomes a criterion of success. A definition of death, for instance, which was psychologically impossible for people to accept, would be inadequate. In this respect bioethics resembles the nontraditional fields for which schools like Michigan State were founded: agriculture, engineering, and so on. (And it's worth noting that bioethics developed early and strongly in schools like my own rather than in more elite institutions with their historical suspicion of the practical. Michigan State University's land-grant heritage is a source of strength, and something of which we are proud.)

As a philosopher, I am refreshed by both the concreteness of the field and the various kinds of knowledge at work in it. During the week of November 7, I both drew from philosophy and felt that I grew as a philosopher; but just as importantly, I drew from many other fields. In particular I remembered anthropology's warning that descriptions of any situation are always suspect. (They sometimes call themselves "natural anti-scripturalists.") When I began each consult that week, I knew that the initial brief case description I had been given would resemble only roughly what an hour's conversation would reveal.

And I learned later how stories live on in hospital folklore, in partial and untrustworthy forms. That painful Friday consult, for instance, is commonly talked about among hospital staff, names and all, and is told entirely from the point of view of the intensivist.

Its interdisciplinarity, its need to be effective, the new forms in which we gain and disseminate knowledge—all of these contribute to making bioethics a distinct practice.

III. EVALUATING BIOETHICS AS A PRACTICE

Thinking of bioethics in this way—as a practice and as a new one—is useful. It presents us with a whole sheaf of questions, ranging from the way we treat one another, through the dangers of interdisciplinary work, and on to the point I first articulated last November: our obligation to be careful about where we direct public attention.

The Implications of Newness

The social contours of a young field may be more readily apparent than those of older, more mystified domains, and possibly more malleable. One of those social facts has to do with entry: People now come to bioethics in various ways, usually after formal training in something more traditional, although a few graduate programs offer explicit bioethics tracks. When someone asks me how to enter the field, and what they want is formal training offering a credential, I am often at a loss. Usually these are people who would not enjoy graduate work in philosophy.

If Master's level work is what they want, I can suggest our interdisciplinary Health and Humanities program. Real entry into the practice of bioethics, however, will come from being actively involved in writing, speaking, and consulting, and may well depend on stumbling into working relationships with those already doing so.

Interdisciplinarity, furthermore, is a kind of country road into bioethics: a slower, richer view of the terrain, but full of potholes. As I explore connections between poverty and health, for instance, I am cautious, painfully aware that I am not trained to evaluate the social science I read. For the same reason, it can be painful to see philosophical concepts misused by those not trained to use them. This must be a problem in many interdisciplinary areas—I know that it is in women's studies, and imagine it is in American studies, medieval studies, and so on. The triangulation provided by different perspectives is essential, but of little value if the perspectives are used haphazardly. Good work, like recognized work (not the same thing!), virtually requires connection with others in the field, in this case people who have mastered methodologies one has not.

Even the relationship to one's field of origin can be a problem. Last November I had little time to read mainstream philosophy, and when I did, I had to clear mental space. There is so much to know and do outside of philosophy that I can anticipate a growing distance from it. Yet this could be a serious loss.

As do virtually all fields, bioethics has an elite, membership in which may arise from merit, from history, or from personal connections. To some extent this is inevitable in any complex human institution. But bioethics is a fresh and still-changing field, and it might be fruitful to cast the light of democratic ideals upon its professional networks.

Internal Goods: Defining Who We Are

MacIntyre points out that a practice has "internal goods," which can only be understood in terms of the practice and which are partly constitutive of it: skillful strategies in chess, for instance [11]. What are they in bioethics?

Roughly speaking, bioethics tries to expand our understanding of what counts as moral health care, and tries to bring it about, especially through education, policy formation, and encouraging moral conversation. These goals have evolved, as they do in every practice. If few would now say that we *apply* ethical theory to medical problems, most would agree that we *use* the skills of philosophy and other disciplines to address ethical problems in health care. The move away from "application" has also been a move away from adjudication and toward discourse. William Ruddick ([14], pp. 20-22) understood that this was an intrinsic goal long before most people did. In 1983 he identified "discursive moral competence" as a goal of ethics education: "the ability to discuss in appropriate moral terminology a variety of routine and rare cases with the variety of people likely to be involved in these cases . . . [:] physicians, nurses, patients, relatives, lawyers . . . ." He thought it was a mistake to describe moral reasoning as "primarily *problem solving* when in fact there may be no way to resolve many medical quandaries." Recently Margaret Urban Walker developed a similar point, portraying ethics consultation as

> a kind of interaction that invites and enables something to happen, something that renders authority more self-conscious and responsibility clearer. It is also about maintaining a certain kind of reflective space (literal and figurative) within an institution, within its culture and its daily life ([16], p. 33).

That, I think, is exactly right, not just about consultation, but about the entire complex of bioethics activities.

All this helps explain why I find the common criticism that bioethics is inordinately "principlist" both false and uninteresting. More than a decade ago Ruddick ([14], p. 31) found "at least *three* modes of moral reasoning, which we might caricature as: 1) The Protestant mode, which stresses conflict of values, dilemmas, personal 'wrestling,' and agonizing decisions; 2) the Catholic mode, which supplies principles (e.g., Double Effect, Omissible Heroic Efforts) to analyze cases and resolve dilemmas (without much agony or personal decision); and 3) the Jewish mode, which uses anecdotes (actual or fictional cases) to pose questions and suggest answers." Philosophy, too, supposedly the source of the evil sterility of "principlism," has been multilingual at least since 1981, with Alasdair MacIntyre [11], Lawrence Blum [2], and Bernard Williams [18], among others, introducing new vocabulary and questions. I also think that "principlist" work, where it exists, has done important and impressive things.

Bioethics is not now, if it ever was, the narrowed discourse that critics present. But my major contention is that published writing is not the proof and the point of our activities, although it is one of the major engines propelling us. The ability to converse, a space in which it can occur: these are public accomplishments, and demand *more* than words.

Each of the activities that make up our days might have its own internal goods, sometimes specifications of Urban Walker's goal, sometimes not. In another essay I would try to pin down the sense in which a consult, for instance, ideally "makes people feel good." I suspect much of the answer lies in the suggestion that a consult should try to foster moral working relationships among the parties involved.

External Goods and Temptation

If internal goods help make a practice what it is, external goods can sometimes strengthen, sometimes destroy it. To the question, "Are the sick better off?" my immediate response was, "I don't know. But we are." The ways in which we benefit from the field need some attention.

MacIntyre explains the concept of external goods by describing a child who has been offered candy for each game of chess she plays [11]. Learning to play for the sake of the reward, she eventually takes pleasure in the game itself. Here external goods do not compete with internal ones. In other cases they do; professional sports[5] are a good example. Rewards—trophies, prizes, fame—ideally help motivate people to the highest kind of physical (and mental) achievement, and serve as a public recognition of what they have done. Competitors—again ideally—spur one another on; each gains from the striving the other makes possible. Big enough prizes, however, can lead competitors to try to cripple one another or to risk serious injury to themselves; payoffs can induce them to throw a game. External goods can be the enemy of internal ones.

All academic fields offer some external goods, typically status, security, and a reasonable income. It's not uncommon for junior faculty to feel torn between these and the internal goods of teaching and scholarship: to feel they need to please students rather than teach them [1], to write for the sake of publication rather than for the sake of knowledge. In bioethics more money is at stake, real fame is possible, and a new set of questions can arise. As a newcomer to the field, I have been struck by its relative affluence. For the "service" components of our work we often receive stipends, five or ten times what a philosopher of comparable stature might ask. (My first year in the field I was invited to speak at a college about an hour away; asked to name a stipend, I asked a colleague on the same program, "Would thirty-five dollars be about right?" "Don't do that! You'll depress the market!" was his astonished, and astonishing, response. I didn't.) Our academic departments—compared to philosophy departments—are rich. The roughest drafts of this essay will be printed on bond paper; photocopying demands as little thought as opening a faucet. Across campus, the philosophy department once held a faculty meeting devoted almost entirely to rules for photocopying—struggling to allocate fairly a minuscule supply budget. Perhaps the contrast only speaks to the deplorable way in which the classic humanities disciplines are treated. What a waste of time and energy for educated, accomplished men and women! On the other hand, perhaps bioethicists are overly comfortable with the disparity because we, in turn, are surrounded by physicians, who earn so much that we can feel relatively impoverished.

I don't know whether there is an ethical problem here or not. I do know that the issue never seems to be raised, and I think it needs attention. Is it true, as it seems to me, that our resources come indirectly from the country's health-care budget? If it is true, does it matter? I am convinced that what we do is important, and that a laborer is worth her hire. On the other hand, I do not believe that whatever a market happens to yield is just.

Fame, or at least a chance to be quoted by the press, is another fringe benefit in bioethics, as unfamiliar to traditional academics as are handsome stipends. Again, there is no necessary tension between this and the internal goods of the field; on the contrary, a chance to be heard is crucial if we are to promote public moral reflection. I hope it is true that issues in medical ethics are seen less simplistically, and conversation about them less often angry, than for other political issues today. If so, we should take some credit for that. On the other hand, the demands of brevity and the competition for an audience—especially in television—create a constant danger that what we say will be distorted. And, flattered to be asked, desiring to be heard, we can succumb. Baruch Brody has suggested one interesting solution. In some cases, he says, refuse to offer "sound bites;" but offer all the help a journalist might want in understanding the issue and framing its presentation [3].

When is a Practice a Good Thing to Have?

Finally, there is a question that MacIntyre does little to answer, but which we must ask. If neurosurgery is a practice, perhaps torture is, too. In asking whether a practice is a good thing overall, we can use a standard, eclectic set of criteria: what are its consequences? What kind of relationships exist within it? Does it respect basic rights? Is it a force toward honesty, fidelity, and freedom? And so on.

Those questions are too large, and some of them too empirical, for this paper. To answer Ron's and Chester's question at last, I hope and I believe, but I do not know, that the sick are better off for our existence. Rather than defend that answer, I want to discuss the issue that crystallized for me during that week in November: bioethics both feeds upon and feeds the American fascination with health care. Henk ten Have tells us that European attitudes are cooler, and suggests that this country will eventually weary of the topic [6].

The fascination has some healthy results; it also does some harm. I see the harm as twofold. First, the valorizing of health and "health behaviors," to which it sometimes seems that every other value is being subordinated. Secondly, the valorizing of medical care, and the consequent obscurity of social contributions to health.

In another paper[6] I look at the first problem. Suffice it to say here that a local hospital will soon host a public session on the health benefits of pleasure, sensuality, and altruism. What greater comment upon American puritanism than our need for reassurance: it's all right to feel good; it's good for your health! And since I think "lifestyle recommendations" promote self-absorption, it's good to know that altruism is again acceptable, since we've discovered it to be healthful.

In this paper I want to concentrate on my second concern. Poverty is a major source of suffering, of sickness, and of early death. We in bioethics use social resources: the media want our words, students want our courses, foundations fund us, part of the health-care budget goes toward our stipends and our travel. There is a feedback mechanism here: we manage so often to be heard because people believe that health care matters enormously. What we say can reinforce that belief.

Yet everything I have learned tells me that the belief is mistaken. Doctors and nurses do help keep people alive and well, but social factors do more. Since last November I have been exploring the relationship between poverty and health. A colleague in epidemiology shared with me John Cassel's classic paper, recently reprinted [4]. Cassel brings together a variety of data suggesting that a disrupted social environment makes us more susceptible to disease. A recent letter in *Lancet* [17] finds a similar, and a particularly poignant, correlation between infant mortality and economic stratification: not the correlation we all know, between *poverty* and perinatal death, but between sheer economic *inequality* and infant mortality. The connection is so strong that if one holds constant the economic status of the lowest class and simply increases the status of the highest, infant mortality will increase. It is not simply lack of resources that makes the poor sick, nor lack of health care; it is lack of meaning, of stability, of self-esteem, of hope.

Health care, in other words, matters much less than people generally think. But in the past twenty-five years the media have begun to treat medical research as headline news, and we have been part of that. My students are fascinated with certain spectacular questions: Should we discard frozen embryos? Should we use anencephalic babies as organ donors? Is Kevorkian a saint or a devil? At least last fall they did not seem very interested in how doctors and nurses should treat one another, nor in whether people were dying for lack of health care—or for lack of fuel.

If bioethics were simply one intellectual discipline among others, such indifference might not matter. Each discipline has its own domain, and none can take on the whole world. Studying pears or polyphony or Aztec ritual is a worthy activity even if other things are more important. And academic writing attracts so little attention that we need not worry about its distracting the general public. If bioethics were simply an academic activity, that defense would be legitimate. It probably *is* a legitimate defense of our scholarly writing and our ethics consultations.

But bioethics is more than that. It helps attract and focus public attention: it tries to help society think more deeply and act more wisely about matters of health. And so we need to ask whether we are making the world better, or dazzling it further into blindness.

Compared with the other fora available to scholars, bioethics supplies a bully pulpit, one which I think we should use more reflectively. We should find ways to say, whenever we speak or write outside the scholarly arena, that both health care and "health behaviors" are relatively insignificant contributors to health and longevity, compared to the circumstances in which people live and work; and that the intriguing puzzles of surrogacy and euthanasia are not the only ones that should engage us, nor the only ones which yield to patient, persistent moral reflection. In a country where political discourse has been bastardized beyond parody, we could do more toward expanding the domain of public moral reasoning.

Center for Ethics and Humanities in the Life Sciences
Michigan State University
East Lansing, Michigan, U.S.A.

NOTES

[1] All names of patients and their families are pseudonyms. Details of the cases have been altered to protect the identity of those involved.
[2] The panel was held during the March, 1995 meeting of the Association for Practical and Professional Ethics, in Washington D.C.
[3] A point I first heard Stephen Toulmin make (Galveston, 1990).
[4] Lisa Parks made this point in conversation, April 1995.
[5] Here I include college football and men's basketball.
[6] "Some Costs of Preventive Medicine," manuscript.

REFERENCES

1. Andre, J.: 1995, 'My client, my enemy', *Professional Ethics*.
2. Blum, L. A.: 1980, *Friendship, Altruism and Morality,* Routledge & Kegan Paul, Boston, MA.
3. Brody, B.: 1996, 'Whatever happened to research ethics?', in this volume, pp. 275-286.

4. Cassel, J.: 1976, 'The contribution of the social environment to host resistance', *American Journal of Epidemiology* 104, 107ff; reprinted 1995, *American Journal of Epidemiology* 141, 798-123.
5. Hart, N.: 1985, *The Sociology of Health and Medicine,* Causeway Press, Ormskirk.
6. ten Have, H.: 1996, 'From synthesis and system to morals and procedure: The development of philosophy of medicine', in this volume, pp. 105-123.
7. Henderson, D. K.: 1994, 'Epistemic competence and contextualist epistemology: Why contextualism is not just the poor person's coherentism', *Journal of Philosophy* XCI: 12, 627ff.
8. Hilfiker, D.: 1994, *Not All of Us Are Saints,* Hill and Wang, New York.
9. Irwin, S. and Jordan, B.: 1987, 'Knowledge, practice, and power: Court-ordered Cesarian sections', *Medical Anthropology Quarterly* New Series 1:3, 319ff.
10. Lukes, S.: 1974, *Power,* The MacMillan Press, London, U. K.
11. MacIntyre, A.: 1981, *After Virtue,* University of Notre Dame Press, Notre Dame, IN.
12. Reiser, S. J.: 1994, 'The ethical life of health care organizations', *The Hastings Center Report* 24, 28-35.
13. Richardson, H. S.: 1990, 'Specifying norms as a way to resolve concrete ethical problems', *Philosophy and Public Affairs* 19, 279-310.
14. Ruddick, W.: 1983, 'What should we teach and test?' *Hastings Center Report* 13, 20-22.
15. Toulmin, S.: 1982, 'How medicine saved the life of ethics', *Perspectives in Biology and Medicine* 25, 736-750.
16. Walker, M. U.: 1993, 'Keeping moral space open', *Hastings Center Report* 23, 33.
17. Wilkinson, R. G.; 1994, letter, *The Lancet* 343, 538.
18. Williams, B.: 1981, *Moral Luck,* Cambridge University Press, New York, NY.

THOMAS R. COLE

TOWARD A HUMANIST BIOETHICS: COMMENTARY ON CHURCHILL AND ANDRE

Sometime in the mid-1960s, the bloom began to fall from the rose of modern medicine. Within a decade it became clear that scientific medicine—a profession that had explicitly detached itself from broader frameworks of meaning and value—was not intellectually equipped to handle the moral and existential questions produced by its own technological power. Hence entirely new fields of academic inquiry and professional practice, known as bioethics and the medical humanities, arose to grapple with problematic issues such as the protection of research subjects, the goals of medicine, definitions of death, the rights of patients, cessation of treatment, the meaning of illness, and the distribution of health care resources.

As H. Tristram Engelhardt, Jr., notes in his essay in this volume, those who contributed to the first Trans-disciplinary Symposium in Philosophy and Medicine (held at the University of Texas Medical Branch in 1974) still enjoyed the intellectual self-confidence and sunny optimism of high modernity in philosophy and the humanities [7]. The stormy waves of French postmodern thinkers like Foucault, Derrida, Lyotard, Lacan, and Baudrillard had not yet reached the shores of American bioethics and medical humanities. Richard Rorty's revival of pragmatism and attack on foundationalism were still five years away.

The chapters by Larry Churchill and Judith Andre reflect the arrival of the postmodern wave, which has brought previously reliable ways of knowing, interpreting, and acting under suspicion, while rendering claims for the unity or universality of knowledge unsupportable. The modernist dream—of decontextualizing and purifying human reason to achieve unified, formal, and unquestioned knowledge—is over.

We are all living through the confusion and uncertainty inherent in a more general reconfiguration of late-twentieth-century thought in which previously

R. A. Carson and C. R. Burns (eds.), Philosophy of Medicine and Bioethics, 173-179.

accepted disciplinary boundaries and unifying ideas are giving way to blurred genres—forms of knowledge that accept (rather than erase or obscure) contingency, contradiction, paradox, irony, and uncertainty in human activity. Yet the death of an older metaphysics does not mean the end of rationality. As Habermas puts it, "Philosophy must operate under conditions of rationality that it has not chosen. . . . It now disposes only over knowledge that is fallible" ([17], p. 171). Churchill and Andre do not allow the postmodern wave to sweep them out to sea. In a sense, they are bearing witness to the dethroning of a certain kind of academic philosophy as the guardian of reason and the privileged arbiter of ethical issues in medicine. Each expresses both the uneasiness and the exhilaration of pluralistic exploration.

One salutary effect of the postmodern attack on the metanarrative of progress is that it has undermined the widespread temporal prejudice favoring the future over the past. Our cultural moment opens the door to appreciating history in general, and to recovering and reconfiguring the humanist tradition in particular. Despite the contemporary attack on the humanist tradition (launched by postmodern skeptics as well as religious fundamentalists), I believe that we are due for a renewal of humanism and for the creation of a humanist bioethics. By emphasizing the concreteness, complexity, and interpersonal nature of moral problems in health care, these elegant chapters demonstrate that the seedling of bioethics is already prepared for the compost of humanism.

To date, the only attempt to link humanism and bioethics is Engelhardt's recent volume, *Secular Humanism and Bioethics.* Yet as Faith Lagay has shown, Engelhardt's outline of a humanist bioethics is unconvincing because it violates core elements of the humanist tradition [6], [11]. Unlike Engelhardt's attempt to salvage a humanist bioethics by aligning it with philosophical doctrine, these chapters (though they do not explicitly allude to humanism) are suggestive of a bioethics that resembles the eclectic, rhetorical humanism of the Renaissance rather than rationalist humanism of the Enlightenment [14]. Ethics, Churchill informs us, is "first and foremost a matter of bumps and bruises encountered as people seek moral direction." Travelling "many miles" and working within "many different institutions," Andre reveals some of the personal bumps and bruises she receives in the course of one week's work as a self-questioning bioethicist.

Churchill's chapter is primarily a critique of the centrality of theory in bioethics and a call for more careful description and attention to the detail and complexity of moral experience. Claiming that human moral activity cannot be grasped by any single theory or discipline, Churchill calls for methodological pluralism in bioethics—for theoretical diversity and inclusion of the humanities and social sciences in moral inquiry.

For Churchill, the goal of bioethics is not attaining philosophically justified knowledge but helping people resolve moral problems. He outlines two scenarios to demonstrate his belief that bioethicists often oversimplify complex moral situations and impose their own preconceived ideas for the patient's experience. "Because moral experience is complex, it requires careful mapping." Unfortunately, academic bioethicists often "bring their own maps and ignore the sketches provided by the natives."

Although I agree with Churchill's concern about academic theory trumping patients' experience, his metaphor of cartography is not rich enough to help us imagine the genuinely dialogical process he envisions. Moral experience cannot be "mapped"—a static, two-dimensional image. It must be brought to language, formulated and negotiated in an ongoing process whereby individuals engage the issues and each other. As Churchill himself puts it, "The idea that the experiences of patients and doctors might . . . be key formative items for ethics . . . has been only a minor note in the bioethics movement to date." Here, I suggest, we can be helped by paying attention to the contemporary recovery of rhetoric [15].

For the most part, "rhetoric" is used today as a term of contempt, to refer to emotionally persuasive language which lacks logical rigor. But it is important to remember that rhetoric originated as a means of teaching free citizens how to deliberate in public. When humane studies first made their appearance in Athens in the Fifth century B.C.E., they did so as a program of adult education designed to produce accomplished, persuasive public speakers ([10], pp. 86-90; [13]). Rhetoric also lay at the heart of Renaissance humanism, an educational movement which recognized, as William Bouwsma puts it, that "forms of thought are part of thought itself, that verbal meaning is a complex entity, like the human organism, which cannot survive dissection" ([2], p. 76; [12]).

The humanist tradition has always contained an unresolved tension between imagination and the arts of language (rhetoric) on the one hand and the search for rational conclusive knowledge (philosophy) on the other. For rhetoricians, the human mind is confined to the subjective and transitory world of experience, and all our truths must therefore be confined to probability and convention. Hence Protagoras's heretical epigram, "man is the measure of all things." Opponents of rhetoric, e.g., Plato or the medieval scholastics, generally believed that language is merely a transparent means for conclusively demonstrating objective reality.

Today, the recovery of rhetoric is one of the joys or consolations (depending on your perspective) arising from the ruined dream of universal reason. Thomas Farrell, for example, urges us to look to Aristotle for a model of rhetoric capable of reinvigorating public communication and deliberation [8].[1] For Aristotle, rhetoric is a legitimate cognitive path (distinguished from analytic and dialectic modes of reasoning), a way of making ongoing sense of appearances by expressing them as proposed themes and arguments inviting decision, action, and judgment. Rhetoric, then, does not aim at definitive knowledge but at contingent, normative understanding necessary for good deliberation.

Despite its tarnished reputation and its potential for degenerating into demagoguery, rhetoric is a promising and untapped intellectual resource for a humanist bioethics. Rhetoric, I think, would not add a competing theory or method for bioethics; understood as a deliberative praxis, rhetoric might help accomplish one of Churchill's goals: encouraging individuals to be comfortable with the many perspectives and theories needed to attend to the varieties of moral experience ([18], p. 255). In other words, rhetoric might be conceived as part of the education of a fully engaged bioethicist, part of the formation of character and competency that has long been a goal of humanist education.

But because deliberative rhetoric aims to integrate knowing and doing, and because it generates dialogical normative knowledge to guide human action rather than factual knowledge, it requires that we relinquish the representational view of reality underlying Churchill's monological map metaphor. Whereas Churchill asks us to place bioethics in a social context as understood through the social sciences, a rhetorically oriented humanist might see bioethics in a public context, understood through the dialogical, deliberative process and the diverse perspectives of the actors involved.

"Life is confused and superabundant," wrote William James in 1904. "And what the younger generation appears to crave is more of the temperament of life in its philosophy, even though it were at some cost of logical rigor and formal purity" ([9], p. 533). Churchill and Andre may no longer qualify as members of "the younger generation," but they are unwilling to sacrifice the "confused and superabundant" quality of moral experience to abstract theory. By emphasizing personal attentiveness and participation, both authors point toward another key element of humanist bioethics: the personal *engagement* of knowledgeable individuals rather than the technical *application* of impersonal knowledge.[2]

Judith Andre's chapter, "The Week of November Seventh: Bioethics as a Practice" epitomizes such engagement. The first section of her chapter—a diary covering one week's personal feelings and thoughts about her work as a bioethicist—is a breath of fresh air in a field where first-person narration and emotion are often excluded as irrelevant. The second section contains a set of more general reflections on bioethics as a social practice. The power of Andre's chapter derives from the interweaving of her personal reflections and her more general claims about bioethics as a practice.

For someone trained in analytic philosophy, Andre's diary-writing is a gutsy gambit. It reflects her own courage as well as a growing cultural awareness of the reflexive nature of all social thought. In genre terms, the diary might usefully be placed in the traditions of spiritual autobiography, women's personal writings, and/or philosophical autobiography. It nicely exemplifies David Tracy's notion that "all theory worth having should ultimately serve the practice of reflective living" [16].

Except for any explicit reference to the humanist tradition, Andre's chapter contains the key elements that I believe humanism can offer bioethics: personal engagement; the pursuit of self-knowledge, self-development, and self-criticism; civic commitment; the integration of humane feeling and knowledge (originally contained in the Latin word *humanitas*); and—here is rhetoric—close attention to the complex verbal meanings that are constitutive of moral deliberation.

Right from the beginning, Andre's diary takes us behind the professional mask, and we learn about some of the pains and pleasures of the bioethicist's personal engagement in her work. Reflecting on the case of Mrs. Bactri, Andre highlights *feelings*: Mrs. Bactri's daughter felt better, the physician probably felt more comfortable, Andre felt uneasy. She wonders whether the expense of the consultation was justified, whether she had unwittingly helped the physician avoid his responsibility to talk directly with Mrs. Bactri.

A central theme of Andre's chapter is how often the "human elements" are more striking than the "philosophical elements" of case consultation. While she acknowledges the lost pleasures formerly derived from philosophy journal readings, Andre has no interest in a theory-driven bioethics. Instead, like Stephen Toulmin in his chapter, "The Primacy of Practice," she wants bioethics to theorize from the human experience of its own practice. "Is there any place in the literature of ethics consults that uses the criterion 'Participants felt much relieved?'" Andre asks. "Probably not; but there should be. I've come to think of it as a necessary condition—not a sufficient one—of success." This call to connect humane feeling with knowledge is a perfect example of *humanitas,* an ideal of personal integration which today might be called spiritual.

Just as Andre strives to connect emotion and understanding, she also balances personal introspection and social justice. Her concern for the poor, her analysis of the social determinants of disease, her worry that bioethics may become merely another self-serving field of expertise, and her belief that bioethicists should expand "the domain of public moral reasoning," are all examples of the civic dimension of humanist educators.

It is not a criticism of these chapters to point out that neither author makes explicit reference to humanism. Contemporary bioethics and medical humanities in general lack any historical understanding of the humanist tradition and therefore have yet to realize how humanist thought can nourish their own self-understanding and practice.

We need not be defensive or intimidated when confronted by overblown, postmodern proclamations of the "end of philosophy," the "end of humanism," or the "death of the subject." Instead, we need to remember that "humanism," "the subject," and "philosophy" are always embedded in the vicissitudes of historical change. Every ending implies a beginning; the only thing (as Gadamer reminds us) that is really certain—that each one of us is finite and mortal—is always available as the terrifying yet reassuring human ground on which to begin again in compassion. The task of connecting bioethics and medical humanities to humanist thought poses an exciting challenge as these fields develop in the *next* twenty years.

Institute for the Medical Humanities
The University of Texas Medical Branch
Galveston, Texas, U.S.A.

NOTES

[1] My thoughts about rhetoric have been shaped by ongoing conversations with Faith Lagay and by her paper, 'Deliberative Rhetoric in 1996: How It Got Here and Where It Might Be Going' (unpublished). See also Renato Barilli (1989) and Thomas Conley (1990).

[2] See Callahan, Caplan, and Jennings, eds., (1985) for a formulation of "applied humanities." The embryonic notion of "engaged humanities" originated with Ronald A. Carson in our jointly-taught graduate seminar "Humanities and the Medical Humanities." It will be more fully elaborated in a future volume of essays, *Practicing the Humanities: Forms of Engagement.*

REFERENCES

1. Barilli, R.: 1989, *Rhetoric*, G. Menozzi (trans.) University of Minnesota Press, Minneapolis, MN.
2. Bouwsma, W.: 1990, 'Changing assumptions in later Renaissance culture', in *A Usable Past: Essays in European Cultural History*, University of California Press, Berkeley, CA.

3. Bouwsma, W.: 1990, 'Socrates and the confusion of the humanities' in *A Useable Past: Essays in European Cultural History* University of California Press, Berkeley, CA, pp. 385-396.
4. Callahan, D., Caplan, A., and Jennings, B. (eds.): 1985, *Applying the Humanities,* Plenum, New York, NY.
5. Conley, T. M.: 1990, *Rhetoric in the European Tradition,* University of Chicago Press, Chicago, IL.
6. Engelhardt, H. T. : 1991, *Secular Humanism and Bioethics: The Search for a Common Morality,* Trinity Press International, Philadelphia, PA.
7. Engelhardt, H. T.: 1996, 'Bioethics and the philosophy of medicine reconsidered', in this volume, pp. 85-103.
8. Farrell, T. B.: 1993, *Norms of Rhetorical Culture,* Yale University Press, New Haven, CT.
9. James, W.: 1904, 'A world of pure experience', *Journal of Philosophy, Psychology and Scientific Methods,* I, 533.
10. Knox, B.: 1993, *The Oldest Dead White European Males,* W. W. Norton, New York, NY.
11. Lagay, F.: 1996, 'Secular, yes; humanism? no: A close look at Engelhardt's secular humanist bioethic' (unpublished paper presented at the conference 'Ethics, Medicine and Health Care: An Appraisal of the Thought of H. Tristram Engelhardt, Jr.', Youngstown State University, September 29 and 30, 1995).
12. Oakley, F.: 1992, *Community of Learning,* Oxford University Press, New York, NY.
13. Stone, I. F.: 1988, *The Trial of Socrates,* Little Brown, Boston, MA.
14. Toulmin, S.: 1990, *Cosmopolis,* Free Press, New York, NY.
15. Toulmin, S.: 1996, 'The primacy of practice', in this volume, pp. 41-53.
16. Tracy, D.: 1987, *Plurality and Ambiguity,* Harper and Row, San Francisco, CA.
17. Van Niekerk, A. A.: 1995, 'Postmetaphysical versus postmodern thinking', *Philosophy Today* 39, 171.
18. Zhao, S.: 1991, 'Rhetoric as praxis: An alternative to the epistemic approach', *Philosophy and Rhetoric* 24, 255.

RONALD A. CARSON

MEDICAL ETHICS AS REFLECTIVE PRACTICE

> Any serious discussion about morality must . . . begin with what people say, spontaneously or reflectively, about what things are worth aiming at.
>
> *J. M. Cameron*

In an effort to regain its moral bearings, medicine is asking: What values should guide healing? What are the virtues of healers? A variety of voices reply. Some say that the profession should look within and reappropriate its own honorable traditions of professional demeanor. Others propose a reanimation of the priestly mode of practice, shorn of the more debilitating elements of paternalism. But the most vocal and insistent reply has come from a bioethics that proffers a methodological remedy for medicine's moral malaise, a way of reasoning about moral quandaries that applies rules derived from principles to conduct.

Bioethics in the "applied" mode has become scholastic, focussed on fine-tuning decisional dimensions of clinical practice. The great pressing issues that prompted the emergence of the field over two decades ago are no less urgent now than they were then. We continue to deliberate and debate about how best to take care of each other in sickness and in dying. Bioethics has contributed appreciably to these discussions, but a repetitiveness has set in, due in part to the prevailing procedural methodology and contractual mind-set of contemporary bioethics.

Thanks in large measure to Alasdair MacIntyre's tour de force more than a decade ago in reintroducing the language of virtue and the concept of practice into discussions of contemporary moral philosophy and theology, bioethics has not been monotonously procedural and contractual. Any book on bioethics that aspires to comprehensiveness now contains a chapter on "virtue ethics," along-

R. A. Carson and C. R. Burns (eds.), Philosophy of Medicine and Bioethics, 181-191.

side obligatory treatments of consequentialist ethics and Kantian ethics. Nonetheless, the bulk of writing in the field continues to be predicated on the vision of a society populated by self-determined individuals whose mutual relations are driven by rational self-interest and governed by negotiated consent. Just how confining and at times misleading that vision is as a guide to reflection on medicine and morality, is becoming more and more apparent. Here I articulate a move beyond calculative decisionmaking to engaged moral inquiry.

Twenty years of overhearing conversations among doctors, nurses, patients and their families, and participating in those conversations myself, has taught me that our actual moral languages are far richer than we typically suppose. This discovery has prompted me to reflect on how this is so and why it is that we are so often tone-deaf to the nuances of these languages. In my experience, sick people almost always bring their own truths with them into their encounters with medical professionals. They may not be able to articulate those truths very clearly, or even identify them with much precision, but they bring their lives and their needs, as well as a sense of what's valuable and what's not, with them into the doctor-patient relationship. Consequently, one of the central moral challenges doctors must take up is that of helping sick people to "find their voices."

Once we concentrate on the voiced quality of conversation about moral matters, questions of interpretation necessarily arise ([6], pp. 1283-1288). Such questions, raised in the context of illness, are fundamentally questions of meaning. The doctor is drawn into a dialogue with the patient and asked to help make livable sense of sickness or injury. This demanding moral work requires not only reasoning skills but reading skills as well. It requires imagination. Metaphor is the preferred language of imagination and one of the chief elements of moral life; it is the vehicle of empathy. A metaphoric capability is the capacity to imagine "what it must be like"—not to know with any certainty how it *is* with another person, but to imagine, to get a provisional working sense of what it must be *like* to suffer in this way or that. To help patients make livable sense of what ails them requires of doctors an enhanced capacity to imagine.

I. THE POVERTY OF PROCEDURALISM

Bioethics was born of strife in the early 1970s. Debate about questions of personal morality had long been a mainstay of moral theology, and thus it is not surprising that much of the early work that accompanied the modern resurgence of interest in morality and medicine was done by theologians. I am thinking here in particular of Paul Ramsey's influential book, *The Patient as Person.* Ramsey believed that whatever our differences, we could surely find common ground in certain basic precepts, such as the sanctity of life and fidelity to others, as a point of departure from which to dispute our disagreements [17].

But such common ground was increasingly hard to come by. Abortion was becoming a contentious social issue, distrust of doctors, especially suspicions about overtreatment of patients near death, was widespread. There was growing public concern that the balance of power over life and death had shifted precipitously from individuals and families to physicians and hospital personnel. Birth and death were being publicly debated under the rubric of control. People were choosing up sides and positions were hardening. Not incidentally, this is the main reason the courts were centrally involved in bioethics from the beginning; medical ethics had become adversarial and people had no other forum in which to adjudicate their differences.

In 1973 a book appeared that radically changed thinking about medical ethics, Tom L. Beauchamp's and James. F. Childress's *Principles of Biomedical Ethics.* The problematic as conceived by the authors is this:

> Many books in the rapidly expanding field of biomedical ethics focus on a series of problems such as abortion, euthanasia, behavior control, research involving human subjects, and the distribution of health care. Rarely do these books concentrate on the principles that should apply to a wide range of biomedical problems. Only by examining moral principles and determining how they should apply to cases and how they conflict can we bring some order and coherence to the discussion of these problems. Only then can we see that there are procedures and standards for deliberation and justification in biomedical ethics that parallel those in other areas of human activity ([2], p. vii).[1]

Coherence was what was wanted, and procedures were to provide it. Right would be distinguished from wrong by the application of principles to practical problems. But is this not to stand human activity on its head? As Michael Oakeshott usefully observed,

> Doing anything both depends upon and exhibits knowing how to do it; and though part (but never the whole) of knowing how to do it can subsequently be reduced to knowledge in the form of propositions (and possibly to ends, rules and principles), these propositions are neither the spring of the activity nor are they in any direct sense regulative of the activity ([15], p. 90).

In the conscientious practice of caring for the sick, doctors confront moral questions, not by applying known principles but by conducting the practice in a morally responsible way: they "practice," they do the thing, using the know-how they have acquired by training and experience—not, however, instrumentally but practically. The moral questions that doctors understand to belong to the practice of responsibly caring for the sick are not known to be such in advance of the activity of trying to answer the questions. The practice itself shapes the questions and the form assumed by their answers. The practice, which amounts to knowing how to engage in the activity of caring for the sick, is "there," and is something into which each novice practitioner gradually must find his or her way.[2]

As the principles-and-applications approach of bioethics came into widespread use, skeptics began to ask how the privileged principles were selected and how they were to be justified. In what sense may such principles be said to be foundational or fundamental? ([9], pp. 219-236.) Several thoughtful participants in the debate about how ethical principles are anchored argue that what is needed is an overarching moral theory or a single unified framework within which to place and parse the dizzying variety of appeals to principle ([4], pp. 1-12). My experience at the bedside of the sick leads me in a different direction. In my view, the principles-and-applications approach is itself flawed because the moral predicaments of medical care are largely impervious to the requirements of logic. Furthermore, I believe that the need for theory to which we can have recourse in cases of conflict can be met, not by further abstracting from experiences of illness and practices of care but by delving more deeply into them, and by engaging in practical moral reflection and analysis there ([1], p. 222).

Medical professionals are distinguished above all by their trained capacity to see beyond the patient's distinctiveness to the regularities by which diseases are recognizable. But morally sound medical practice requires that the subtraction of the patient and the search for universals, though essential to the diagnostic hunt and instrumental to patient care, be temporary. The movement from diagnosis to prognosis and treatment remains incomplete until the doctor enters a relationship with the patient. For this, interpretive methods different from those appropriate to diagnosis are required, methods sensitive to subjectivity, attentive to particularity, and alert to meaning ([5], pp. 51-59).

Implicit in the practice of medicine is a vision of people in relation and occupying certain roles or positions. The practice requires that relations that occur within it be seen in light of this vision and its accompanying norms of fittingness and appropriateness. Practices are modes of social relation, forms of shared tacit understanding, virtuous ways of interacting with each other toward certain ends (health, the assuaging of suffering), which carry within them their own meanings and norms ([12], pp. 187-196; [19]). Norms need not be imported, but in situations of moral predicament and conflict, neither are they simply there for the asking, obvious to any right-thinking person; if that were so, no problems would arise. Instead, in practice, one proceeds, not certainly but provisionally, not unilaterally but dialogically—discerning, sizing up, discovering, constructing, engaging other parties to a morally vexing case to arrive at an understanding of what the case calls for by way of moral response. Michael Walzer puts it this way:

> the most interesting parts of the moral world . . . have to be "read," rendered, construed, glossed, elucidated, and not merely described. All of us are involved in doing all of these things; we are all interpreters of the morality we share. That does not mean that the best interpretation is . . . the product of a complicated piece of survey research—no more than the best reading of a poem is a . . . summing up of the responses of all the actual readers. The best reading is not different in kind, but in quality, from the other readings: It illuminates the poem in a more powerful and persuasive way. Perhaps the best reading is a new reading, seizing upon some previously misunderstood symbol or trope and re-explaining the entire poem. The case is the same with moral interpretation ([20], pp. 29-30).

Indeed it is, in medicine as in every other moral practice. The moral practice of medicine is one in which doctors engage patients and other professionals in the work of interpreting experiences of illness and injury with a view to making livable sense of them. What are the terms of this engagement?

II. CONSCIENTIOUSNESS AND MORAL INQUIRY

In the turn from applied ethics to reflective moral practice, the principal guidance we need emerges from our interpretive encounters with experience aided by provisionally settled opinion about previously controversial moral matters. By provisionally settled opinion, I refer to maxims, the articulable form of probable certainty on a particular matter. Maxims guide us as long as they stand up to the scrutiny of collective practical judgement.

Maxims are, of course, a kind of principle but they are illuminative rather than prescriptive. By throwing the light of provisionally settled opinion on a concrete situation, they enable us to interpret what direction subsequent events ought to take in order to maintain and promote the good and to select or devise the action most appropriate to that end. Maxims aid moral reflection by pointing or inclining the reflecting person toward what is fitting as he or she contemplates a novel situation demanding decision. The distinction is neatly drawn by James Gustafson.

> In the illuminative use of principles the center of gravity is on the newness, the openness, the freedom that is present, in which the conscientious man seeks to achieve the good and do the right. In the prescriptive use of principles, the center of gravity is on the reliability of traditional moral propositions and their reasonable application in a relatively open contemporary situation ([10], p. 191).

When principles in the form of maxims are used to illuminate circumstances, lawfulness displaces legalism. Legalism is a univocal concept in that it applies accepted principles to relevant situations. Legalistic thinking is conservative in that it accommodates the new to the tried and true. Lawfulness is not univocal but analogical in that it brings to bear upon ethical decisions the guidance, though not the imperative, of a generalized compendium of moral experience. Moral experience resists codification, but of course the perplexities of moral experience cannot be dealt with situationally without some steady guides. Maxims, understood as moral rules of thumb hammered out historically by people of conscience, are such guides to practical reasoning in the craft of casuistry [11].

In *Conscience and Its Problems,* Kenneth Kirk observed that "the exigencies of language force us often enough to speak of conscience as a distinct entity; but we must continually remind ourselves that it is no such thing. When conscience speaks, it is my own best self that speaks . . . conscience is my self in so far as I am . . . moral . . . . Conscience is not a discoverer of the unknown, but a craftsman working on the known" ([13], pp. 57, 103). Casuistry extends the maxims that encapsulate received wisdom to unforeseen cases and new problems. This is no application of the known to the unknown but an extension, in which light is thrown forward on the situation to be interpreted as well as backward upon the maxim so that the received wisdom is adjusted to take the new, heretofore unimagined, situation into account.

Casuistry is the exercise of conscientious judgment. It begins in the recognition that no principle, no matter how laudable, can avoid running up against another equally worthy principle and that no theory, no matter how overarching, can cover all the problematic particularities of lived experience. We begin with the things that have a felt rightness about them and, where necessary, examine that felt rightness in the light of our broader experience and the experience of others with whom we agree and disagree, asking how what we take for granted measures up when we no longer take it for granted but instead hold it up to scrutiny.

Given a morally problematic medical case, and disagreement about it, one wants to know which, if any, among alternative ways of handling the case is the right one. What is being asked for here? What does one want to know when asking which interpretation is the right one? Moral practices are not the object of assessment in the light of some independent standard of judgement, but the means to rightness [16]. When moral deadlock is reached, as often occurs in contemporary medical ethics, what commonly happens is that a definitive judgement is rendered. This is sometimes achieved by invoking the law as trump, or by appealing to ethical principles. When principles conflict, a further appeal is made to overarching theories within which the principles may be reconciled with each other, either by some lexical ordering or by reference to some scale on which the significance of conflicting principles can be weighed. When there is disagreement about the lexical ordering or further conflict over what weight to assign to this or that principle, a move into yet further abstraction is advised in what is surely a futile search for an ultimate theoretical foundation. As W. J. T. Mitchell has astutely observed,

> Theory is monotheistic, in love with simplicity, scope and coherence. It . . . always places itself at the beginning or the end of thought, providing first principles . . . or . . . schematizing practice in a general account. It is unhappy with the middle realm of history, practical conduct, and business as usual and so tends to seek a final solution, a utopian perspective, which presents itself as a point of origin ([14], p. 7).

What does this desire for moral finality amount to? If you and I have a moral disagreement, and I have carefully thought through the position I have taken, it would be odd for me to ask, "I wonder which of us is right?" If I ask this question I must be unsure of where I stand on the matter in dispute. Of course I may be impressed by the power of your convictions or swayed by the cogency of your arguments and thus be led to reconsider my position. But if I have reflected on my own moral convictions and am able to articulate why I hold them, what further question about rightness remains? Furthermore, it is redundant, and all too often dismissive, to say that a person's moral perspective is only his own point of view. What moral values could one have which were not one's own?

The fact that one has a moral point of view is not a sign of bias, as is often implied in the semantics of suspicion. It is, on the contrary, a minimum condition of moral seriousness.

The objection to this way of thinking seems to be that one person is in no better position than another to say who is right. But what does this reference to a better position amount to? It sounds like a factual claim but it is not. It sounds like one person's view of the circumstances is obscured, and his judgement thus stilted, while the other person's view is clear and her judgement therefore unimpaired. There is an element of truth in this claim. There are occasions and circumstances in which, by virtue of the character of a person's experience, one person may be in a "better position" than another, perhaps by virtue of having had more relevant experience, to size up a situation in dispute.

But in ethical discourse the claim of occupying a better position is usually bolstered by appeal not to experience but to something abstract, something beyond experience. It is reference to something "beyond" that is believed to render the claimant's position "better." This way of thinking about moral values as requiring regulation by something beyond themselves and of demanding that every moral judgement be shown to be an instance of a principle that in turn derives its authority from a moral theory is confused. This is not to say that when people differ they are both right from their different points of view; that would be a formula for deadlock or truce. The point is that the values adhered to by the parties to a conflict reside in practices that shape attitudes and actions.

A recognition of the multiplicity of moral practices should not prompt despair or skepticism, but should be considered an invitation to conversation. Our lives are affected by diverse moral influences. We feel the attraction of those different influences and are sometimes torn between them. Moral views develop in relation to one another. In order to understand why one stands where one stands, it is probably necessary to carefully consider the views of others. This is, after all, how virtues, understood as practice-shaped dispositions or inclinations, are cultivated.

Once we appreciate our interlocutors' views, we may still differ with them but we will now likely not be as quick to judge their beliefs and actions as we were before we discovered their meaning for them. Our judgements of another's view of what is required of him must be suspended until we understand that view. Whatever our judgements, they must not distort what the other person believes he must do by ignoring the context within which his life makes sense to him. To the extent that contemporary bioethics takes the form of normative theory, it obscures the importance of moral practices and the values and virtues constitutive of them in people's lives.

To sum up, reflective practices are derived from earlier acts of appropriation and criticism. Confronted with moral perplexity, they serve as a provisional aid to practical reasoning which moves simultaneously into a consideration of the particulars of a puzzling case and up to a level from which to judge the case. This latter is not an ascent into theory but a search for a vantage point from which to see the case whole and in context—that is, a point from which to view larger areas of particularity. Thus does reflective moral practice not desert particularity, but take in more of it, with the aim of finding a vantage from which the relevant particulars of the case are discernible and construable. Medical ethics in a cases-and-interpretations mode is less like problem-solving than it is like moral inquiry. Moral inquiry in medicine bears little resemblance to applied ethics but looks instead more like practical, critical reflection on the moral features of relationships between patients and doctors.

The tools for the moral work required in medicine are at hand in the practice and social context of medicine. These need to be adapted as new situations arise, they need to be whetted and honed and taught anew to every generation, but they are there, already in use, albeit often clumsily, and at times even beneath the awareness of the practitioners who use them. The task of medical ethics is to construe medical practice in a morally meaningful way, with a view to making it more deliberate, more reflective, more critical. This work of construal or interpretation is part discovery, part construction, and it aims at cultivating and refining the myriad ways that doctors have of responsibly attending sick people and taking good care of them.

Institute for the Medical Humanities
The University of Texas Medical Branch
Galveston, Texas, U.S.A.

NOTES

[1] Now in its fourth edition, this popular book, with its cogent discussion of the guiding principles of autonomy, nonmaleficence, beneficence, and justice and their application continues to shape the field.

[2] Oakeshott acknowledges that "It is, of course, not impossible to formulate certain principles which may seem to give precise definition to the kind of question a certain sort of activity is concerned with; but such principles are derived from the activity and not the activity from the principles" ([15], p. 97). I agree but prefer the language of "maxims" to

that of "principles" for reasons that should become clear later in the paper. See a related discussion of "embodied understanding" by Charles Taylor, "To Follow a Rule . . . ." ([18], pp. 45-60).

REFERENCES

1. Baier, A.: 1985, 'Theory and reflective practices', in *Postures of the Mind: Essays on Mind and Morals,* Methuen, London, U. K.
2. Beauchamp, T. L., and J. F. Childress: 1979, *Principles of Biomedical Ethics,* 4th ed., Oxford University Press, New York, NY.
3. Beauchamp, T. L.: 1995, 'Principlism and its alleged competitors', *Kennedy Institute of Ethics Journal* 5, 181-198.
4. Brody, B. A.: 1988, 'Introduction', *Moral Theory and Moral Judgments in Medical Ethics,* Kluwer, Dordrecht, The Netherlands, pp. 1-12.
5. Carson, R. A.: 1990, 'Interpretive bioethics: The way of discernment', *Theoretical Medicine* 11, 51-59.
6. Carson, R. A.: 1995, 'Interpretation', in W. T. Reich (ed.), *Encyclopedia of Bioethics,* 2nd ed., Simon & Schuster, New York, NY.
7. Childress, J. F.: 1994, 'Principles-oriented bioethics: An analysis and assessment from within', in duBose, E. R., Hamel, R. P., and O'Connell, L. J. (eds.), *A Matter of Principles? Ferment in U. S. Bioethics,* Trinity Press International, Valley Forge, PA, pp. 72-100.
8. Childress, J. F.: 1994, 'Ethical theories, principles, and casuistry in bioethics: An interpretation and defense of principlism', in P. F. Camenisch (ed.), *Religious Methods and Resources in Bioethics,* Kluwer, Dordrecht, The Netherlands, pp. 181-201.
9. Clouser, K. D. and Gert, B.: 1990, 'A critique of principlism', *The Journal of Medicine and Philosophy* 15, 219-236.
10. Gustafson, J. M.: 1965, 'Context versus principles', *Harvard Theological Review* 58, 191.
11. Jonsen, A. R. and Toulmin, S.: 1988, *The Abuse of Casuistry,* University of California Press, Berkeley, CA.
12. MacIntyre, A.: 1984, *After Virtue,* 2nd ed., University of Notre Dame Press, Notre Dame, IN.
13. Kirk, K. E.: 1933, *Conscience and Its Problems: An Introduction to Casuistry,* Longmans, Green, London, U. K.
14. Mitchell, W. J. T.: 1985, 'Introduction', in W. J. T. Mitchell (ed.), *Against Theory: Literary Studies and the New Pragmatism,* University of Chicago Press, Chicago, IL.
15. Oakeshott, M., 1962: *Rationalism in Politics,* Methuen, London, U. K.
16. Phillips, D. Z. and Mounce, H. O.: 1970, *Moral Practices,* Schocken Books, New York, NY.
17. Ramsey, P.: 1970, *The Patient as Person,* Yale University Press, New Haven, CT.

18. Taylor, C.: 1993, 'To follow a rule . . . .', in Calhoun, C., LiPuma, E., and Postone, M. (eds.), *Bourdieu: Critical Perspectives,* University of Chicago Press, Chicago, IL.
19. Turner, S.: 1994, *The Social Theory of Practices,* University of Chicago Press, Chicago, IL.
20. Walzer, M.: 1987, *Interpretation and Social Criticism,* Harvard University Press, Cambridge, MA.

ANNE HUDSON JONES

FROM PRINCIPLES TO REFLECTIVE PRACTICE OR NARRATIVE ETHICS? COMMENTARY ON CARSON

In his essay "Medical Ethics as Reflective Practice," Ronald A. Carson makes a valuable contribution to this anniversary volume with his scrutiny and critique of the principle-based paradigm that has dominated American bioethics during the past twenty years. Carson's critique derives from his conversations during these years with doctors, nurses, patients, and families, as well as from the emerging intellectual dissent best represented for him by Alasdair MacIntyre (virtue ethics) [6] and Albert R. Jonsen and Stephen Toulmin (casuistry) [5]. The routinized application of principles to quandaries in medical ethics has led, Carson maintains, to a "contractual mind-set," the "poverty of proceduralism," a prescriptive legalism, and a scholastic irrelevance to the world of medical practice in which doctors and patients struggle to make meaning of illness in moral languages that are far richer than principlism acknowledges. These are strong claims, and the proponents of principlism must be able to answer them persuasively if the dominant paradigm is to hold its ascendancy into the next century.

Carson is at his best in making this critique. His proffered remedy—attend to the world of practice and the lived experience of doctors and patients—is, in general, correct, but I fear that the specific steps he recommends—recovering illuminative maxims to guide dialogical interpretation of moral experience—will not take us as far from the pitfalls of principlism as he hopes. Maxims are too much like principles to escape the risks of misuse that have impoverished principle-based bioethics. Like principles, maxims can be used illuminatively or prescriptively. Ultimately, it's the user, not the tool, that determines how the tool will be used. The question then becomes whether the process of dialogical interpretation that Carson describes can ensure that a principle or maxim be

R. A. Carson and C. R. Burns (eds.), Philosophy of Medicine and Bioethics, 193-195.

used illuminatively. For Carson, dialogical interpretation means something like tacking back and forth between the received wisdom of maxims (drawn from the world of medical practice) and the particularity of an individual patient's circumstances, in a process akin to, but not exactly the same as, the analogical reasoning of casuistry. The interpreter is presumably the practitioner—doctor or ethicist. Whether an interpretation is truly illuminating depends entirely upon the interpreter's skill. Rita Charon [1] and others [2] have called such interpretive skill narrative competence.

Narrative competence constitutes an important strand of narrative ethics and is usually identified with Charon's position [3]. In this strand, narrative competence serves as an adjunct to other ethical approaches. The presumption is that doctors and ethicists who have skill in listening to (or reading) patients' and families' stories, and who know how to recognize and interpret metaphor, symbol, and other figurative aspects of language, are better prepared to carry out traditional ethical analyses. At its best, this kind of narrative competence helps avoid ethical quandaries by encouraging effective dialogue between doctors and patients before they reach conflicts that must be resolved by analytic ethicists, ethics committees, or the courts. Nonetheless, this strand of narrative ethics serves as handmaiden to other ethical approaches, principlism among them. Thus, I believe that despite apparent similarities between his approach and this form of narrative ethics, Carson intends to propose something more sweeping.

Another strand of narrative ethics makes more assertive claims about offering an alternative to principle-based ethics ([4], pp. 267-286). In this version, narrative ethics presumes a nonhierarchical narrative paradigm that empowers patients and families—those in whose lives the consequences of medical and ethical decisions will be lived out—to make decisions for their own lives, whenever possible, relying upon doctors for their expert knowledge of medicine more than for their expertise in ethics. This paradigm recognizes and respects what Carson has observed: "... sick people almost always bring their own truths with them into their encounters with medical professionals. ... they bring their lives and their needs, as well as a sense of what's valuable and what's not, with them into the doctor-patient relationship." And as Carson acknowledges, "Consequently, one of the central moral challenges doctors must take up is that of helping sick people to 'find their voices.'" As part of doing so, Carson encourages doctors and nurses to engage in conversations with patients and families. In these ways, Carson's approach and this more radical strand of narrative ethics are syntonic. They diverge when Carson turns away from the patient and asks the doctor to carry out a kind of dialogical reasoning that will presumably resolve and conclude the case. Both his privileging of the doctor (or ethicist) and the kind of dialogical interpretation he recommends veer away from true empowerment of the patient and family and from a dialogical process that, ideally, encourages everyone involved in a particular case to become part of the chorus of voices that seeks its best resolution. The hope, of course, in this second form of narrative ethics, is that through a nonhierarchical dialogical process, over time, a consensus will emerge that is satisfying to all those who are involved with the case and who will have, in one way or another, the consequences of its resolution forever embedded in their lives.

Institute for the Medical Humanities
The University of Texas Medical Branch
Galveston, Texas, U.S.A.

REFERENCES

1. Charon, R.: 1994, 'Narrative contributions to medical ethics: Recognition, formulation, interpretation, and validation in the practice of the ethicist', in E. R. DuBose, R. P. Hamel, and L. J. O'Connell (eds.), *A Matter of Principles? Ferment in U. S. Bioethics*, Trinity Press International, Valley Forge, PA, pp. 260-283.
2. Charon, R., Banks, J. T., Connelly, J. E., Hawkins, A. H., Hunter, K. M., Jones, A. H., *et al.*: 1995, 'Literature and medicine: Contributions to clinical practice', *Annals of Internal Medicine*, 122, 599-606.
3. Churchill, L. R.: 1996, 'Bioethics in Social Context', in this volume, pp. 137-151.
4. Jones, A. H.: 1996, 'Darren's case: Narrative ethics in Perri Klass's *Other Women's Children*', *The Journal of Medicine and Philosophy*, 21, 267-286.
5. Jonsen, A. R., and Toulmin, S.: 1988, *The Abuse of Casuistry: A History of Moral Reasoning*, University of California Press, Berkeley, CA.
6. MacIntyre, A.: 1981, *After Virtue*, University of Notre Dame Press, Notre Dame, IN.

CARL ELLIOTT

HEDGEHOGS AND HERMAPHRODITES: TOWARD A MORE ANTHROPOLOGICAL BIOETHICS

I. INTRODUCTION

"The fox knows many things," goes a fragment from Archilochus made famous by Sir Isaiah Berlin, "but the hedgehog knows one big thing." Berlin wanted to make a distinction in the world of ideas between hedgehogs, who understand the world in terms of a single central principle or concept (Plato, Dante, the early Wittgenstein) and foxes, who pursue a plurality of ends, sometimes unconnected, and seek a variety of experiences and ideas each for itself without attempting to relate them to a unitary vision (Aristotle, Shakespeare, the later Wittgenstein.) It is an artificial distinction, of course, but useful nonetheless, and with a bit of imagination (and considerably more artifice) one could expand it to a whole range of human activities: say, theology, physics, and golf as the territory of hedgehogs; psychiatry, poetry, and decathlons the territory of foxes.

What I would like to suggest, if I may extend this metaphor just a little further, is that there are often times when bioethicists would do well to imagine themselves less like hedgehogs and more like foxes. By this I mean not just that bioethical problems can be usefully approached using a plurality of visions and methods, but that those visions and methods may be culturally variable, and that at least in some cases the answer one gets will depend on the time and place one asks the question. To say that some problems in bioethics are most fruitfully examined in the light of local knowledge, to use Clifford Geertz's well-chosen phrase [9], is to say that the problems do not answer well to a single, universal, timeless solution but rather depend upon local, culturally contingent moral visions. Not one big thing, but many little things.

Suggesting that bioethics should look less like metaphysics and more like cultural anthropology is by no means a new idea, of course. It is an old idea to which little attention has been paid.[1] Partly, I think, this is because it is not clear just how a more anthropological bioethics might look, or in many cases what

R. A. Carson and C. R. Burns (eds.), Philosophy of Medicine and Bioethics, 197-211.

difference it would really make. Calling for a new way of doing bioethics is generally easier than actually doing it. So here I want to explore at least one kind of difference a more anthropological bioethics might make, and in a way appropriate to the fox: by looking at a particular ethical problem, the resolution to which is dependent on particular, local, culturally contingent frameworks of understanding.

II. THERAPIES OF THE SELF

The problem in question is the matter of identity, who we are and how we think of ourselves—an old philosophical problem, of course, but important now for bioethics because of what we might call "therapies of the self," a range of medical interventions that change human identity in certain fundamental ways. My aim here will be to look at these therapies in the light of another much older issue of identity—sexual identity—in order to show how an appropriate ethical stance to these therapies may depend on a very different range of concerns than bioethicists are accustomed to addressing.

When I use the term "therapies of the self" I have in mind three specific interventions that are currently controversial, though certainly more of them will emerge over time: first, the debate over providing growth hormone for short children who do not have growth-hormone deficiency; second, the debate over the so-called "cosmetic" use of the antidepressant fluoxetine, or Prozac; and third, the debate over the conditions for which it would be ethical to use gene therapy. What these therapies have in common is the potential to alter human characteristics closely bound up with individual identity: physical appearance, personality, intelligence, genetic constitution. The growth-hormone debate, for example, has focused on whether it is ethical to administer synthetic growth hormone to short children who do not have growth-hormone deficiency—the condition for which the treatment was developed—but who might be short for other reasons (or, in fact, who may not even be short, but whose parents would like them taller). The Prozac debate has revolved around what Peter Kramer calls "cosmetic psychopharmacology," or the use of the antidepressant Prozac by patients who are not clinically depressed [12]. Kramer describes how, in an admittedly small minority of cases, Prozac has transformed the personalities of patients: a patient who is uptight and obsessive becomes easy-going and relaxed, one who is shy and withdrawn becomes outgoing and assertive. Kramer and others have wondered whether there is anything morally troubling about these cosmetic personality transformations. My third example is the gene therapy debate, which has evolved along lines roughly similar to the debates over synthetic growth hormone and Prozac [7], [11]. Everyone seems to agree that gene therapy is ethically acceptable in limited cases—somatic cell gene therapy for circumscribed genetic illnesses, such as ADA deficiency or cystic fibrosis—but many worry that in the future, enthusiasts will want to use gene therapy to improve intelligence, personality, or physical appearance.

In each case the ethical worry concerns a person's desire to alter himself (or his children) in ways that are harmful neither to himself, nor to other people. Quite the contrary, in fact: the worry is that people will want to become taller, smarter, more self-assured, better-looking. And why not? Don't we all? It is awkward, to say the least, to try to argue against becoming more intelligent and beautiful, yet I suspect more than a few of us are troubled by the prospect.

An ethical stance toward these therapies will depend on how we identify just what is at stake, which will in turn point—or at least gesture—towards the reason many people find them troubling. What is at stake, I suggested earlier, is the effect of these therapies on individual identity: what people are trying to do is change themselves. The potential for change is most profound in the case of gene therapy, but it is most immediately striking in Kramer's cases of cosmetic psychopharmacology, where a patient on Prozac might even describe herself as becoming a "new person," or "becoming myself again." [12].

There is also a sense, however, in which these transformations of identity, for all their apparent novelty, are old problems: old in the sense that medicine is about illness, which can itself be profoundly transformative of human identity (Alzheimer's disease, stroke, or schizophrenia), but also because bioethicists have seen these types of issues before. I am thinking particularly about psychosurgery, once a widely-used procedure that seemed to relieve a patient of symptoms of mental illness but also changed the patient into a different type of person. Yet these transformations are also an old problem in a more general way. In ordinary life we maintain the fiction that while circumstances change, character is constant—that whatever we do, our essential, core identity remains the same. That fiction has a lot of truth in it, of course, but it is also true that over time our characters can change dramatically. Anyone who has been to a high school reunion knows this. And we do have some control over the manner and direction in which we change, although often in an indirect way. I realize that I might have become quite a different person had I gone to war, gone to business school, or gone to prison. In some ways, changing our identities is merely an extension of what has always been a part of our lives [19].

The way that bioethicists have generally come to approach these identity-transforming therapies, of course, is to distinguish between interventions that are meant to *cure* a condition, and those that are used as *enhancement*. The distinction is used to mark off those interventions that are ethically acceptable from those that are not: the acceptable being curative interventions such as growth hormone for growth-hormone deficiency, or gene therapy for metabolic disorders, and the unacceptable being interventions for enhancement, such as gene therapy for one's looks, or Prozac for shyness. Enhancing oneself, changing one's identity, is what is thought to be ethically questionable.

Identity, then, is what is at stake here, and the conditions under which it might be transformed. The next question is whether we have a parallel to these therapies, and I think we do. That parallel, at least in some important ways, is the question of sexual identity. Like personality, or intelligence, or physical appearance, one's sexual identity is an essential constituent of who one is. Moreover, sometimes sexual identity can be changed. What I want to turn to now is how, in certain types of cases, decisions about altering sexual identity are made. I want to suggest that an ethical stance towards these decisions about sexual identity depends on local, culturally contingent frameworks of understanding, and further, that these frameworks may undercut the distinction between therapy and enhancement that we have grown accustomed to using.

III. SEXUAL IDENTITY AND ILLNESS

If there is any aspect of our identity that seems—at least to common sense—fixed, it is sex. There are men and there are women, and nature assigns to us one category or the other. So we are used to thinking. In fact, however, as Anne Fausto-Sterling has pointed out, these arrangements belong more to us than to nature, whose sexual arrangements are arguably much more complicated. Fausto-Sterling argues that biologically, there are many degrees of maleness and femaleness, among which are at least five sexes, probably more [6]. The medical literature uses the term "intersexual" to designate individuals with both male and female sexual characteristics; the more common term, of course, is hermaphrodite, derived from the Greek mythological figure Hermaphroditus, son of Hermes and Aphrodite, who became half female when his body fused with that of a nymph.

Medicine classifies intersexuals into three main subgroups: true hermaphrodites, male pseudohermaphrodites, and female pseudohermaphrodites, or to use Fausto-Sterling's abbreviations, "herms," "merms," and "ferms." There are important anatomical differences between the three groups, but all have genitalia, reproductive organs, and sex chromosomes in combinations unlike ordinary males and females. For example, a male pseudohermaphrodite is chromosomally XY and has testes, like ordinary men, but also has a vagina and a clitoris, and at puberty may develop breasts. A true hermaphrodite, on the other hand, often has the gonads and chromosomal composition of both a female and a male: both XX and XY chromosomes, and both a testis and an ovary, which may grow separately or together. The genitalia of a true hermaphrodite are often ambiguous—for example, a phallus that resembles more a penis than a clitoris, with a urethra that runs near or through it, but through which, at puberty, menstrual blood exits.

What is most striking about intersexuals, however, is not so much the range of anatomical variation they possess as the ease with which we treat these naturally occurring variations as deformities, as fit objects for medical and surgical intervention. Conventional treatment is to alter these anatomical variations during infancy through surgery and hormonal therapy so that intersexuals can later pass for men or women. Or, as one intersexual puts it, with considerably more flair: "Intersex specialists are busily snipping and trimming infant genitals to fit the procrustean bed that is our cultural definition of gender" [2].

The point here is that we move very easily between the province of identity and that of illness, and it is not always so easy to see just where we are. If, as Fausto-Sterling writes, intersexuals constitute up to two to three percent of live births, one might well start to question why intersexuality is considered a deformity in need of surgical correction and not a normal variation in human identity. But sexual variation is not unique in this respect, of course; we are equally at ease in other areas of medicine with these shifts between the languages of illness and identity. A person with diabetes quickly becomes a diabetic, one with schizophrenia a schizophrenic—one's identity, for better or worse, is wrapped up in one's disease.

This is particularly true, of course, for genetic variation. We used to have a very rich vocabulary, albeit a somewhat backward one by today's standards, for all manner of genetic variation: one spoke of dwarves, lunatics, imbeciles, mongoloids—a vocabulary that has now been transformed into one of illness. A person with three copies of chromosome 21 is no longer a mongoloid; she has a genetic disease, Down's syndrome. Whereas we used to think of her as a different type of human being, now we think of her as sick. And of course, we also slide easily in the other direction, from illness to identity. A person who is sexually attracted to others of his own sex is not considered mentally ill, as he once was. Homosexuality is simply part of a person's identity, a constituent part of the way some people are.

It is this easy travel between the provinces of illness and identity that makes the distinction between cure and enhancement so seductive, but also so difficult to sustain. We are nervous with the idea of enhancing the self, and would like to mark off that territory with glaring caution signs, but the territory is constantly changing under our feet, often so slowly so that at first we do not even notice it. Frequently, as Willard Gaylin points out, it is a new technology that changes the boundaries of illness [8]. Before various reproductive techniques such as artificial insemination were developed, infertility was simply a fact of nature; now that it can be treated, it is a medical problem. Before the invention of the lens, poor vision was simply a consequence of getting old. Now it is something to be treated by a medical specialist. And by virtue of knowledge, skill or, in some cases, mere happenstance, doctors have also come to treat a broad range of conditions that no one considers illnesses, that would more easily be called enhancements, but which no one seems especially bothered by: minoxidil for baldness, estrogen for post-menopausal women, cosmetic surgery for people unhappy with their looks, acne treatment for self-conscious teenagers. It may seem obvious to us now that shortness, shyness, and ugliness are not fit objects for medical intervention, but this may not always seem so obvious.

It is precisely when we move closer to aspects of identity that the line between enhancement and cure becomes fuzziest—things like physical appearance, intelligence, sexual identity, and personality. Psychiatry is a striking example. Before the development of psychotherapy, mental illness was limited to psychotic disorders; now it includes phobias, obsessions, compulsions, personality disorders, and the like. Today it is disarmingly easy to speak of any disagreeable personality trait as if it were an illness—and even some that are not so disagreeable. Consider the way this question is phrased, for example: "Have shy people made reasonable efforts to overcome the condition—participating in social events, asking others for tips on socializing, taking public speaking classes and so on?" [13], p. 11). It comes from a recent discussion of third-party insurance payments for the medical treatment of shyness.

Now while I have been emphasizing the fluidity of our concepts of identity and illness, it is also true that these concepts often seem fixed, part of the very nature of things. This aspect of identity—simultaneously anchored and transient—is what makes it so slippery: we realize that our concepts are contingent, that they could have been otherwise, but nevertheless they are all we have.

Contrast, for example, today's attitudes towards intersexuals with those of Navahos in the 1930s, who, according to W. W. Hill, believed them to be divinely blessed: a cause for reverence and awe [10].[2] "They know everything. They can do the work of both a man and a woman," says one. "They are responsible for all the wealth in the country," says another. "If there were no more left, the horses, sheep and Navaho would all go." For the Navaho, I imagine, the notion of surgically fixing an infant intersexual would seem very odd indeed: foolish, irrational, morally perverse—not unlike, perhaps, the way we would see the idea of surgically "fixing" a boy.

Contrast the Navaho with yet another example, that of the Pokot in Kenya, for whom intersexuals are not objects of reverence, but simple errors [3]. For the Pokot, intersexuals are useless—unable to reproduce, incapable of bringing a bride-price. "I can only sleep, eat and work," says one Pokot intersexual. "What else can I do? God made a mistake." For the Pokot, it seems, intersexuals are not monstrous, and not blessed, but just botched.

Yet while the conceptual location of intersexuals can vary dramatically from one culture to the next, within each culture that location seems fixed, so fixed that it seems like part of the very nature of things. And it remains fixed, though to a lesser degree, even when we realize that it is culturally variable. Having the genitalia of both a man and a woman seems hardly less a reason for horror and compassion for many of us even when we realize that the Navaho and the Pokot do not see things the same way.

At least part of the reason for this sense, I suspect, is that concepts like these do not float freely and in isolation; they anchor (and are anchored by) a whole range of cultural practices. The Navaho make intersexuals the heads of the family and give them control over family property. And why not, since intersexuals are divinely blessed? The Pokot treat them with utter indifference—perhaps they are killed at birth, perhaps they are simply ignored and live on the margins of society. And why not, since they are botched products? They are useless in a culture for whom usefulness is a source of high value. And for us, intersexuality is a medical problem that deserves to be fixed. And why not? We live in a society that looks to medicine for its salvation, where novels are far outsold by diet manuals and self-help books. It seems fitting, somehow, that intersexuality appears in the pages of medical texts.

My broader point here is that like intersexuality, the debates over self-transforming therapies are culturally located in particular places, and we will be unable to say how best ethically to approach them without looking very closely at that cultural location. Just as it matters crucially how local frameworks of understanding accommodate intersexuality (divinely blessed or anatomically deformed), it matters crucially how they accommodate aspects of identity such as intelligence and beauty and self-confidence. But to understand how they are accommodated requires a hard look at one's own culture, a task more difficult than it appears.

IV. THE SCOPE OF MORALITY

A common complaint about bioethics, at least outside the borders of the United States, is that its methods and vocabulary have been dominated by Americans. Often this complaint is not directed so much at the methods and vocabulary themselves as at the assumption, often implied rather than argued, that they are (or should be) universal. The most common target is the language of rights and autonomy; nowadays, it is criticized not only by non-Americans who question its universality, but also by Americans who find it conceptually thin.

Yet for all the current dissatisfaction with talk about rights and autonomy, it is a paradigm that, like Wittgenstein's flybottle, is maddeningly difficult to escape. It is not so much that we Americans choose to construct our ethical problems in this way; they appear to come ready-made, and the materials are not negotiable. Some problems just seem to us to be, in their essence, a matter of respect for autonomy, and that is that. Here we are not so much bewitched by language as trapped by it.

Which is not surprising, given where we stand. These are not just the tools that we are accustomed to using; they are the materials from which our identity is constructed. We cannot, by simple force of will, start to see moral problems the way a Vietnamese Confucian would, or for that matter, an eighteenth-century Scot. What we can do, though, is step back from our own particular moral and cultural location and try to understand the relationship between where we are located and what we are seeing. To do this, however, requires a broader understanding of morality than that to which Anglo-American philosophers have become accustomed.

I have in mind an understanding of morality like that described by Charles Taylor, who configures moral thinking along three axes ([14], pp. 14-17). Taylor's first axis concerns questions about respect for others and our obligations toward them. These are the moral beliefs that cluster around questions such as the value of human life and the respect it is due, freedom and its limits, and the avoidance of suffering. In bioethics one might add questions such as those surrounding the concept of personhood, truth-telling, and confidentiality. These kinds of questions have been thought by some, in fact, to constitute the whole of morality, or something close to it.

Taylor, however, broadens the notion of morality to include two further axes. One is a range of notions concerning dignity. By dignity Taylor means "the characteristics by which we think of ourselves as commanding (or failing to command) the respect of those around us"—those attributes in respect of which others think well or badly of us ([14], p. 15). We are shaped from the time we are children, says Taylor, by the knowledge that "we stand in public space, and that this space is potentially one of respect or contempt, of pride or shame" ([14], p. 15). We may think of our dignity as bound up with our power, or our self-sufficiency, or being liked or admired. Such beliefs may vary widely across cultures. But all would recognize that there are standards in respect to which one's dignity (or something like it) is evaluated by others, and that self-respect is tied up with this.

Taylor's final axis concerns beliefs about what is often called the meaning of life. These are notions about what kind of life is worth living, what constitutes a full life or an empty one, the ways in which a life can fail to fulfil its promise. This axis, like the other two, is very likely a constituent of moral thinking in any culture, but one variation of it is probably unique to the modern West. As Taylor points out, a person in another time and place will often worry about meeting the demands made on him by the moral framework that defines his life. For example, he may fail to live a Christian life and face the loss of his soul; or he may fail to meet the demands incumbent on a person of his social station and lose his honor. But the predicament of the modern Westerner is different. The question is often not whether he will meet the requirements that his moral framework demands, but whether there exists any framework that has uncontested authority. It is a predicament not of failing to meet given standards, but of questioning what standards, if any, are the right ones [4]. The result here is not so much a concern about failure as about meaninglessness, or as Taylor puts it: "the world loses altogether its spiritual contour, nothing is worth doing, the fear is of a terrifying emptiness, a kind of vertigo . . . " ([14], p. 18).

The first of Taylor's three axes has dominated recent bioethics and probably Anglo-American moral philosophy as well. We have been far more concerned with questions of respect and obligation than with questions surrounding meaning, dignity, what sort of person to be and how to live a life. I want to argue, though, that an approach to bioethics limited to the first axis will be incomplete. It will be incomplete because how one answers questions along the first axis depends on how one answers them along the second two: questions about respect and obligation depend on often unstated assumptions about dignity and the meaning of life.

A striking example concerns the evolution of a particular conception of social justice and what Michael Walzer calls "the cure of bodies and the cure of souls" ([17], p. 87; [18], p. 28). The medieval Christian world was organized, says Walzer, so as to make the "cure of souls" universally available. Christians had a system of what we might call socialized salvation: a system backed by public funds (tithes) and ecclesiastical law that ensured that every parish had access to a priest, confession, catechism, and communion. This elaborate ecclesiastical machinery cannot be properly understood, of course, without some idea of the commonly agreed upon cultural understanding of the meaning of life, or more specifically, of the reality of heaven and hell. The good for distribution was eternal salvation, and the system served the imperatives of distributive justice. The "cure of bodies," on the other hand, was a much less important matter. There was no such elaborate distributive system, and no one seems to have been very concerned that the rich and powerful had access to health care unavailable to anyone else.

Today, I hardly need mention, matters are reversed. As people came to question the reality and significance of eternal salvation, a just system for access became a far less pressing concern. But as medical knowledge made health and longevity more easily achievable, a just system for access to health care became a much *more* pressing concern. A particular conception of social justice evolved along with a particular moral and cultural understanding of life's purpose.

A more mundane but still pressing example is the current debate surrounding the notion of medical futility. That debate is most often focused on the question of when the medical treatment of a comatose or moribund patient can be considered medically futile and thus discontinued over the objections of the patient's family—overriding their autonomy or liberty rights. But since the concept of futility has its full meaning only in relation to a given aim (if the aim is to cure, ventilator treatment of a comatose patient may be futile, but not futile if the aim is prolongation of life), a fruitful discussion of the moral question can be undertaken only if it attends to the question of which aims are worthwhile. Consequently, the issue of medical futility cannot be resolved unless it is openly recognized that these aims will vary from one culture to the next, and in fact from one person to the next within a culture, and that they are interwoven with a variety of conceptions about human dignity and what kinds of lives are worth living.

V. BIOETHICS AND THE GOOD LIFE

If such questions are as important as many of us believe, then why do we pay them so little attention? One reason, I believe, comes from a widely-held picture about how these sorts of questions fit into moral thinking. This is a picture of a pluralistic society whose members each have different visions of the good, who may all have different goals and desires, and in which the function of moral thinking is largely to preserve and arbitrate between the liberty of each person to realize his own particular goals and desires. Just what those particular goals and desires are—what kind of life one considers worth living, for example—does not become a matter for our concern unless they interfere with the well-being of others.

This picture of morality, whose most important exemplar has been the political philosophy of John Rawls, has taken root and flourished particularly in bioethics, I believe, because the earliest agenda of bioethics was to encourage, or even enforce, a sense of respect for individual patients whose particular values and desires had been often overlooked. This has been a valuable model for bioethics, and as a model for political philosophy, it may well be the best that liberal democrats can hope for. But it has also had the effect of pushing substantive visions of the good life to the fringes of moral thinking ([14], pp. 15-20). As long as the liberty of individuals to pursue their vision of the good life is preserved, just what that good life might be is a matter of secondary importance. Hence the thin and impoverished discourse on these questions that we have tended to produce.

However, when it comes to questions about the nature and meaning of life, this picture of moral thinking is far from neutral. Underlying it, as Michael Walzer points out, is a historically and culturally particular conception of life. It is a conception of life as a project, a planned undertaking "in which we ourselves are the undertakers, the entrepreneurs, the managers and organizers of our own activities" ([18], p. 23). The notion of life as a project that we plan, control and are responsible for is by no means universal. Think, says Walzer, of other possible conceptions: 1) a spontaneous life, unplanned and governed by circumstance; 2) a "divinely ordained" life, where the plan for one's life is God's, not one's own, and where each person's job is to determine and follow God's will; 3) an inherited life, where one takes over the position and accomplishments of one's parents; 4) a socially regulated life, where a person receives exactly what is due him in consequence of his birth or his virtue. Plainly, these differing conceptions of the form and significance of life will affect quite profoundly how one believes life ought to be lived, and what counts as a better life or a worse one.

I want to argue that our modern ambivalence about therapies of the self arises from our notions about dignity and life's meaning, and perhaps more crucially, from the shaky foundation upon which those concepts stand. Take, for example, the notion of life as a project. If a person's life is her own project, then it seems she should be given the liberty to pursue the project to its successful completion in whatever way she wants. If that project includes gene therapy or growth hormone or Prozac, then so be it—as long as their use does not harm others or interfere with their liberty to pursue their own projects. Of course, the notion that life is a project has the corollary that it can be a successful project or a failed one. Unlike, say, a life that is planned by God, this life is one whose trajectory is largely determined by the person who is living it, and with whom the responsibility for that life lies. Under this vision of life, self-transforming therapies are tools to help a person ensure a good result, a successful project.

Here is one source of the deep disquiet that many of us feel about these therapies. We do indeed tend to see our lives as projects, but the criteria for their successful completion is up for grabs. We have no uncontroversial, commonly agreed understanding of what counts as a successful life. Rather, we have a plurality of visions of the good life, none of which has uncontested authority. Given this context, it is natural to feel a great deal of ambivalence about therapies of the self. If the criteria for a good life are unclear, then the value of tools to achieve those criteria will be even less clear. The value of these therapies is dependent on whether they help to achieve a result the value of which is itself up for debate. My worries, for example, about a drug that produces an extroverted self-confidence are worries about whether an extroverted self-confidence is itself a virtue, or whether it serves a vision of the good life to which I do not subscribe.

The distinction between cure and enhancement often used to mark off these therapies can be understood in light of another, more troubling variant of this conception of life as a project—that is, life as a competitive project. For many people the most persuasive argument for using these therapies is that without them some people are at a competitive disadvantage. Short people, shy people, people who are less attractive or less intelligent—these people do worse, it is argued, at getting good jobs, attracting partners, or succeeding in school or work. Therapies that are "curative" bring them up to the level of others in society and make them competitive on a fair level. Thus they are ethically acceptable. "Enhancement therapies" are unacceptable, however, because they give people an unfair advantage. Rather than levelling the playing field, they tilt it, and thus using them is cheating. Given this particular framework of what counts as a good life, and perhaps more importantly, this vision of what gives a person dignity and respect in life, this distinction between cure and enhancement can make a lot of sense. It is the framework itself that is shaky.

Given other visions, of course, the distinction between cure and enhancement would make less sense. Many of us tend to frame our lives around what Taylor calls "ordinary life," which endows meaning on a life of work, family, householding, and loving relationships [15]. Therapies of the self have a much different (and much deeper) significance in this kind of framework, where people see the therapies as playing an important role in how they find meaning and dignity in life, than in a framework where meaning is drawn, say, from one's relationship with God, or in a culture where dignity is defined not by personal achievement but by inherited social station. In such a framework these therapies might well seem trivial, because no one would see them as making their lives more meaningful, or in contributing to their dignity.

Finally, I believe that yet another widely-held moral ideal stands at the root of many people's worries about therapies of the self. This ideal is what Taylor calls the ethic of authenticity, the notion that each person has the right to determine and pursue what he thinks is valuable, and more importantly, that personal fulfillment comes by discovering this for oneself [15]. It is a familiar and powerful ideal, and it is part of what we mean when we talk about things like "being true to oneself" ([15], p. 14). To do anything else would be—to use a familiar and telling phrase—to betray oneself, in the sense of failing to fulfil the promise that is uniquely one's own.

Yet this ethic of authenticity is directly challenged by therapies that fundamentally alter the self. This ethic calls on one to discover oneself, not to change or create oneself. To the extent that these therapies truly change a person, they run squarely against an ethic that implores a person to look inward and find himself, to come to terms with who he is. The ethic of authenticity is so subtly ingrained in American culture that even Peter Kramer, Prozac's most thoughtful analyst, argues for Prozac's "cosmetic" use by turning the authenticity argument on its head [12]. He argues that, rather than altering the self, as he suggests at length elsewhere, Prozac actually *restores* the self. That is, it brings to the surface the "true" self that was previously hidden by depression, or obsessiveness, or hypersensitivity. In this way, Prozac becomes a way of heeding the call to self-discovery, rather than ignoring it, and the ethic of authenticity can be preserved.

VI. CONCLUSION

How we answer questions about the proper use of therapies of the self will depend on when and where those questions are asked. I have suggested that how we answer these questions in North America will be contingent upon cultural strands that are often overlooked. In looking at these therapies, the relevant questions to ask are not only those such as "Under what conditions can the state restrict a person's liberty?" but questions such as "What kind of person should I be?", "What should I want?", and "How should I live my life?" For these latter questions especially, it is far from clear that an answer can be given that will be the same for all people, at any time, in any culture, under any circumstance. Rather, how a person sees the answer to these questions will depend on local frameworks of understanding, in which are embedded a whole variety of concepts and practices that are, as Walzer puts it, "richly referential, culturally resonant, locked into a locally established symbolic system or network of meanings" ([18], p xi). To paraphrase Wittgenstein, a moral language is tied to a form of life. Understanding how this is so, and how things might have been otherwise, is the natural territory not of the hedgehog, but of the fox.

McGill University, Centre for Medicine, Ethics and Law, and
Montreal Children's Hospital
Montreal, Canada

This research was supported financially by Fonds pour la Formation de Chercheurs et l'Aide à la Recherche, Québec.

NOTES

[1] This is at least part of what I understand Stephen Toulmin to be suggesting when he says philosophers need to look at the timely, the local, the oral, and the particular (1988). I believe this approach is also compatible with the view of interpretive bioethics set out by Ronald Carson [1]. See also Elliott [5].

[2] I first learned of these varied cultural attitudes towards intersexuals through Clifford Geertz's wonderful essay, 'Common Sense as a Cultural System,' in *Local Knowledge* [9].

REFERENCES

1. Carson, R.: 1990, 'Interpretive bioethics: The way of discernment', *Theoretical Medicine* 11, 51-59.
2. Chase, C.: 1993, 'Intersexual rights' (letter), *The Sciences,* July/August, 3.
3. Edgerton, R. B.: 1964, 'Pokot intersexuality: An East African example of the resolution of sexual incongruity', *American Anthropologist* 66, 1288-1299.
4. Edwards, J. C.: 1990, *The Authority of Language: Wittgenstein, Heidegger, and the Threat of Philosophical Nihilism,* University of South Florida Press, Tampa, FL.
5. Elliott, C.: 1992, 'Where ethics comes from and what to do about it', *Hastings Center Report* 22, 28-35.

6. Fausto-Sterling, A.: 1993, 'The five sexes: Why male and female are not enough', *The Sciences,* March/April, 20-24.
7. Fletcher, J. C. and Anderson, F.: 1992, 'Germ-line gene therapy: A new stage of debate', *Law, Medicine and Health Care* 20, 26-39.
8. Gaylin, W.: 1993, 'Faulty diagnosis', *Harper's,* October, 57-64.
9. Geertz, C.: 1983, *Local Knowledge: Further Essays in Interpretive Anthropology,* Basic Books, New York, NY.
10. Hill, W. W.: 1935, 'The status of the hermaphrodite and transvestite in Navaho culture', *American Anthropologist* 37, 273-279.
11. Kahn, J. P.: 1991, 'Genetic harm: Bitten by the body that keeps you?', *Bioethics,* 3(4), 289-308.
12. Kramer, P.: 1994. *Listening to Prozac: A Psychiatrist Explores Antidepressant Drugs and the Remaking of the Self,* Fourth Estate Limited, London, U. K.
13. Sabin, J. E., and N. Daniels: 1994, 'Determining "medical necessity" in mental health practice', *Hastings Center Report* 24, 5-13.
14. Taylor, C.: 1989, *Sources of the Self: The Making of the Modern Identity,* Harvard University Press, Cambridge, MA.
15. Taylor, C.: 1991, *The Malaise of Modernity,* House of Anansi Press Limited, Concord, Ontario, Canada.
16. Toulmin, S.: 1988, 'The recovery of practical philosophy', *The American Scholar,* 37-54.
17. Walzer, M.: 1983, *Spheres of Justice: A Defense of Pluralism and Equality,* Basic Books, New York, NY.
18. Walzer, M.: 1994, *Thick and Thin: Moral Argument at Home and Abroad,* University of Notre Dame Press, South Bend, IN.
19. Williams, B.: 1981, *Moral Luck: Philosophical Papers 1973-1980,* Cambridge University Press, Cambridge, U. K.

GERALD P. MCKENNY

AN ANTHROPOLOGICAL BIOETHICS: HERMENEUTICAL OR CRITICAL? COMMENTARY ON ELLIOTT

Calls for a hermeneutical model for bioethics have appeared with some frequency over the past decade and have usually levelled two criticisms at standard forms of bioethics in the name of two different kinds of particularity. One is that abstract principles are too indeterminate to resolve cases, whose particularity demands something like Aristotelian phronesis. The other is that bioethics has not done justice to the particularity of cultures, whose moral and relevant nonmoral beliefs may differ significantly from those held by Western philosophers. Since I believe standard bioethicists have responded to some extent to the first form of particularity, I will begin this response by sorting out what is legitimate in Carl Elliott's criticism against standard bioethics from what I believe is a confusion of these two types of particularity. However, I will also show how Elliott voices a third hermeneutical criticism of standard bioethics that is less frequently heard: that standard bioethics ignores the thicker visions of human dignity and the meaning of life that "therapies of the self"—those medical interventions that have the capacity to alter basic features of individual identity—call into question. Having set up Elliott's basic position vis-à-vis standard bioethics, I will then point out what I believe is at stake in his central argument and will suggest both a revision in Elliott's hermeneutical model and its completion by a critical approach. I will conclude by pointing out some difficulties in determining what role the anthropological bioethicist would play and by suggesting a possible role.

I

Elliott does not specify which theoretical position(s) his call for "a more anthropological bioethics" opposes, but since he is clearly challenging what he believes to be the reigning position(s) it seems reasonable to assume that he has

R. A. Carson and C. R. Burns (eds.), Philosophy of Medicine and Bioethics, 213-220.

principlism and/or casuistry in mind. If so, then his characterization of the difference between the dominant approaches to bioethics and his own approach as a difference between hedgehogs, who view the world in terms of one central principle or concept, and foxes, who seek an irreducible plurality of ends and experiences, is mistaken. For neither principlism as practiced, for example, by Beauchamp and Childress nor casuistry as practiced, for example, by Jonsen and Toulmin is committed to a single principle or concept. To the contrary, Beauchamp and Childress recognize an irreducible plurality of principles and, in the most recent edition of their collaborative work, the need to specify and balance those principles in order to negotiate the irreducible particularity of cases ([2], pp. 28-37, 100-109). Similarly, Jonsen and Toulmin recognize a multiplicity of paradigm cases that can not be reduced to a single principle or general theory [3]. The foxes, it appears, are everywhere in bioethics while the hedgehogs are everywhere in retreat.

Nevertheless, Elliott's criticism is not entirely off the mark, and his plea to bioethicists to deal explicitly with beliefs and convictions could overcome a shortcoming in both principlism and casuistry. In order to determine which principle or specification of a principle to bring to a case (principlism) or which paradigm case is relevant to a controverted case and what similarities and differences qualify this relevance (casuistry), both principlism and casuistry must appeal to a set of beliefs and convictions that neither Beauchamp and Childress nor Jonsen and Toulmin ever make explicit. Beauchamp and Childress assert an overall coherence or reflective equilibrium of specified and balanced principles, on the one hand, with a set of moral norms and considered judgments, on the other hand, while Jonsen and Toulmin assume a widespread set of judgments and convictions about cases. But in both cases the relevant norms, judgments, or convictions are never articulated.

Of course, as Elliott knows, when the beliefs and judgments that govern the specification and balancing of principles or the mapping of paradigm cases in relation to new cases are made explicit, there is often widespread disagreement. This brings us to the second form of particularity and to Elliott's insistence on the need for bioethics to recognize the culturally contingent character of the relevant beliefs and convictions. Both principlism and casuistry assert (largely on faith) the existence of a universal common morality. Beauchamp and Childress argue that their principles derive from shared moral beliefs and therefore serve as universal standards that can be used to criticize or correct merely customary moralities insofar as the latter do not acknowledge these principles ([2], pp. 100-101). Jonsen and Toulmin seem to recognize with Elliott that the resolution to many bioethical problems depends upon contingent, culturally variable moral visions. But the variation they are willing to concede is limited: they insist that the basic categories and methods of internal self-criticism are largely common to the casuistical traditions of different cultures ([3], pp. 285, 325-326).

However, as I noted above, Elliott's most significant contribution involves a third kind of hermeneutical criticism: that while standard forms of bioethics primarily address issues on Charles Taylor's first axis of morality (namely, issues involving obligations and respect for others), many of the most important issues surrounding therapies of the self fall along Taylor's second and third axes [5]. These axes address, respectively, ideas about what constitutes dignity and the meaning of life. There is no questioning the importance of reaching beyond the first axis, for as Elliott points out, many of these therapies seem to pass the standard bioethical hurdles involving self-determination and harm to oneself or others (fairness is perhaps a higher hurdle), and yet anxieties and uncertainties about the use of these therapies remains.

II

At the risk of oversimplification, the core of Elliott's argument could be schematized into three claims: that some crucial ethical worries about therapies of the self are worries about identity; that identity (at least in our culture) is best understood in terms of ideas about dignity and the meaning of life; and that some crucial ethical issues regarding therapies of the self therefore revolve around these ideas. The first claim is Elliott's strongest. The view that what troubles many people about therapies of the self is that they change, or have the potential to change, our identity in profound ways is highly plausible. One of its virtues is that it helps explain both why we so readily seek to draw a sharp line between cure and enhancement, and why any such line is so unstable. If the line between cure and enhancement distinguishes between treating an illness and altering identity, then it is clear that it must be a fuzzy and shifting line, since our beliefs about what belongs to the categories of illness and identity are fuzzy and constantly changing—and this within a single (albeit highly diverse) culture. However, it may also be that those who worry to this extent about altering identity are simply misguided. As Elliott points out very briefly, alterations of our identity have always been a part of our lives. He mentions crucial life decisions we make, and later points to minoxidil, estrogen, and cosmetic surgery (though he does not seem to think these change our identity). But social engineering, psychotherapy, and body building, for example, also change our looks, behavior, and fundamental attitudes in profound ways, and most people have little trouble accepting these. Perhaps, then, it is not the therapies of the self themselves that trouble people, but only the rapidity, thoroughness, and perhaps irreversibility that technology makes possible—a matter of degree and magnitude rather than of kind. I suggest, however, that it may not be the alteration of identity that is the problem, but that we are not sure what kind of identity we should cultivate and therefore do not know whether it is a good or a bad thing to use these therapies to become more beautiful, intelligent, self-assured, or tall. (Elliott makes a very similar suggestion later in his essay.)

The second claim, that these identity issues are best understood in terms of ideas about dignity and the meaning of life, is debatable because it is not clear from Elliott's brief summary of several of these ideas how they entail the conclusions regarding therapies of the self he attributes to them. Elliott gives a rather confusing description of the idea of life as a project. If, as it appears, it is a contentless idea that must be filled in with whatever project an individual chooses to fulfill—I waive concerns about whether an idea without substantive moral content can qualify as a view of dignity or the meaning of life—then it cannot rule out the use of therapies of the self. But it is an error to conclude from this, as Elliott does, that for this vision of life, therapies of the self ensure the success of the project. For surely many of these individual projects would reject therapies of the self as incompatible with a successful life. The ideal of authenticity is likewise ambivalent on the role of therapies of the self, and it is not clear why such therapies play "an important role" in the ideal of ordinary life. All of this suggests that ideas of dignity and the meaning of life cannot play a very significant role in determining moral evaluations of therapies of the self, unless they are made much more specific. I will offer another suggestion: that ideas about dignity and the meaning of life must be formulated to include views about the place of the body in a morally worthy life. Views of dignity and a meaningful life have, since antiquity, often reserved a prominent place for understandings of bodily integrity or perfection, and notions of what kind and degree of attention to devote to the body. Therapies of the self may be rejected because they are judged to violate the integrity of the body or devote the wrong kind of attention to it, or welcomed because they help to realize an ideal of bodily perfection. A hermeneutical bioethics, I think, must attend to such ideas.

There is a deeper question to be raised regarding the understanding of identity in terms of ideas of dignity and the meaning of life. The question is whether this is all that an anthropological bioethics would say about identity. It seems likely that a genuinely anthropological bioethics would pay attention not only to our ideas but also to those practices by which we form our identities. The argument here would be that we form our identities not only by interpreting our lives in terms of certain ideals but also by problematizing certain aspects of our selves (including our bodies) as in need of attention or change; monitoring our thoughts, feelings, desires, bodies; acting upon ourselves to change ourselves; and so forth, all in accordance with certain societal norms or standards against which we measure ourselves and others. This question of what we might call practices of the self goes to the heart of Elliott's project, since it raises the question of whether a bioethics calling itself anthropological can be exclusively hermeneutical or must also be critical, exposing the roles of medicine and society in our formation as subjects.[1] It is important to know whether we are to become ethnographers who study moral and other beliefs in their complex connections with institutions and practices or (like Taylor) philosopher-historians of ideas. Elliott has clearly chosen the latter, but there are good reasons, even on his own grounds, for choosing the former as well. One is that it enables us to do justice to a common complaint made by intersexuals, which Elliott records but strangely ignores. He quotes an intersexual who complains about intersex specialists who "are busy snipping and trimming infant genitals to fit the procrustean bed that is our cultural definition of gender." But in the analysis that follows, Elliott misses the point of this complaint: It is not just that intersex specialists have a different view of identity than intersexuals but that intersexuality is being brought under the domain of medical intervention for the purpose of enforcing societal norms for gender. An ethnographic bioethics would have been able to point out similar phenomena with regard to the therapies of the self. In a society that, as Michel Foucault argued, has learned to control by stimulation rather than repression of desire, therapies of the self (along with the personal disciplines, dissemination of information, and techniques of persuasion that ensure that we will desire and obtain them for ourselves and our children) play a major role in producing the bodies and dispositions that characterize the optimally useful and productive subjects our society needs. Such an approach also helps us understand why those who reject such therapies do not simply live out an alternative ideal but meet the scorn, derision and often cruelty of others (as intersexuals, those who "fail" to make their bodies beautiful and productive, and those who make the "wrong" genetic decisions can all testify). And finally, this approach helps us understand how by acting upon our bodies in these ways, we form a certain moral identity—one, for example, whose compassion toward others is expressed by helping them to meet societal norms (and eliminating them to save them from suffering when they can not meet these norms) rather than by caring for them as they are.[2]

III

My final point involves an issue Elliott does not consider at all, but one which anthropologists nowadays spend a great deal of time and energy considering. The issue is the location of the anthropological bioethicist. This issue in turn is directly related to the role of the bioethicist. To argue that bioethicists ought to become cultural anthropologists (or philosopher-historians of ideas) is not to say anything about the role of the bioethicist in resolving moral disputes. Neither cultural anthropologists nor Taylor himself seek to resolve actual moral disputes. Cultural anthropologists attempt to understand the role of moral beliefs in the community; the practices, forms of authority, and lines of transmission in which they are upheld, applied, revised, and passed on; the tensions and negotiations that accompany them; and the like. No cultural anthropologist who was not also a member of the community would claim authority to settle moral disputes in that community. Taylor attempts to articulate the frameworks (i.e., visions of a worthy or meaningful life) that constitute the transcendental horizon of certain moral beliefs and commitments that are central to the moral identity of the modern West but which, under the pressure of reductionist theories of human behavior and motivation, are in danger of being lost. While his method could, if developed in the way I suggested above, enable one to show how acceptance of these various ideals would lead one to evaluate therapies of the self, that method cannot tell modern Westerners which of these visions, or which combination of them, they ought to adopt.[3]

There is nothing wrong with limiting anthropological bioethicists to a descriptive role while substantive moral disputes about dignity and the meaning of life are resolved by particular communities, traditions, or individuals. But since we expect bioethicists to resolve moral disputes, this would mean that most actual bioethical work on matters involving dignity and the meaning of life would be done by, say, Navajo and Pokot bioethicists (and their counterparts elsewhere) rather than by anthropological bioethicists. What role the latter would play aside from the roles cultural anthropologists and philosopher-historians already play is unclear.

Once again I will make a suggestion that befits the hermeneutical-critical gloss I have inscribed into Elliott's text. Recognizing that normative bioethics must be done within particular communities or traditions that hold substantive views of the moral significance of the body and practices in which these ideas are imbedded, it is plausible to expect that many of these views will conflict with the view of the body and the practices of modern, normalizing societies.

Members of such communities would presumably welcome an anthropological bioethics that enables them to understand and resist that dominant view. They might also welcome a deeper understanding of the discourses and practices by which their own communities and traditions form them. While the anthropological bioethicist will not play a direct role in formulating their alternative convictions and practices or carrying them out, she may play an important indirect role by making them aware of other alternatives, through which they may gain a deeper understanding of themselves and engage in dialogue and form alliances with other communities and traditions. Perhaps most important, the anthropological bioethicist is in a position to bring to our attention the ways in which nearly everyone in the modern West has been formed by the dominant discourse and practices of the body and, by making us aware of alternatives, to make it possible for us to question the inevitability of our modern identity.

To conclude: an anthropological bioethics 1) distinguishes itself from current models of bioethics by its attention to cultural particularity rather than the particularity of foxes, 2) must in its hermeneutical activity understand notions about the role of the body in a morally worthy life, 3) carries out an interpretation and critique of the practices that form our identities in the society under investigation, and 4) takes its place relative to the particular communities and traditions in which normative judgments are binding. I am not sure whether Elliott would accept my hermeneutical-critical amendment to his proposal as a friendly or a hostile one, nor do I know whether he would accept my suggestion regarding the location and role of the anthropological bioethicists relative to those in particular communities or traditions. But he has nevertheless made an intriguing proposal that deserves to be pondered in relation not only to therapies of the self but to bioethics more generally.

Department of Religious Studies
Rice University
Houston, Texas, U.S.A.

NOTES

[1] The reader should not conclude that I think beliefs or ideas are reducible to or merely the products of social institutions and practices. Like Charles Taylor, I believe that it is possible (and useful for certain purposes) for historians to discuss the attractiveness or plausibility of an idea apart from the concrete conditions that make it effectual or that harness it to other more material ends. But I also believe with Taylor that a complete

account of social reality must take account of these latter factors (1989, pp. 202-203). In effect I am arguing that bioethics cannot really be anthropological (as opposed to cultural-historical or hermeneutical) unless it does the latter. Tamal Asad criticizes Geertz, and by implication Elliott, for reducing the anthropology of religion to a hermeneutical enterprise, though I do not fully agree with him (1993, pp. 27-54).

[2] The argument of the preceding paragraph is presented in more detail in my forthcoming book.

[3] The debates in the field of comparative religious ethics in the 1970s and 1980s between those who used concepts and methods of moral philosophy to reconstruct the moral reasoning of non-Western moral traditions and those who insisted on viewing those traditions in their own cultural and historical context are instructive in this regard. Those who used the philosophical approach often did so in the interest of securing a normative cross-cultural basis for, say, human rights, while adherents of the cultural-historical approach were less interested in normative ethics.

REFERENCES

1. Asad, T.: 1993, *Genealogies of Religion*, Johns Hopkins University Press, Baltimore, MD.
2. Beauchamp, T. and Childress, J.: 1994, *Principles of Biomedical Ethics*, 4th ed., Oxford University Press, New York, NY.
3. Jonsen, A. and Toulmin, S.: 1988, *The Abuse of Casuistry*, University of California Press, Berkeley, CA.
4. McKenny, G.: *To Relieve the Human Condition: Bioethics, Technology and the Body*, SUNY Press, Albany, New York, NY (in press).
5. Taylor, C.: 1989, *Sources of the Self*, Harvard University Press, Cambridge, MA.

LORETTA M. KOPELMAN

MEDICINE'S CHALLENGE TO RELATIVISM: THE CASE OF FEMALE GENITAL MUTILATION

Our families, communities, and institutions increasingly include people from different cultural groups. Many of us have parents of different religious and ethnic origins. This diversity enhances our lives as we learn the pleasures of different foods, arts, music, and views. Being open-minded about various attitudes and beliefs challenges and benefits us. We do not want to rule out any unexamined options or suppose our preferences are absolute standards. Consequently, openness is a practical way to test what is true or meritorious in our views. Being open to others' views also shows that we respect and care about the people expressing those views and are interested in what they think and feel ([21], p. 1; [11]). Being receptive to diversity of thought, however, means that different views should be heard and debated on their merits, not that they should all be accepted.

Medicine in the last half of the twentieth century has also experienced the benefits and challenges of cultural diversity. In addition to the growing differences within our nation, people from around the world increasingly seek medical care in the United States. While we ought to be respectful and receptive to other customs, some seem wrong. On what basis do we rationally establish that another culture's practices should be stopped? For example, Abdalla writes a ". . . custom practiced in Southern Yemen and along the Persian Gulf is to put salt into the vagina after childbirth . . . [because practitioners believe this] induces the narrowing of the vagina . . . to restore the vagina to its former shape and size and make intercourse more pleasurable for the husband" ([1], p. 16).

In what follows, I argue that a clear example of interculturally shared values and methods may be found in medicine. These values and beliefs sometimes effectively and rationally challenge deeply embedded moral and cultural beliefs.[1] That is, medicine cannot define or establish our moral values, but it can help evaluate them and the means we use to attain these goals. A consequence of this, however, is the implausibility of those versions of ethical relativism hold-

R. A. Carson and C. R. Burns (eds.), Philosophy of Medicine and Bioethics, 221-237.

ing that something is right means it is approved in a person's culture and something is wrong means that it is disapproved. To demonstrate how shared goals and values can rationally challenge entrenched cultural attitudes and beliefs, thereby showing the implausibility of ethical relativism, I discuss the rite of female circumcision/genital mutilation. To introduce these issues, consider the following case:

I. A CASE

Mr. and Mrs. A immigrated from a country in East Africa to North America. In their original homeland, the surgical rite of female circumcision and infibulation is performed on about ninety percent of the girls, and parents are thought to fulfill an important variety of duties to the child by having this done. These duties include religious and parental obligations to make it possible for daughters to marry and to foster their health, well-being, and beauty. This surgery involves removal of most of the girl's genitalia, including the clitoris, internal labia, and most of the external labia. Because of the risk of infections, this family prefers a medical facility over traditional practitioners, so they go to Dr. B. Dr. B refuses to perform the surgery, denying it is a Muslim religious requirement, as they believe; moreover, doing such surgery violates physicians' duties to do no harm and prevent unnecessary suffering because it causes increased likelihood of mortality and morbidity, including sexual dysfunction, infections, scarring, incontinence, and maternal-fetal jeopardy during labor and delivery. Mr. and Mrs. A return to East Africa and ask their cousin Dr. C to perform the surgery. They find Dr. C also refuses, saying this practice is wrong despite its cultural approval, for similar reasons as given by Dr. B.

While this case is imaginary, situations like it are increasingly common as people from Africa and Southern Arabia engaging in such rites immigrate. They find people outraged over their practices that concern how to promote people's well being and fulfill important duties [12], [27], [1], [14].[2] Also characteristic is that doctors and nurses in these countries actively try to stop these ancient practices of genital mutilation.[3] According to some versions of ethical relativism, there is no basis for morally authoritative cross-cultural criticism because the right or dutiful action is one that is approved by the person's society or culture, and the wrong action is one that is disapproved by the person's society or culture; there are moral truths, but they are determined by the norms of the culture. In our example, this theory means that Dr. B's judgments have no moral authority to people of a different culture, and Dr. C is mistaken about this surgery being wrong, given its cultural approval.

Female circumcision/genital mutilation powerfully demonstrates both the role of medicine in evaluating the reasonableness of some entrenched cultural and moral beliefs, and in assessing the plausibility of ethical or cultural relativism. First, female circumcision/genital mutilation has wide approval within the countries where it is practiced and wide disapproval outside these cultures. In Southern Arabia and Africa,[4] it is commonplace for varying amounts of the women's external genitalia to be removed. About eighty million living women have had this surgery, and an additional four or five million girls undergo it each year [16], [18]. Many international groups, including UNICEF, the International Federation of Gynecology and Obstetrics [22], the World Health Organization [31], and the American Medical Society [2] have condemned these rituals. A second reason these rites are a potent test case is that these practices are supposed to promote well-being and health and fight disease, goals that have approval in all cultures; consequently if cultures share values and methods about how to fulfill such goals, then these aims can be evaluated in a way that has moral authority.

II. SHARED VALUES AND METHODS

Medicine incorporates many values and methods having wide intercultural acceptance, including agreement about the evils of causing unnecessary pain, the goods of promoting personal and public health, the duty to try to relieve suffering and avoid disease and pain, the importance of enhancing people's opportunities, and the need to help children thrive. For example, Mr. and Mrs. A, as well as Drs. B and C, want to promote the child's health; but there is a disagreement over how to achieve this end. We need not value all things similarly with people in another culture, or our own, to have coherent discussions with them about whether certain means will achieve ends. In addition, we share methods of discovery, evaluation, and explanation. These include methods for translating, debating, deliberating, criticizing, analyzing, negotiating, and evaluating data or technology. To do these things, however, we must share some consensus about them [10].

There are also internationally shared values and methods that make possible scientific research between cultures, and these are incorporated into medicine. These include the duty to evaluate information based upon evidence and merit, to be skeptical of results until there is sufficient evidence to rule out alternative explanations, and to be disinterested in collecting and weighing evidence. I will argue that medical practice illustrates that we can determine what is right or wrong in terms of means other than cultural approval or disapproval, because we share intercultural values and methods that enable us to evaluate our reasons. As moral beings, we guide our behavior in part by reasons. Examination of reasons given by men and women who practice female circumcision/genital mutilation reveals many ways of entering the debate, using these shared values and methods.[5]

III. REASONS FOR FEMALE GENITAL MUTILATION

Defenders of female circumcision/genital mutilation do not claim that this practice is a moral or religious requirement and end the discussion there; they are willing to give and defend reasons why they believe it is a good tradition. Five studies conducted by investigators from countries where female circumcision is widely practiced [6], [4], [18], [14], [1] establish that the primary reasons given for performing this ritual surgery are it 1) fulfills religious duties, 2) preserves group identity, 3) helps to maintain cleanliness and health, 4) preserves virginity and family honor and prevents immorality, and 5) furthers marriage goals including greater sexual pleasure for men. Each of the reasons given for this practice has been contested as unfounded or inconsistent with other important beliefs held by the communities that practice these rites [6], [14], [1], [13].

Most of those practicing female circumcision/genital mutilation are Muslim, and regard it as a religious requirement; yet it is not practiced in Saudi Arabia, the spiritual center of Islam, nor is it required in the Koran [6], [28], [18]. Many Islamic leaders vehemently deny it is a Muslim tradition [15]. Koso-Thomas ([14], p. 10) writes: "None of the reasons put forward in favor of circumcision have any real scientific or logical basis." To illustrate how medicine can effectively challenge some cultural beliefs, I will focus on the claim that female circumcision prevents illness and maintains cleanliness and health.

IV. HEALTH CONSEQUENCES

Beliefs that female circumcision or infibulation promotes health and hygiene are incompatible with evidence from surveys done within cultures where these rites are practiced. It has been linked to mortality or morbidity such as shock, infertility, infections, incontinence, maternal-fetal complications, and protracted labor. The tiny hole generally left for blood and urine to pass is a constant source of infection. It causes painful intercourse and menstruation [6], [14], [1], [4], [18].

These surveys show that pain and complications vary with the extent of the mutilation. Female circumcision/genital mutilation is somewhat arbitrarily viewed as taking three forms. Type 1 circumcision involves removal of the clitoral hood, or prepuce. This is the least mutilating type and might not preclude or diminish sexual orgasms in later life, unlike other forms. When this surgery is performed on infants and small children, however, it may be difficult to avoid removal of additional tissue, because infants' genitalia are small, and the tools commonly used are razors and knives. In the southern Arabian countries of Southern Yemen and Musqat-Oman, Type 1 circumcision is commonly practiced. In African countries, however, Type 1 circumcision is often not regarded as a genuine circumcision [14], [1]. Only about three percent of the women in one East African survey had this type of circumcision [6].

Type 2, or intermediary, circumcision, involves removal of the clitoris and most or all of the labia minora. In Type 3 circumcision, or infibulation, the clitoris, labia minora, and parts of the labia majora are removed. The gaping wound to the vulva is stitched tightly closed, leaving a tiny opening so that the woman can pass urine and menstrual flow. There is evidence that Type 3, also known as Pharaonic circumcision, has been done since the time of the pharaohs [1]. In some African countries most young girls between infancy and ten years of age have Type 3 circumcision [1], [18], [4]. Traditional practitioners often use sharpened or hot stones, razors, or knives, frequently without anesthesia or antibiotics [22], [1], [6]. In many communities thorns are used to stitch the wound closed, and a twig is inserted to keep an opening. The girl's legs may be bound for a month or more while the scar heals [1], [6].

Types 2 and 3, both of which diminish or preclude orgasms,[6] are the most popular forms. More than three-quarters of the girls in the Sudan, Somalia, Ethiopia, and other north African and southern Arabian countries undergo Type 2 or Type 3 circumcision, with many of the others circumcised by Type 1 [6], [18], [4], [14], [19]. One survey by Sudanese physician Asma El Dareer shows that over ninety-eight percent of Sudanese women have had this ritual surgery, twelve percent with Type 2 and eighty-three percent with Type 3 [6]. These rites are popular with one study reporting ninety-two percent of Somali women surveyed favor continuing Type 3 (seventy-six percent) or Type 2 (twenty-four percent) for their daughters [18].

Almost all girls experience immediate pain following the surgery [22], [6]. El Dareer found other immediate consequences, including bleeding, infection, and shock correlating with the type of circumcision: Type 1, 8.1 percent; Type 2, 24.1 percent; and Type 3, 25.6 percent. Bleeding occurred in all forms of circumcision, accounting for 21.3 percent of the immediate medical problems in El Dareer's survey. She writes, "Hemorrhage can be either primary, from injuries to arteries or veins, or secondary, as a result of infection" ([6], p. 33). Infections are frequent because the surgical conditions are often unhygienic [22], [6]. The inability to pass urine was common, constituting 21.65 percent of the immediate complications [6]. El Dareer found 32.2 percent of the women surveyed had long-term problems, with 24.54 percent suffering urinary tract infections and 23.8 percent suffering chronic pelvic infection.

Published studies by investigators from the regions where these rituals are practiced uniformly find that women had similar complaints about and complications from female genital mutilation: at the site of the surgery, scarring can make penetration difficult and intercourse painful; cysts may form, requiring surgical repairs; a variety of menstrual problems arise if the opening left is too small to allow adequate drainage; fistulas or tears in the bowel or urinary tract are common, causing incontinence, which in turn leads to social as well as medical problems; maternal-fetal complications and prolonged and obstructed labor are also well-established consequences [16], [22], [6], [14], [1], [20], [19], [27]. El Dareer's studies lead her to conclude that immediate and long-term complications are "almost invariable . . . especially at childbirth. Consummation of marriage is always a difficult experience for both partners, and marital problems often result. Psychological disturbances in girls due to circumcision are not uncommon" ([6], iiiiv). The operation can also be fatal because of shock, tetanus, and septicemia [22]. Despite this evidence, some practitioners of these rituals when interviewed insist that these rites are neither painful nor harm their patients [28].

Morbidity and mortality may even be higher than these studies indicate because, first, investigations are often conducted in cities and the complication and death rate is probably higher in rural areas [18]. Second, many of the countries where these rites are practiced have unenforced laws prohibiting these surgeries (some are remnants of colonial days); consequently, some people are reluctant to discuss with investigators technically illegal actions and their consequences. Third, some women do not attribute the complications to the surgery that they believe promotes their health and well-being [6].

Thus, a series of studies from these regions by investigators from these cultures documents that female genital mutilation has no benefits and is harmful in many ways, with both short- and long-term complications. The shared goals to promote cleanliness and health and prevent disease and disability are not fulfilled by the practices studied. One could argue that since one goal of medicine is to reduce morbidity and mortality, doctors and nurses should cooperate with traditional practitioners toward this end. This narrow understanding of medical goals and values as technical efficacy in reducing morbidity and mortality, however, would also justify doctors and nurses assisting in the torture of political prisoners. The condemnation of such practices by international medical organizations shows that medicine's values extend beyond technical expertise. They include promoting practices that prevent morbidity and mortality and help the health, well-being, and flourishing of members of our communities.[7]

V. ETHICAL RELATIVISM

There are many forms of ethical relativism, and I criticize and reject the version holding that to say that something is right means that it is approved of in the speaker's culture, and to say that something is wrong means that it is disapproved. In contrast, other versions of relativism may be noncontroversial. *Descriptive relativism,* for example, is the noncontroversial view that people from different cultures *do* act differently and have distinct norms. Descriptions about how or in what way we *are* different do not settle the moral issue of how we *ought* to act. Often relativism is presented as the only alternative to clearly implausible views entailing absolutism or cultural imperialism; sometimes it is used to stress obvious points that our own preferences are not absolute standards, or that different rankings and interpretations of moral values or rules by different groups may be justifiable; sometimes "relativism" is the term employed to highlight the indisputable influence of culture on moral development, reasoning, norms, and decisions. It may also be used to show how decisions about what we ought to do depend on the situation—for example, that it may not be wrong to lie in some cases.

These points are not in dispute herein or even controversial, so my comments do not apply to these versions of relativism. The controversial position under discussion herein, called *ethical* or *cultural relativism,* is that to say an action is right means it is approved of in a person's culture and to say it is wrong means it is disapproved. If this view is correct, then there is no basis for establishing that one set of culturally established duties or moral values is right and another wrong. On this view, moreover, it is incoherent to claim that something is wrong in a culture yet approved, or right yet disapproved by the culture. According to ethical relativism, when people make moral judgments about things done in other cultures, they are expressing only their cultural point of view, not one that has moral authority in another culture; positions taken by other countries or international groups merely reflect particular societal opinions having no moral standing in another culture [29], [9], [23], [25].

In contrast to such versions of ethical relativism, other traditions hold that to say something is morally right means that the claim can be defended with reasons in a certain way. Saying that something is approved does not settle whether it is right, because something can be wrong even when it is approved by most people in a culture. Moral judgments do not describe what is approved but prescribe what *ought* to be approved. Thus people's belief that female circumcision/genital mutilation is right because it promotes health and cleanliness does not make this opinion true, or their subsequent behavior to fulfill these goals right.

What we believe and do can be rationally evaluated using some shared values and methods. Because medicine is practiced in many similar ways all over the world, medicine's values and beliefs can be a powerful means to challenge some entrenched moral and cultural beliefs such as those about female genital mutilation. Consequently, the version of ethical relativism considered herein seems an implausible view for this and other reasons I will now consider.

VI. DIFFERENTIATING AND COUNTING CULTURES?

According to ethical relativism, the final determination of what is good or bad, right or wrong, is cultural approval and disapproval. People who are in doubt about what they ought to do need to find out what is approved in their cultures. To develop a useful theory, therefore, we must know how to distinguish one culture from another. How big or old or vital must a culture, subculture, or group be in order to be recognized as a society whose moral distinctions are self-contained and self-justifying? In short, how exactly do we count or separate cultures? A society is not a nation-state, because some social groups have distinctive identities within nations. If we do not define societies as nations, however, how do we distinguish among cultural groups, for example, well enough to say that an action is child abuse in one culture but not in another?

Consider this issue of differentiating cultures in relation to our example. Do Drs. B and C belong to the same culture? One might cogently argue that they do because they both have similar training in medicine, read the same journals, and attend international medical meetings that condemn female genital mutilation, as well as work for other important goals such as world peace and preventative health care. On the other hand, one might also plausibly argue that Drs. B and C are not members of the same culture since they belong to different religions, races, are citizens of different countries, share no common language, live half a world apart, and differ in many of their customs and experiences. Another problem is determining Mr. and Mrs. A's culture because they, while Muslim, live in predominantly Christian North America, and cling to cultural practices from East Africa not shared by most Muslims. Do they even share a culture with their cousin Dr. C who grew up with them, yet disagrees about this and other cultural practices?

One difficulty, then, is that it seems implausible to say most of us live in one culture. Consider a real case. An Islamic scholar from Gambia, Baba Lee, committed to eradication of female genital mutilation, left his daughter with his mother. She disagreed with him, and had her granddaughter circumcised while he was gone [28]. It is rare to find a place in the world where people live under the umbrella of a single culture.

Another difficulty for those claiming ethical relativism is a useful theory for establishing the meaning of right and wrong, or learning of our duties, concerns how we know what is approved or disapproved within some culture. Not only is there a problem about distinguishing cultures, but in any community there can be passionate disagreement, ambivalence, or rapid changes about what is approved or disapproved. Approval ratings for political leaders, laws, or wars, for example, may change weekly. According to ethical relativism, where there is no significant agreement within a culture, there is no way to determine what is right or wrong. But what agreement is significant? As we saw, some people in cultures practicing female genital mutilation, often those with higher education, actively work to stop it [6], [14], [28], [1].

If defenders of ethical relativism mean it to be a useful theory, they cannot say that people's cultures are an overlapping patchwork of similarities and differences, as many of us would be inclined to say. Defenders of ethical relativism must offer some cogent means to distinguish between cultures and establish what is approved or disapproved within them; otherwise, the theory is useless as a means to find out what traits are virtuous or vicious, what things are right or wrong, or how people ought to act. To say that people may belong to various cultures that overlap and have many variations fails to give a way to determine this. To summarize, it seems implausible to say most of us live in one culture or that cultures are clearly separated. If we cannot identify the relevant culture or what counts as sufficient approval or disapproval to be judged right or wrong in a culture, then it is not helpful to put forth the theory that to say something is right means it has cultural approval and wrong if it has cultural disapproval.

VII. WORKING FOR IMPROVEMENTS

Another cluster of problems arises for this version of ethical relativism when we consider the important task of working for improvements within our cultures or communities. We often make claims like, "This is approved in my culture, yet wrong." The activists working to change practices of female genital mutilation in their countries, for example, do so because they believe what is approved sometimes is wrong.

Relativists who want to defend the meaningfulness of saying that something is approved but wrong, or there are sound cross-cultural moral judgments (for example, about the value of freedom and human rights in other cultures) seem to have two choices. On the one hand, if they agree that some cross-cultural norms have moral authority, they should also agree that some intercultural judgments about female circumcision/genital mutilation also may have moral authority. Some relativists such as Sherwin [24] take this route, thereby abandoning the version of ethical relativism being criticized herein. On the other hand, if they defend this version of ethical relativism yet make cross-cultural moral judgments about the importance of values like tolerance, group benefit, and the survival of cultures, they will have to admit to an inconsistency in their arguments. For example, anthropologist Scheper-Hughes [23] advocates ethical relativism and tolerance of other cultural value systems; however, she fails to acknowledge that she is saying that tolerance between cultures is *right* and that this expresses a cross-cultural moral judgment about tolerance. Similarly, advocates of ethical relativists who say it is *wrong* to eliminate rituals that give meaning to other cultures are also inconsistent in making a judgment that presumes to have genuine cross-cultural moral authority about what is right and wrong.

Thus, it is not consistent for defenders of this version of ethical relativism to make intercultural moral judgments about tolerance, human rights, group benefit, intersocietal respect, or the value of cultural diversity. The burden of proof, then, is upon defenders of this version of ethical relativism to show why we cannot do something we think we sometimes do very well, namely, meaningfully criticize what is approved or disapproved in our own cultures, engage in intercultural moral discussion, cooperation, or criticism, or support people whose welfare or rights are in jeopardy in other cultures in a way that has moral authority. In addition, defenders of ethical relativism need to explain how we can justify the actions of national or international professional societies that take moral stands in adopting policy. For example, international groups may take moral stands that advocate fighting a pandemic, stopping wars, halting oppression, promoting health education, or eliminating poverty, and their views have moral authority in some cases. Some might respond that our professional groups are themselves cultures of a sort. But this response raises the already discussed problem of how to differentiate cultures.

The version of relativism criticized herein holds that cross-cultural moral judgments have no moral force. Its distinctive feature is its skeptical position about the limits of knowledge, claiming that one can *never* make a cross-cultural moral judgment that has moral authority.[8] Williams [29], Ginsberg [9], and Shweder [25] hold forms of this position. This view is false if people from one culture can *sometimes* make judgments about actions in another society that have moral authority.[9] That is, opponents need not show *all* our judgments, values, and methods are similar to defeat this position. For example, whatever other differences they may have, if Drs. B and C, who are from different cultures, make the same judgments about female circumcision in a way that has intercultural moral authority, then ethical relativism cannot be true. In short, this version of ethical relativism cannot give a plausible account of something we regard as an important part of our lives, working for improvements in our communities.

VIII. DO WE SHARE VALUES AND METHODS?

Some might object to my arguments saying that we do not really share values and methods as I have presupposed, and thus there is no real basis for meaningful cross-cultural moral evaluations. We cannot really understand another society well enough to criticize it, the argument might go, because our feelings, concepts, or ways of reasoning are too different; our so-called ordinary moral views about what is permissible are determined by our upbringing and environments to such a degree that they cannot be transferred to other cultures. For example, critics might argue that we may think that Drs. B and C make the same moral judgments; but, we cannot know this given our cultural barriers.

There are two ways to understand this objection [26]. The first is that nothing counts as understanding another culture except being raised in it. If that is what is meant, then the objection is valid in a trivial way. But it does not address the important issue of whether we can comprehend well enough to make relevant moral distinctions or engage in critical ethical discussions about, for example, aggression, oppression, theft, or cooperation. In addition, the problem remains about what it means to be raised in the *same* culture when, as we discussed earlier, it is implausible to suppose most people are raised in or belong to just one culture.

The second, and nontrivial, way to understand this objection is that it means our experiences have shown that we cannot understand another culture well enough to justify claiming to know what is right or wrong in that society. Yet our experiences suggest we can do this very well. We ordinarily view international criticism concerning, for example, human rights violations, aggression, torture, and exploitation as important ways to show that we care about the rights and welfare of other people, and in some cases these responses have moral authority. In addition we can translate, debate, criticize each other's reasoning, and so on, and by these means show our shared values and methods. This does not mean that all disputes can be resolved by these shared values and methods within or across cultures. Some disagreements, such as those over abortion and euthanasia, resist solution because reasonable and informed people of good will rank important values differently. Our experiences, then, suggest we share many values and methods between cultures.

IX. AVOIDING CULTURAL IMPERIALISM

The version of ethical relativism we have been considering does not avoid cultural imperialism, as defenders such as Scheper-Hughes [23] sometimes suppose. To say that an act is right on this view means that it has cultural approval. A culture, however, may approve acts of war, terrorism, oppression, enslavement, aggression, exploitation, racism, or torture; and, on this view, that means those acts are right in that culture. The disapproval of other cultures is irrelevant in determining whether these acts are right or wrong; accordingly, the disapproval of people in other cultures, even victims of war, oppression, enslavement, aggression, exploitation, racism, or torture, does not count in deciding what is right or wrong except in their own culture. This view not only entails agreement that female genital mutilation is right in cultures where it is approved but the affirmation that anything else with wide social approval is right, such as slavery, war, discrimination, oppression, racism, and torture.

If defenders of the version of ethical relativism criticized herein are consistent, they will dismiss any objections by people in other cultures as merely an expression of their own cultural preferences, having no moral standing whatsoever in the society that is engaging in the acts in question. Defenders of ethical relativism must explain why we should adopt a view which, when consistently defended, leads to such atrocious conclusions. It leads to the conclusion that we cannot make intercultural judgments with moral force about societies that start wars, practice torture, or exploit and oppress other groups; that is, as long as these activities are approved in the society that does them, according to this view, they are right. Yet the world community believes that it makes important cross-cultural judgments with moral force when it criticizes slavery, apartheid, aggression, repression of dissidents, or denial of equality of opportunity to women. Representatives from the criticized society usually try to defend themselves by saying that their practices are morally justified in their own cultures because of their special traditions, even if they would not be in another society. If ethical relativism is convincing, such responses ought to be as well.

X. OPPRESSION

Even where female circumcision/genital mutilation is popular with the majority of people, a minority would like to stop these rites. For example, Baba Lee, an Islamic scholar from Gambia, denies that female genital mutilation is an Islamic tradition, but is a bad tribal practice: "It is a means of suppressing women . . . [and] to teach women how to be obedient, how to be subdued with men, how to carry on traditions that matter" ([28], p. 326). Investigators from these regions have found other men who join women in these countries by condemning these rites. Abdalla writes that they agree this custom " . . . serves no purpose; it has ill effects on health; it lessens sexual desire (enjoyment) for women; it is based upon male chauvinism" ([1], p. 97). Unless adults understand the harms, it is still questionable whether they make a voluntary choice. Children, of course, cannot assess the harms and have no opportunity to choose.

Some men and women in these regions, particularly those with more education, would prefer not to perform these rites, Abdalla contends, but many of them "do not have the guts to totally abandon circumcision of women" ([1], p. 94). Studies indicate many men and women would not want to have this surgery for themselves or their daughters if they believed it would not hurt their marriage prospects or other opportunities. This suggest that their approval is manipulated by an institutional framework they do not entirely accept. People who oppose these practices, especially outsiders, run risks. Lightfoot-Klein points out some have even been mutilated or killed for their efforts [17]. Koso-Thomas [15] attributes the popularity of these rites to ignorance.

Some argue that we should use the same sort of international sanctions to oppose genital mutilation as those employed to challenge other human rights violations; while others favor education as the best way to change these cultural practices [28]. I have neither addressed how best to persuade people to stop these rites, nor discussed the proper limitations of tolerance in a just society. I have only tried to show that we have means for evaluating these practices, attitudes, and beliefs in a way that has moral authority across cultures.

XI. CONCLUSION

An effective way to debate the merits of our different but entrenched cultural beliefs and values is to employ a common ground of shared values and methods. Despite our cultural differences, we share many values and methods of discovery, evaluation, and explanation. These enable us, sometimes correctly, to judge other cultures, and they us. Moral judgments can be evaluated at least in terms of their consistency, factual presuppositions, and their coherence with stable evidence, like medical or scientific findings. By this means, certain moral claims can be challenged across cultural boundaries in a way that has moral authority. As values and methods become more and more integrated worldwide, they increasingly provide inroads to evaluate the rational basis of some deeply held cultural and moral beliefs.

The common ground used here to show these points is that of medical practice. Medicine does not embrace, define, or establish all our moral goals and values. Cultures around the world share enough of medicine's values and methods, however, for them to serve as the basis of some genuine cross-cultural examination and intercultural moral criticism of goals and means to those ends. Defenders of female circumcision/genital mutilation, as we have seen, claim it prevents disease, promotes health and cleanliness, as well as maternal and child well-being. The stable evidence shows that this is false. Moreover, Ntiri [18], Abdalla [1], and Koso-Thomas [14] find many women in these cultures, when interviewed, attribute health problems to circumcision and infibulation such as their keloid scars, urine retention, pelvic infections, puerperal sepsis, and obstetrical problems. Studies document that these rituals deny women sexual enjoyment or orgasms, cause significant morbidity or mortality among women and children, and strain the overburdened health-care systems in these developing countries. Beliefs that female genital mutilation enhances fertility and promotes health, that women cannot have orgasms, and that allowing the baby's head to touch the clitoris during delivery causes death to the baby are incompatible with medical information gathered in these cultures using methods having approval therein.

Furthermore, outright inconsistencies of reasoning appear when people within these cultures defend female circumcision/genital mutilation. For example, on the one hand, they believe these practices deprive women of nothing important because they do not think that women can have orgasms or that sex can be directly pleasing to women; their pleasure comes only from knowing they contribute to their husbands' enjoyment. On the other hand, one justification for the surgery is that women without it have uncontrollable sexual appetites [14], [6],

[1]. Moreover, on the one hand, they believe these rites promote health and prevent morbidity and mortality. On the other, the rites are known to cause scarring, infection, incontinence, morbidity, and mortality [14].

The version of ethical relativism under consideration can be consistently held. Ethical relativism seeks to explain the meaning and use of moral terms and judgments with a single-dimension, cultural approval and disapproval. It leads to conclusions so far from how we use moral terms and judgments, however, that the theory is very implausible. When consistently defended, the version of ethical relativism under consideration leads to the implausible doctrine that we live in separate and uniform societies rather than in families and communities that draw from many identifiable groups. It denies that we can take the first step to moral improvements, namely judging that the status quo is wrong, or what our culture disapproves is right. In addition, rather than avoid cultural imperialism or tyranny, this version of ethical relativism seems to promote it, giving a license to some cultures to ignore international condemnation of their aggression, wars, suppression or torture, as long as what they do has approval within their own culture. Finally, it seems to belie the experiences we have of working internationally on such projects as peace, world health, and environmental safety. Thus, while this view can be consistently defended, it leads to implausible and even abhorrent conclusions.

Department of Medical Humanities
East Carolina University School of Medicine
Greenville, North Carolina

NOTES

[1] This thesis should not be confused with medicalization of moral problems by mistakenly turning "bad" behavior into "sick" behavior. I also reject the view that moral judgments can be "reduced" to descriptive or scientific claims.

[2] It is practiced, for example, by Muslim groups in the Philippines, Malaysia, Pakistan, Indonesia, Europe, and North America [12], [27], [1], [14]. Parents may use traditional practitioners or seek medical facilities to reduce the morbidity or mortality of this genital surgery. Some doctors and nurses perform the procedures for large fees or because they are concerned about the unhygienic techniques that traditional practitioners may use. In the United Kingdom, where about 2,000 girls undergo the surgery annually, it is classified as child abuse [27]. Other countries have also classified it as child abuse, including Canada and France [12].

[3] For example see the work of Abdalla [1], Calder [4], Rushwan [22], El Dareer [6], Koso-Thomas [14], Kouba and Muasher [16], and Ntiri [18].
[4] This tradition is prevalent and deeply embedded in many countries, including Ethiopia, the Sudan, Somalia, Sierra Leone, Kenya, Tanzania, Central African Republic, Chad, Gambia, Liberia, Mali, Senegal, Eritrea, Ivory Coast, Upper Volta, Mauritania, Nigeria, Mozambique, Botswana, Lesotho, and Egypt [1], [18], [4], [22], [6], [14]. Modified versions of the surgeries are also performed in Southern Yemen and Musqat-Oman [1].
[5] Portions of this paper were adapted from Kopelman [13].
[6] Some people in these cultures deny that it precludes orgasms for women, but the author who reports this doubts their claims [17].
[7] See codes in the Appendix to the *Encyclopedia of Bioethics,* Volume 5, 1995.
[8] Many attacks have been made on the skepticism underlying such ethical relativism [3], [10], and my remarks are in this tradition.
[9] As I pointed out earlier, one can also question if it is meaningful to speak of most people as belonging to one culture.

REFERENCES

1. Abdalla, R. H. D.: 1982, *Sisters in Affliction: Circumcision and Infibulation of Women in Africa,* Zed Press, London, U. K.
2. American Medical Society: 1991, 'Surgical modification of female genitalia', House of Delegates Amended Resolution 13 (June).
3. Bambrough, R.: 1979, *Moral Skepticism and Moral Knowledge,* Routledge and Kegan Paul, London, U. K.
4. Calder, B. L., Brown, Y. M. and Rac, D. I.: 1993, 'Female circumcision/genital mutilation: Culturally sensitive care', *Health Care for Women International* 14: 3, 227-238.
5. Dirie, M. A., and Lindmark, G.: 1992, 'The risk of medical complication after female circumcision', *East African Medical Journal* 69, 479-482.
6. El Dareer, A.: 1982, *Woman, Why Do You Weep? Circumcision and Its Consequences,* Zed Press, London, U. K.
7. *Encyclopedia of Bioethics,* 2nd ed., 1995, W. T. Reich, (editor-in-chief), Simon & Schuster-Macmillan, New York, NY.
8. Fourcroy, J. L.: 1983, 'L'eternal couteau: Review of female circumcision', *Urology* 22, 458-461.
9. Ginsberg, F.: 1991, 'What do women want? Feminist anthropology confronts clitoridectomy', *Medical Anthropology Quarterly* 5, 17-19.
10. Hampshire, S.: 1989, *Innocence and Experience,* Harvard University Press, Cambridge, MA.
11. Kane, R.: 1994, *Searching for Absolute Values in a Pluralistic World,* Paragon House, New York, NY.

12. Kluge, E.-H.: 1993, 'Female circumcision: When medical ethics confronts cultural values' (editorial), *Canadian Medical Association Journal* 148, 288-289.
13. Kopelman, L. M.: 1994, 'Female circumcision/genital mutilation and ethical relativism', *Second Opinion* 20, 55-71.
14. Koso-Thomas, O. A.: 1987, *The Circumcision of Women,* Zed Press, London, U. K.
15. Koso-Thomas, O. A.: 1995, 'Circumcision: Female circumcision', in W. T. Reich (editor-in-chief), *Encyclopedia of Bioethics*, 2nd ed., Simon & Schuster-Macmillan, New York, pp. 382-387.
16. Kouba, L. J., and Muasher, J.: 1985, 'Female circumcision in Africa: An overview', *African Studies Review* 28, 95-109.
17. Lightfoot-Klein, H.: 1989, *Prisoners of Ritual: An Odyssey into Female Genital Circumcision in Africa,* Haworth, New York, NY.
18. Ntiri, D. W.: 1993, 'Circumcision and health among rural women of southern Somalia as part of a family life survey', *Health Care for Women International* 14, 215-216.
19. Ogiamien, T. B. E.: 1988, 'A legal framework to eradicate female circumcision', *Medicine, Science and the Law* 28, 115-119.
20. Ozumba, B. C.: 1992, 'Acquired Gynetresia in Eastern Nigeria', *International Journal of Gynecology and Obstetrics* 37, 105-109.
21. Rawls, J.: 1993, *Political Liberalism,* Columbia University Press, New York, p. 1.
22. Rushwan, H.: 1990, 'Female circumcision', *World Health,* April-May, 24-25.
23. Scheper-Hughes, N.: 1991, 'Virgin territory: The male discovery of the clitoris', *Medical Anthropology Quarterly* 5, 25-28.
24. Sherwin, S.: 1992, *No Longer Patient: Feminist Ethics and Health Care,* Temple University Press, PA.
25. Shweder, R.: 1990, 'Ethical relativism: Is there a defensible version?' *Ethos* 18, 205-218.
26. Sober, E.: 1991, *Core Questions in Philosophy,* Macmillan, New York, NY.
27. Thompson, J.: 1989, 'Torture by tradition', *Nursing Times* 85, 17-18.
28. Walker, A., and Parmar, P.: 1993, *Warrior Marks: Female Genital Mutilation and the Sexual Blinding of Women,* Harcourt Brace, New York, NY.
29. Williams, B.: 1985, *Ethics and the Limits of Philosophy,* Harvard University Press Cambridge, MA.
30. World Health Organization: 1991, *Female Genital Mutilation,* Geneva.
31. World Health Organization: 1992, *International Journal of Gynecology and Obstetrics* 37, 2, 149.
32. World Health Organization: 1993, Press release WHA/10. International Planned Parenthood Federation. *Statement on Female Genital Mutilation*, Developed by IPPF International Medical Advisory Panel, IPPF, London, U. K.

GRANT GILLETT

"WE BE OF ONE BLOOD, YOU AND I": COMMENTARY ON KOPELMAN

When Mowgli learns the greeting "We be of one blood, you and I," he has learnt the master words of the jungle law. It is, however, not just the words themselves that he must learn, he must also learn to say them in the language of the creature to whom he is talking. These points about identification and communication contain a deep lesson about cross-cultural ethics and the proper response to cultural relativism. It is also congenial to some of the things that Loretta Kopelman wants to say in her forthright paper about female circumcision.

Kopelman correctly identifies the challenging type of cultural relativism as that which baldly asserts that being right is and can be no more than being approved in one's proper cultural setting. Her argument against this view has a number of strands.

The main strand of argument is that medicine, in and of itself,

> incorporates many values and methods having wide intercultural acceptance, including agreement about the evils of causing unnecessary pain, the goods of promoting personal and public health, the duty to try to relieve suffering and avoid disease and pain, the importance of enhancing people's opportunities, and the need to help children thrive ([5], p.).

This sounds undeniably right and good until one heeds the skeptical voices and wonders whether we can fill in the crucial value terms. Terms such as "health," "suffering," "enhance," "opportunity," and "thrive" surely admit of ambiguities. One can systematically relativize all such terms to a conception of what constitutes a "worthwhile, culturally embodied life"[1] in which one allows a strong reading of the cultural aspect to submerge basic facts about human embodiment. If we do not underplay our embodiment, we are able to attend to some fairly universal

R. A. Carson and C. R. Burns (eds.), Philosophy of Medicine and Bioethics, 239-245.

features of our function as rational social beings so as to ground some relatively robust conceptions of goods and harms. Arguably, such things as the experience of nurturing or being encouraged on the one hand, and pain and suffering or rejection on the other, all, typically, have unambiguous values no matter what cultural context they occur in and no matter who it is that experiences them. Kopelman develops moral considerations based on the biological and therefore species-wide aspects of our human function to make some very plausible claims about medical grounds on which one could say that genital mutilation causes pain, suffering, and real physical harm to the women who undergo it. These effectively overturn the spurious health-based reasons adduced in support of the practice but also give independent reason to believe that the practice is harmful per se. She then develops the other strands of her argument.

Kopelman argues that locating a person in one univocal and distinctive culture is a difficult task, indeed it is difficult even to individuate cultures. In this she is surely right. However, this thought never quite engages with the felt need to belong evident in works such as Alice Walker's *Possessing the Secret of Joy* [8]. The need to identify provides Walker's character with a powerful reason to be circumcised and, ultimately, a reason to kill the woman who did it to her. In my view this need to belong can be very powerful but could also be outweighed by the kind of systematic violence and oppression suffered in a ritual such as female circumcision, and therefore can be seen as a Trojan Horse in which violence and betrayal make their way into the believing or needy heart.

Kopelman next argues that the existence of shared voices from within and without a culture provides an argument against hermetically sealed and diverse moral discourses. This argument builds to some extent on a fact evident but not made explicit in the previous discussion of cultural overlap and interference. Such blurring of cultural boundaries results in any given individual occupying a position in more than one symbolic and moral discourse. These differing, and in areas conflicting, subjectivities must interact with each other if they meet in the same person, and thus there is created internal reflection and criticism within some of the subjects inhabiting any given discourse. The only subjects unable to experience such a multitude of voices are those who are kept insulated from alternative discourses. In fact, it is increasingly becoming clear that the voice of African women is emerging from that isolation and beginning to be heard. Where it is being heard it is often being heard to agree with the extracultural voices condemning genital mutilation. Kopelman also cites other voices from within the cultures concerned (those of health-care workers, for instance) that agree with her condemnation of the practice.

Where there is unrest within a culture one begins to suspect that there exist protesting minorities and moral reformers who are being silenced because of their resistance to the dominant discourse. We, as ethical commentators, are therefore bound to develop a theory in which the voices of the oppressed and silenced subjectivities can emerge and inform our moral reflection. Such a theory cannot accept dominant cultural norms as a final arbiter of moral praxis nor can it accept that the oppressed within a culture have necessarily rejected their belief and value system and become cultural aliens.

Kopelman's claim for an intellectual basis on which one can engage with a strong and culturally legitimated voice is one that commends itself to most moral commentators. Many, however, would be suspicious of the aspirations of Western medicine as that basis [2]. Western medicine too often finds itself at odds even with those whose best interests it is supposed to have at heart. Medicine as a profession has more than its share of elitism, paternalism, and the trappings of a culture of privilege and these features tend to distance it from colonized peoples. It is by nature imperialistic and to look to it for any informed identification with the voice of the oppressed is to live in a world of misplaced romanticism.

It is equally romantic to buy into the myth of the economically independent, rational, autonomous agent; a myth ascribed to by those who reject shared ethical values and who are only committed to an ethic of contract and informed consent. Contractarians, for example, overlook our real lived experience and rely on a philosopher's understanding of the moral world as a domain where arguments are built on moral axioms. It is clear that in such a world a difference in basic axioms would predispose different moral subjects (both agents and patients) to different substantive moral views. However, if one were to hold that morality is based on shared encounters and experiences between persons, then one would have something apart from premises on which to base moral convergence or the hope thereof. Consider, for instance, the person who falls and hurts her arm—our instant response is to soothe, to try to offer some balm or comfort. This works because her plight is so universal to the experience of being human that in the typical case we are not in doubt about what she needs [1]. [2] It is plausible that if we examine some of the things that are basic to human life—the actual experiences of nurturing, being welcomed, having one's hurts tended, having one's greeting snubbed, pain, loneliness, and so on—we can erect a common ground in lived experience for some of our moral intuitions. These experiences, as Nel Noddings argues, are not prone to the widespread and systematic distortion that can be applied to commandments and abstract absolutes that rest heavily on interpretation and discourse [6].

The abstraction of propositional morality from the interpersonal reality in which morals live and grow and have their being makes it an unlikely starting place for grounded moral knowledge. Nussbaum [7] talks of being attentive to a sense of life and Williams [9] refers to what in personal and social life counts as something. This general train of thought suggests that there is a kind of perception or sensitivity and a set of related dispositions that are very basic in the human interactions typically experienced in being cared for and raised to adulthood. These features of living with others as a being-in-relation therefore may inform our actions towards others at a very deep level. I believe that the interactions of living-in-relation create a primitive awareness of experiences that are negative and positive, albeit heavily overlain and transformed by the cultural milieu in which they take shape. As we begin to reflect on our experiences they become linked to certain symbols and significations, that are culturally shaped and that influence us in certain ways. At this point our moral thinking can begin to go in a number of divergent directions even though its roots may be almost universal.

Given the almost inevitable emergence of significant cultural divergence in moral discourse, how do we come to share the sensitivities and perceptions of other moral subjects? That we do is clear to anyone who reads and immerses themselves in a variety of historical and cultural narratives. For instance, it is quite possible to understand the moral desolation of Oedipus; to feel something of the mix of outrage, grief, and frustration reported by Alice Walker; or to be moved by the cry of post-colonial writers from within the experience of New Zealand Maori. In each of these we identify the narratives and subjectivities of fear, rejection, belonging, care, and so on. The premises that these are based on are often unclear. That there are such premises seems also to be unclear and, in fact, singularly irrelevant to the implicit moral understanding involved. It remains true that one cannot always see the pain or joy, or hear the silenced voice when one continues steadfastly to occupy the position of privileged agent and abstract commentator but that does nothing to detract from the moral force of the lived experience which someone is trying to convey. Provided only that one has engaged in some sufficiency of the relevant human relationships, it should be possible to appreciate, to some extent, the cry of the elusive voice. To do so requires a certain descent from the level of abstraction to the level of situated narrative and personal subjectivity.

This experiential, or phenomenological, view of the nature of moral knowledge undermines the claim to inaccessibility on the basis of divergent premises. The premises can come along later, they are constructions of a situation which may or may not serve well the purposes of inhabiting that situation with some empathy for the moral subjects exposed to it. Some constructions silence others and speak past them and so must be modified, engaged with, and brought to a fuller grasp of the various positions in which one might find oneself. The constructions that emerge will, of course, inform subsequent experience but, as Hans-Georg Gadamer long ago realized in relation to texts other than that of lived experience, the hermeneutic circle is not closed, it is open to surprise. If we regard our moral discourse as a framework within which we construct our moral experience we must equally realize that here the circle is not closed but that moral knowledge arises in the presence of imminent surprise through interpersonal encounter and shared experience. To lose this is to lose all and submit to the never-ending distraction of "thin" moral concepts [9]. "Thin" moral concepts are high level and abstract; they allow the moral thinker to avoid the encounter with the human reality underpinning his words. That is why moral life is meant to be "thick," full of content, and therefore engaging, resistant, intractable, and challenging. Kopelman's morals are thick.

Kopelman speaks to us as a concretist and as such places her faith in medical praxis and its caring values. This has much to commend it [3]. Her position cannot help but recall Hegel's image of the master and the slave [4]. Hegel argues that the master, although apparently powerful and knowledgeable, is actually helpless and ignorant because the slave is his point of actual contact or encounter with the domain he depends on for his very being. The slave is the one with the actual knowledge of how the world is and what makes it work because he is engaged there and not removed from it at a level of derivative experience and indirect operations. We can liken it to the story in which the wife remarks to her distinguished guests that her husband makes all the important decisions in the household and she merely looks after the details. He it is who ponders and decides about whether the world had a beginning, whether God is dead, what is the real nature of the atom, and whether there are objective truths. She takes over the little things like where they will live, what career paths they will follow, whether they will have children, and how they will spend their money.

For me, the real level of moral knowledge is the level at which we engage with one another as beings-in-relation who are both vulnerable and powerful, needy and satisfied, dependent and independent. In all these modes we experience mutual situations, speak to one another in them, and are changed by them. This is perhaps why Aristotle formulates a theory in which no amount of abstract knowledge can accomplish phronesis or practical wisdom. Phronesis is gained by virtuous practice in which one lives through decisions and their consequences and acquires those hexes which result in good intuitions. To learn in this way requires a certain "openness to the text" of lived human life and the voices and silences that speak to us about subjectivities not our own.

My own view of the possibility of a shared morality clearly sides much more with Kopelman than against her. She observes that the cultural or ethical relativist claims "that we cannot understand another culture well enough to justify claiming to know what is right or wrong in that society" and comments, correctly in my view, "Yet our experiences suggest we can do this very well." There are two ways of taking this claim. First, we might claim we can achieve sufficient moral understanding to tell what would count as right or wrong to a person within that culture. Second, we might claim to understand *absolute* right and wrong sufficiently to be sure that something within that culture is absolutely wrong. In discussing female circumcision, Kopelman claims both. To do this she brings us to a level of embodied experience and imaginative identification that becomes vivid when we read the narratives of the women and health workers concerned. We are not claiming that we will necessarily agree with a dominant or official view held in that culture (sometimes by an oppressive elite). In her argument, Kopelman returns us to the Aristotelian hunting ground of lived moral experience and phronesis or wisdom. On that ground we find a level of moral assurance that allows us to say to those who are vicious, "If any argument or belief can make you treat human beings like that, then there is something wrong with you." This, I believe, is the implicit moral knowledge in the saying, "We be of one blood, you and I."

Otago Medical School
University of Otago
Dunedin, New Zealand

NOTES

[1] This phrase arose in the context of a multicultural group of scholars from Africa, America, Scandinavia, Britain, Eastern Europe, Turkey, Lebanon, Israel and Southern Asia, who were at the time meeting in Hungary.

[2] She may, of course, be atypical and resent help as paternalistic, but the occasional albino tiger does not defeat generalizations about the species.

REFERENCES

1. Gillett, G.: 1993. "'Ought' and wellbeing," *Inquiry* 36, 287-306.
2. Gillett, G.: 1994, 'Beyond the orthodox: Heresy in medicine and social science', *Social Science and Medicine* 39, 1125-1131.

3. Gillett, G.: 1995, 'Virtue and truth in clinical science', *Journal of Medicine and Philosophy* 20, 285-298.
4. Hegel, G. (trans., A.V. Miller): 1977, *Phenomenology of Spirit,* Oxford University Press, Oxford, U. K.
5. Kopelman, L.: 1996, 'Medicine's challenge to relativism: The case of female genital mutilation', in this volume, pp. 221-237.
6. Noddings, N.: 1989, *Women and Evil*, University of California Press, Berkeley, CA.
7. Nussbaum, M.: 1987, 'Perceptive equilibrium: literary theory and ethical theory', *Logos* 8, 55-83.
8. Walker, A.: 1993, *Possessing the Secret of Joy,* Random House, London, U. K.
9. Williams, B.: 1985, *Ethics and the Limits of Philosophy* ,Collins, London, U. K.

COURTNEY S. CAMPBELL

MUST PATIENTS SUFFER?

I. SUFFERING AND THE IMPERATIVE OF MEDICINE

One of contemporary medicine's central purposes is to relieve pain and suffering. Substantial policy funding and clinical efforts are directed towards pain research, clinical trials for pain management, and programs of palliative care. While we accept some pain as a necessity in the performance of certain medical procedures, from needlepricks to invasive surgeries, this relative tolerance for pain makes sense only in the context of broader aims to achieve a pain-free existence for patients and thereby enhance patient control and quality of life.

While pain may be tolerated as a medical necessity in some circumstances, it is difficult to discern a medical purpose for suffering. Rather, in the ideology of contemporary medicine, suffering is an unmitigated evil, worse than death. The relative weighting of these concepts in medicine is reflected in the standard moral formulation that it is permissible to administer pain medication to relieve suffering, even if this method hastens death. Suffering, not death, is the ultimate or absolute "enemy" of modern medicine, while death is a relative or qualified enemy, which may be transformed into a dignified "exit" under appropriate circumstances. It is possible to say, then, that medicine has added to its goal of relieving suffering an aspiration to banish suffering from the human condition.

It follows from this imperative that chronic, protracted pain and unremitting suffering experienced by patients is a symbol of medical failure. Medical research on pain control, however far advanced, and greater willingness to use palliative care by professional caregivers, is still inadequate to the task; reliance on technology has not proven to be an efficacious mediator of human presence. Moreover, recent right-to die legislation has *presumed* this failure in claiming that competent patients have a right to physician assistance in suicide on the grounds of a compassionate response to patient suffering [2].

R. A. Carson and C. R. Burns (eds.), Philosophy of Medicine and Bioethics, 247-263.

This failure of medicine to meet the imperative to relieve pain and suffering confronts us with very different kinds of options. We might attribute this failure to a biomedical research agenda that historically has given a low priority to pain control, and to inadequacies in institutional and ongoing medical education that leave caregivers with insufficient knowledge and skill in palliative care. Three implications follow from this understanding of medical priorities and limits:

> 1) The professional community accepts the validity of the imperative to relieve pain and suffering;
> 2) The community recognizes that medicine's failure to fulfill this alternative gives to pain and suffering great meaning and direction for the identity and integrity of medicine;
> 3) The profession commits itself to a more dedicated effort at the research and clinical levels to relief of pain and suffering as a goal of medicine in general and in the treatment of particular patients.

These collective commitments, which presuppose that pain and suffering are remediable *medical* conditions, are constitutive characteristics of the contemporary medical paradigm.

A second option is available, however, namely to question the overriding status of the medical imperative. Although acceptance of pain and suffering may strike many caregivers and patients as "cruel," the point of this approach is to suggest that pain and suffering are inevitable features of the *human* condition. Thus, the effort to eradicate pain and suffering may appear misguided and even dehumanizing. I want to draw on comments from two student journals to illustrate this perspective.

In one entry, a woman who occasionally experiences arthritis and back pain comments: "I actually get to the point where I welcome the pain and suffering and revel in it because it is at such times that I know I'm really alive and a finite being." In short, pain and suffering bear witness to the reality of our existence, and to both the possibilities and limits of our mortality. A medicine devoted to uncompromising fulfillment of its imperative may deprive us of mediums of self-knowledge that are vital to making our way in the world.

A second entry, reflecting on the aging process, is equally revelatory: ". . . even though I have a low threshold for pain and sickness, I also feel the most alive when I'm in pain or suffer. This is because I'm aware that I'm a body and not just a mind and I get a reality check that I need to take better care of my body if I'm going to have a healthy old age. . . . I just don't want to have to rely on medicine and doctors to make up for what I should have been doing my whole life." An embodied self is a self that bears and is re-membered by pain and suffering; this is the existential "reality check" to which medicine must conform. By contrast, the imperative of the medical paradigm seeks to transform this reality.

I have put these alternatives rather starkly—in full knowledge that there are gradients of pain and suffering, and a distinction between pain and suffering—in order to suggest that the identity and integrity of medicine is inextricably oriented by its understanding of pain and suffering. We clearly do not aspire to a medicine that is cruel, insensitive, or dehumanizing; caregivers, patients, and the public alike seek a medicine that is technically proficient, current in its knowledge base, and that embodies caring in its practices. Nonetheless, I want to argue in this essay that the grasp of contemporary medicine exceeds its reach. It seeks to transform basic features of the human condition into a medical problem and thus inevitably sets itself up for failure. Ultimately, this must mean that a competent, contemporary, and caring medicine will embody more modest ambitions. The imperative of medicine must then be revised so that it encompasses relief of pain, but recognizes that much suffering is medically intractable. How we respond to suffering is much more a problem of human presence than a lack of medical presence.

II. THE ORDEAL OF SUFFERING

Why is it thought that an existential dis-ease can be transformed into a repairable medical disease? I want to examine four philosophic and medical assumptions that support the view that suffering is an eliminable experience.

A. The Concept of Suffering

When caregivers and ethicists ask whether patients *must* suffer, they assume that there are medical alternatives that will spare patients from the burden of suffering. However, this in turn presupposes that we have a well-articulated and widely shared understanding of what suffering is.

It is doubtful that this latter assumption can be defended (which calls into question the cogency of the related medical projects). An inherent ambiguity attends the term "suffering." Suffering may involve "undergoing" or "enduring" an experience or process ([12], p. 608). The sufferer is one that "allows" and "lets the experience happen." This is the sense of "suffer" that occurs in a poignant biblical narrative in which Jesus blesses little children, chastising his over-protective followers: "Suffer the little children to come unto me, and forbid them not, for of such is the kingdom of heaven" (Mark 10:13-15). Moreover, a sense of suffering as "undergoing" experience is embedded in the etymological rootage of "patient" (*pati*, meaning "to suffer") and suggests the meaningfulness of the virtue of patience in the moral life.

This first understanding clarifies why suffering is so discomforting within our contemporary medical paradigm and cultural ethos. "Suffering" on this first rendering implies a rejection of the ideologies of mastery and control over contingency. The sufferer embodies an acceptance of the limits and finitude of the self, as well as its interrelational character. Yet, this acceptance should not be confused with a passive resignation or fatalism. Religious traditions speak of *active* responsiveness as the mark of patience; patience presupposes self-control, goodness, and understanding, and in turn is the foundation for developing the virtues of devotion, kindness, and love. This mark of patience as an active virtue is illustrated in William F. May's rejection of the resigned fatalism often attributed to the person who endures suffering: "Patience requires waiting, receiving, willing; it demands a most intense sort of activity; it requires taking control of one's spirit precisely when all else goes out of control, when panic would send us sprawling in all directions" ([13], p. 133).

A second understanding of suffering bears a radically distinct, and to some extent, conflicting meaning. The unifying feature of these meanings is the sufferer's experience, actual or perceived, of a loss of control, or a feeling of powerlessness in the face of overwhelming forces beyond personal control. The radical difference consists in the place of self; in the first interpretation, the active responsiveness expressed in patience presumes self-control, while on the second account of suffering, the very capacity to sustain a "self" over time is seriously challenged. Indeed, suffering is characterized by an actual or perceived threat of *dissolution* to the identity and integrity of the self. The experience of suffering calls into profound question the continuity of the "I" by which we refer to ourselves.

This suffering self seeks first understanding, not patience, because it is experiencing some catastrophe that erodes coherence, meaning, and purpose to one's place and being in the world.[1] "Such universe altering events . . . as severe illness, death, violence, or war call into question both the individual's very capacity to maintain itself as a whole and the idea of the whole according to which that organization takes place" ([17], p. 48). This ordeal of suffering seems to deprive the sufferer of all that is most humanly meaningful in life—relationships, memory, creativity, communication, the body as a means of self-revelation, autonomy. The threat of dissolution is very particularized, private and subjective; the catastrophe—whether in the form of the onset of terminal illness, or the death of a loved one, or a relational rupturing such as a divorce—diminishes and challenges *my* identity, loyalties, and sense of self. The prospect of threat strikes to the very core of who one is, and the challenge of patience and self-control must then give way to a primary need for self-reconstitution in the midst of the brokenness of life.

The characteristics of suffering experienced in the face of catastrophe can be delineated further. A first distinguishing mark may be designated by the metaphor of *magnification.* Faced with the multitudinous losses entailed by threatened self-dissolution—losses of relationship, historical memory, creativity, body, and autonomy—the claims and interests of the self become more present or magnified in our consciousness. We do not attribute suffering to patients in persistent vegetative state because they have already experienced these losses, and have minimal organic capacities and functioning. For the suffering self threatened with the loss of integrity and identity, however, self-awareness is magnified to the point of becoming all-consuming.

The sufferer also experiences *alienation* or *fragmentation.* The self becomes estranged from its embodiment through the diminishment of disease; the body is transformed into a hostile other that frustrates life plans and aspirations rather than facilitating their achievement ([22], pp. 129-131). Moreover, the self may be estranged from its life projects and their achievement. As Rawlinson contends, "suffering names the experience of and alienation from or disruption in one's own ends or purposive activity and an inability to maintain the ordered wholeness of one's world" ([17], p. 50). Or, the self becomes a stranger-in-relation to others, who for various reasons find themselves repelled by suffering, critical of the sufferer ("others have it worse than you"), or engage in victim-blaming, as in the paradigmatic biblical account of the suffering of Job.

A loss of relation and creativity is also attributable to a third mark of suffering, a profound *shattering of language* and a diminishment of voice. What Elaine Scarry has observed of pain is no less true of embodied suffering: "Physical pain does not simply resist language but actively destroys it, bringing about an immediate reversion to a state anterior to language, to sounds and cries a human being makes before language is learned" ([19], p. 4). Similarly, Warren T. Reich contends the ordeal of suffering is in part constituted by the "search for a language for suffering—through the sufferer's struggle to discover a voice that will express his or her search for the meaning of suffering" ([18], p. 86). As with pain, screams and moans rather than silence may be the medium of expression, but the general point is that the suffering self is initially unable to communicate about his or her suffering.

A fourth mark of suffering may be identified as the *contraction* of the world, even as the self becomes an enlarged presence in what remains of that world. Tolstoy's *The Death of Ivan Ilyich* reveals a two-fold dimension to contraction. First, the phenomenal, physical world contracts or turns inward on the self. A person with a chronic terminal condition, for example, may experience the contraction of his world from a natural to an entirely social environment, and then, as a "shut-in" or "home-bound" person, to the home environment, and, finally, to his bedroom and perhaps confinement to bed.

Contracted in spatial mobility, the suffering self may find solace and companionship through memory. Yet, the process of re-membering may actually drive home the experience of dissolution of the self's identity and integrity more poignantly: "I'm not the person I once was (or remembered myself to be)." This realization invites a second feature of contraction, a turning inward of the self as it undertakes an inner dialogue. The fragmented self cannot but avoid asking questions: Why is this happening? Why me? What is the point of this suffering? and so on.

These two perspectives on suffering, which I shall call "existential" and "particular" suffering, comprise foundational understandings of suffering. Existential suffering is experienced by all human beings as a condition of life, while particular suffering is experienced more personally and contingently through catastrophe. Neither one of the experiences of suffering, I would argue, is medically malleable. Medicine certainly can prevent or remedy the occasions out of which particular suffering arises—at least for a time, since we cannot escape the inevitability of our mortality. However, this is different than saying that suffering can be remedied, or that patients need not experience suffering. The alleviation of suffering requires the presence of a compassionate community and practices that support patient acceptance and purposive endurance of existential suffering and/or the location of particular suffering within a theological or philosophical world-view or narrative that offers coherence and understanding. To suggest that medicine has the answers for suffering is profoundly arrogant, for such answers will deprive the person of virtues and narratives needed in the journey of moral life. This is why a medicine, with cultural connivance, that aspires to the elimination of suffering risks dehumanization.

B. Medicalized Suffering

The only plausible way that this imperious claim for the scope of medicine's ends can be substantiated is for suffering to be conflated with pain. Suffering is understood medically as a sub-set of pain, rather than a distinctive human phenomenon.

We can refer to this understanding as *medicalized* suffering. This paradigm is certainly related to the ever-expanding horizons of the medical model for almost all human pathologies. There is, moreover, an inexorable logic to the medicalization of suffering. Since medicine can, or at least arguably ought (once professional skills, knowledge, and caring consciousness are sufficiently refined), to bring pain under its domain, continued research and clinical application of pain control methods will, in due course, bring medicine to the end of the pain continuum. Insofar as suffering is understood as a marker on this continuum, then suffering will as well be subsumed under the biomedical model, and its mastery and control will become a medical possibility. In that case, suffering should turn out to be no less susceptible to therapeutic intervention and cure as pain and disease. The medicalization paradigm, in short, transforms a human existential condition into a manageable technical problem.

It is surely the case that, especially in clinical or institutional care settings, pain and suffering can be reciprocally related. However, the reduction of suffering to a form of pain is not justified in all circumstances; the relationship is thus contingent, not necessary. We can experience pain without suffering, such as a person who dies immediately subsequent to a cardiac arrest. Indeed, for many persons this is a preferable way of dying, since it allows the person to avoid a protracted process of dying and its accompanying particularized suffering.

Alternatively, it is no less the case that a person can experience suffering without incipient pain. One of the pioneering narratives in biomedical ethics, the story of Dax Cowart [8], illustrates this situation powerfully. The daily tubbings in an antiseptic bath to which Mr. Cowart was subjected clearly were an occasion for great pain and suffering. However, during his lucid interviews with a psychiatrist, Mr. Cowart's request to die is motivated by particularized suffering, while pain recedes into the background of the narrative.

A more recent example of the experience of suffering sans pain is displayed in the story of Ms. Janet Adkins, who in 1990 was the first "medicide" victim of Dr. Jack Kevorkian. Adkins had been diagnosed with early stage Alzheimers, with a prognosis of a "good" quality of life for ten years. This diagnosis clearly induced particularized suffering in Ms. Adkins, but, at least at the time she sought out Dr. Kevorkian, she was not afflicted with pain, and on the contrary had been very physically vigorous (e.g., playing tennis) within her family.

On these narratives, the medicalization paradigm loses credibility. It is conceptually and clinically mistaken to reduce all particular suffering and personal anguish, even that induced by health-related conditions, to some physiological pathology. This is even more the case for common suffering and the irritants of everyday life. As it is currently constituted, and driven by its set of present presuppositions, medicine cannot readily generalize from its expertise in somatic illness to answer questions of meaning and purpose. At the very least, a wholistic orientation to care, such as that embodied in the hospice movement, must be integrated fully with professional practice for medicine to begin to make good on its promise to relieve pain and suffering.

C. Persons as Patients

In asking whether patients must suffer, we not only presuppose understandings of suffering and its contingency, and of the relationship of pain and suffering, but we also suggest a conception of what it means to be a patient. "Patient" language has rightly been criticized for its positing of a passive and vulnerable role for the sick that has and can be exploited by a paternalistic ethos.

There is, nonetheless, something of significance in the moral clustering of terms rooted in *pati:* patient, patience, and suffering. The "patient" is a person who is given moral *sanction* to suffer and who is also *responsible* to cultivate the virtue of patience to persevere through the ordeal of suffering. Illness brings to patients inescapable dependency on caregivers for competency, knowledge, skill, and care, and re-naming patients as "consumers" or "clients" will not change this fact.

The emphasis on consumer rights and client advocacy in biomedical ethics, however necessary as a remedy for the abuses of paternalism, can ironically actually be disempowering. The demanding consumer or litigious client is a person armed with rights and discharged from duties, including any responsibility to cultivate patience as an excellence of character. Yet, this moral asymmetry is no less diminishing to personhood, for "duty-free" patients are treated as lacking full moral agency ([16], pp. 199-214). My point is that respect for persons must include space and sanction for the person to mourn and suffer and accountability for the kind of person one is as patient, themes that are obfuscated by consumer and client care models. Indeed, the conferral of moral immunity, that is, an ethic of rights without duties and accountability, can breed subtle forms of discrimination, and reflect social marginalization and abandonment of the moral community ([13], p. 130). Thus, when we ask whether patients must suffer, we are inquiring about whether we will allow patients to be patients.

D. The Epistemology of Suffering

A fourth concern in the management of patient suffering raises epistemic issues. Since problem-seeing precedes problem-solving, a professional imperative to relieve suffering presumes caregiver capacities in recognizing suffering. This capacity to "see" or know the suffering of the patient would seem to be facilitated by the proximity to patients and immediacy of experience that caregivers possess. This point is illustrated poignantly in Albert Camus's *The Plague*, in which a physician fighting an outbreak of bubonic plague receives moral support over the radio from a world-wide community: ". . . kindly, well-meaning speakers tried to voice their fellow-feeling, and indeed did so, but at the same time proved the utter incapacity of every man truly to share in suffering that he cannot see." Camus describes an "unbridgeable gulf" between speaker and caregiver, which can be crossed only by embodied solidarity and presence, that is, a commitment to "love or die together" ([3], p. 131). Thus, while an intolerance toward pain and suffering surely has roots in the technological ethos of modern medicine, its sources also lie in the proximity of professional caregivers to suffering patients.

There are sound reasons to think that the very assumptions of modern medicine distort caregivers' vision of suffering. The experience of suffering is, by all accounts, irreducibly subjective. The sufferer knows of his or her suffering even when others do not; this may place the caregiver outside the realm of access to suffering just as much as the well-meaning but remote community. Indeed, one consequence of the shattering of language that occurs in particularized suffering is that the sufferer may be unable to name his or her experience as suffering. Lacking conventional patterns of communication, the sufferer can experience a social death, the death of abandonment and isolation, antecedent to biological death.

Moreover, the subjectivity of suffering makes it conceptually problematic within the scientific (i.e., "objective") presuppositions of contemporary medicine. The scientific quest for objectivity is displayed in the daily ice-breaker between caregiver and patient: "So, on a scale of one to ten, how do you feel today?" Somehow, the patient is to interpret and translate the meaning of his or her qualitative experience of suffering into a quantitative scale that holds interpretative meaning for the professional. Such an inquiry into the world of suffering not only reaffirms the power of medicalizing suffering, but disempowers the patient by displacing control over the meaning of his or her experience.

Eric Cassell suggests further reasons for questioning the validity of medicine's epistemological privilege. While Cassell acknowledges that the relief of suffering is the enduring test of medicine's success, "[t]he central assumptions on which 20th century medicine is founded [particularly, objectivity, quantification, and perception of the person as an organic machine] provide no basis for an understanding of suffering" ([5], p. Vii; [6], p. 1900). Such methodological commitments create a profound tension between the epistemological and ethical foundations of medicine, for if Cassell's claim is correct, then medicine lacks the capacity to comprehend what it is professionally committed to relieve. Indeed, these epistemological presuppositions not only preclude recognition of suffering, but make *denial* of its reality a compelling clinical alternative. As Cassell perceptively observes, "The dominance and success of science in our time has led to the widely held and crippling prejudice that no knowledge is *real* unless it is scientific—objective and measurable. From this perspective, suffering and its dominion in the sick person are themselves unreal" ([5], p. xi). It is then little surprise that a common complaint of suffering selves is that they receive impersonal care.

The conclusion I wish to draw from these four avenues of inquiry—the concept, medicalization, and epistemology of suffering, and the idea of patient—is that suffering is an intractable problem from a medical perspective. No matter how technically proficient and technologically advanced medicine becomes, the professional imperative to relieve suffering must be radically modified and rendered appropriate to medicine's means. Absent such modesty, the presence of suffering patients will continue to be viewed as a failure of medicine and its caregivers. Moreover, if medicine cannot deliver on its promise to patients, increased attention will be given to a certain "solution" proposed by right-to-die advocates: Ending the suffering by ending the life of the sufferer through physician assistance in suicide or active euthanasia.

If suffering is intractable to the medical community (at least under its current paradigm), it does not follow that we simply let the sufferer suffer. The defining features of suffering—submission to powers beyond our control and a quest for coherence and meaning—suggest that practices of understanding and relieving suffering are best situated within a context of spirituality, despite the fact that religious responses to the problem of suffering present their own insidious temptations.

III. TWO TEMPTATIONS OF THEODICY

Within religious communities, theological discourse may seek to give suffering a purpose and meaning by situating it within some larger design for human beings. The "problem" of suffering is re-located from the medical to the metaphysical realm. In Western faith traditions, the experience of suffering may assume spiritual significance either as the working out of divine designs of retribution for sin, or the formation of character through education, or the refinement of the soul for salvation. Unlike medicine, religious theodicies (justifications for evil) do presume an objective reality to suffering. What is unclear is whether this epistemic claim is itself sufficient to "relieve" suffering.

The intellectual explanatory power of a theodicy can be limited because it may induce a mentality of *victimization*. In a religious community in which all suffering is interpreted as a crisis or test of spiritual devotion, the suffering survivors may respond with a passive resignation or even fatalism. That is, the ordeal of suffering is simply accepted as part of the divine ordering of life.

Passivity before suffering can become especially problematic in the context of medicine. While medicine sees in suffering a manifestation of failure, the religious community often will perceive suffering as a manifestation of divine presence. The result of this conflict may be either (a) restraining the practical arts of medicine to enable the working out of the divine purpose, or (b) subjecting the divine will to the contingent state of medical technology. For example, in the case of Helga Wanglie, the family's motivation for continuation of life support was a perceived divine mandate to prolong life with all available medicinal means ([15], p. 513). However, such medical means as ventilators, feeding tubes, antibiotics, and transplants are human and social artifacts that are developed, tested, and refined in a trial-and-error process over time. This would mean that the divine will is contingent on the status of human knowledge, such as whether proposed treatment is nonvalidated or established. This seems, at best, an odd theological position to affirm.

My argument here is that a theodicy of victimization must guard against a temptation to abandon human responsibility for decision-making, whether in the realm of medicine or in the broader moral horizons of life. The distinctiveness of humanity resides in part in the possession and exercise of moral agency, notwithstanding the mortal limits of finitude (incomplete information or lack of control) and fallibility (a propensity for mistakes that favor self-interest). In the face of suffering, moral responsibility requires us to distinguish natural suffering (that which is induced through the processes of disease, diminishment, and death) from suffering brought on by human agency and choices (poverty, discrimination, violence), occasions of suffering that are theoretically preventable. While natural suffering is an uneliminable aspect of the human condition, the moral agent responds with purposive patience, not passivity. Persons who are victimized by suffering should not use this experience as an excuse for abdication of moral responsibility, and in particular, the responsibility to ameliorate those human decisions and choices that contribute to suffering.

A second temptation is that constructing punitive, educational, or redemptive theodicies involves what David Smith refers to as the "domestication" of suffering ([20], pp. 255-261). Domesticated suffering is suffering that we control, not through immunizing ourselves from the experience, but rather through the interpretation of its meaning. This grounds a common distinction in both medical and religious discourse between "unnecessary" and "necessary suffering." On this distinction, unnecessary suffering should be susceptible to remedy through medical intervention, while necessary suffering may be beyond medical amelioration, but is bestowed significance and coherence within a *Weltanschauung* of some kind.

Several kinds of issues emerge with this distinction, some professional, others theological. First, some practitioners and biomedical ethicists do not accept a concept of "necessary" suffering [11]. *All* suffering is definitionally unnecessary and in principle medically manageable. This approach is a manifestation of the commitment to medicalize suffering critiqued previously; it likewise should be rejected because its denies the reality of medically intractable suffering.

A theology of domesticated suffering, however, may not be found personally validating by the suffering self. The theodicies of domestication, first of all, speak more directly to the issue of justifying the ways of God than to the very personalized experience of suffering. Moreover, such theodicies tend to be "macro" in scale, articulating the purposes of suffering for human beings in general. It is unclear that these generic purposes apply directly to a particular patient with a certain disease, such as bone cancer. This absence of personal validation within a theodicy of domestication is eloquently expressed by C. S. Lewis. The very same theologian who domesticated suffering in the memorable phrase, "pain is God's megaphone to rouse a deaf world" ([9], p. 93) would, some years later, following the protracted dying of his wife and deep in the throes of grief, seek God and find only silence. "Talk to me about the truth of religion and I'll listen gladly. Talk to me about the duty of religion and I'll listen submissively. But don't come talking to me about the consolations of religion or I shall suspect that you don't understand" ([10], p. 26).

Still, medicalized suffering and domesticated suffering may possess some common features. As Stanley Hauerwas contends: "Our questions about suffering are asked from a world determined by a hope schooled by medicine—a world that promises to 'solve' suffering by eliminating its causes" ([7], p. 35). We would not be so insistent on asking for explanatory causes of the necessity of suffering were we not in the grip of a medical science that relies on cause-effect methodology to explain the occurrences of the material world, including disease pathologies. In short, the questions embedded in contemporary religious theodicy are parasitic on a medical ideology that seeks to "legitimate the Enlightenment project of extending human power over all contingency" ([7], p. 48). Suffering, however, is problematic for medicine precisely because it resists reduction to the contingent. The full range of human suffering, from lamentation and grief over the death of a child, to the loss of identity that ensues when parents divorce, to the diminishment of self that accompanies the aging process, to the dis-ease over one's purpose in being in the world, cannot be so easily dismissed as "unnecessary," that is, contingent, suffering.

This presents a serious conundrum, however. Medicine offers to the sufferer a compassionate community, but since its practical response is oriented by the medicalization and mastery of contingency, it necessarily ignores the credibility of a good deal of the reality of human experience. A religious theodicy that domesticates suffering may be intellectually compelling, but it is an intellectual, not practical, expression of a communal voice, and it is formulated, typically, by a person or community other than the person who is suffering. However well-considered this domestication of suffering may be, an alien construction can inhibit the sufferer's own quest to give voice to his or her own suffering. A form of religious paternalism is thereby added to the burdens carried by the suffering self.

IV. EMPATHETHIC PRESENCE

In this world, we suffer. That is a "fact" of the human condition. This existential given means that medicine must draw back from its ambitious aspirations to master contingency, medicalize suffering, or quantify its experience. Theological accounts must resist the temptation of offering global explanations that deprive persons of finding personal validation and meaning through suffering. This medical and theological modesty does not entail a prescription for indifference to the suffering self. It means instead that expressions of care and compassion must assume different forms and practices. I shall refer to these in the language of "embodied presence."

A necessary beginning point is to recognize and appreciate the particularity of suffering. An account of the marks of suffering does not yet tell us about the meaning of this experience *for* the suffering self. Each of us will make of suffering what we will, and/or, suffering will make of us different persons. This particularity requires personalization of care, compassion, and embodied presence, in contrast to the one-size-fits-all generic response of both medical and theological models.

The particularity of suffering does indeed present a difficult epistemic issue about the capacity to know of the suffering of another, but this is not insurmountable ([4], p. 27). Embodied presence to the suffering self requires the cultivation of the faculty of imagination and the virtue of empathy. It involves little imagination or empathy, only a great deal of pretension, to hear the cry of suffering and respond, "I know how you feel." Such cruel words isolate the sufferer, who is rightly skeptical, if not cynical, about such an intrusive claim to personal, intimate knowledge. Moreover, the "I know" phraseology deprives the suffering self of an occasion to express the experience of suffering on his or her own terms, and thereby only perpetuates a feeling of powerlessness.

The imaginative presence involves a disposition to learn from the suffering self. The caregiver thereby acknowledges his or her own limitations, rather than asserting explanatory knowledge or power. Even if caregiver and patient have undergone a similar ordeal of suffering, the *meaning* of that experience will nonetheless be very particularized and personalized. Nor, for the same reason, will it be legitimate to generalize from patient to patient. The imaginative disposition instead offers an invitation to the sufferer grounded in our inadequacy, interdependence, and solidarity. The statement of embodied solidarity with the sufferer is thus expressed as: "I don't know how you feel, but I am here with and for you, so that you may tell me of your suffering." This invitation reflects a subverting of conventional power relationships in medicine, in which knowledge and authority reside in the professional. The subversion created by suffering means the professional assumes a position of dependency and limited knowledge. The authoritative voice of suffering is that of the patient. A recognition of the patient's narrative authority can, especially when particular suffering threatens discontinuity of the self, be an important means of retaining or transforming control for the sufferer. The imaginative disposition to learn engenders a willingness to listen. Thus, while theodicies often portray suffering as a pedagogical ordeal for the sufferer, my claim is that it is no less a learning occasion for the caregiver, who must also learn patience and be taught by the patient.

Imaginative presence, the disposition to learn, and the patience to listen are the traits of character necessary to the embodied expression of empathy. We typically understand empathy to be a virtue that consists in putting oneself in the place of another person, understanding and sharing the emotional burden, and seeing the world through his or her eyes. The cultivation of empathy is arduous and time-consuming; it is a learned rather than instinctive virtue. First, putting oneself in the place of the other requires sensitivity to *context and story*. The caregiver's awareness of the present "place" of the patient will require knowledge of the life narrative of the person; only this contextualized experience of suffering will render insight on both why this particular ordeal is experienced as "suffering," and the resources the person might bring to bear to make the experience meaningful or coherent within his or her life. The question to ask is, "Where is this person now in his or her life journey such that this is an ordeal of suffering?"

The *affective and emotional* aspects of the moral life must also direct the embodied expression of empathy. We commonly attribute the virtue of compassion to the person who "suffers with" the sufferer, which means that at a fundamental level, the experience of suffering is one of "passion," intense emotion, and deep affectivity, to the very ground of one's being. This dimension of suffering explains in part why suffering is intractable to medical technology and also cannot be rationalized through a religious theodicy, for neither gives adequate attention to the emotional construction of suffering.

This emotional depth also bears its influence on patterns of communication. If one of the marks of suffering is diminishment of voice, this is in part because we find it generally difficult to describe or even name our emotions. Just as the experience of love often defies language, so also, I contend, with the language of suffering. A person who is able to name the emotions of his or her suffering may still not achieve a ready match between language and experience.

It is unquestionably problematic whether a person can share the suffering of another to the same degree. The Christian doctrine of atonement, for example, presumes that only a divine being can achieve such solidarity in suffering. However, it is possible to share suffering in kind. We can appreciate with some depth, for example, those emotions connected with abandonment, isolation, loss of social relations, or a shattered world. These experiences become the basis by which the caregiver can reach out to the sufferer with authenticity and genuineness, offering embodied presence and empathy. For as Leon Bloy suggests, "suffering disappears, but the fact of having suffered remains always with us" ([1], p. vii). The caregiver can, with honesty and integrity, express to the sufferer, "I only know the smallest part of that which you feel."

The method of analogical approximation may also make possible the empathetic envisioning of the world through the eyes of the sufferer. At some level, this attempt is bound to fail because the world of the sufferer is in the process of what Scarry calls "unmaking," or what is put in more popular discourse as "my world is falling apart." A person whose world-view is reasonably stable, coherent, and purposeful will be at a disadvantage in sharing the fragmented, ruptured, and alien world of the suffering. Nonetheless, "memoriative knowing" ([14], p. 615) of one's own suffering may enable the caregiver to have indirect or proximate access to the sufferer's world. The caregiver's memory of what it means to live in a fragmented and hostile world—even if that world is not de-constructed precisely the same as the world experienced by the sufferer—can offer a memorialized and embodied suffering presence to the sufferer. Such presence can enable him or her to remember (build again) their world and begin the process of self-reconstitution.

Another way of making this point suggests itself. All suffering is embedded in a narrative context. This context includes a particularized life narrative of the person, and a more general narrative that, because it is grounded in some world-view, gives interpretive meaning and direction to the particularized narrative. For example, a Christological story about the suffering of Jesus can provide profound orientation for a particular narrative self who is experiencing suffering. The basis for an empathetic embodied presence to the suffering may then be achieved through sharing of those narratives that provide the context for the crucible of suffering. A sharing of narratives in which suffering is embedded can be a means to the renewing of relationships and the overcoming of social alienation ([6], pp. 19011903). The shared narrative enables the creation of a community of the suffering.

As articulated by Camus, the commitment of community is expressed in a statement of solidarity to the sufferer to live and die together. The moral task of both the medical profession and religious traditions is to provide contexts of community within which such commitments can be made. A story of suffering does not itself impart meaning, but it does provide a context for a common quest for meaning and commitment. Those commitments express our human interdependence, patience, and embodied empathy in the face of suffering.

Department of Philosophy
Oregon State University
Corvallis, Oregon

NOTE

[1] The language of catastrophe to convey an experience of suffering is borrowed from Metz [14].

REFERENCES

1. Berdyaev, N.: 1935, *Freedom and the Spirit,* Centenary Press, London, U. K.
2. Campbell, C.: 1995, 'When medicine lost its moral conscience: Oregon Measure 16', *BioLaw* II:1, S:1-16.
3. Camus, A.: 1972, *The Plague,* Vintage, New York, NY.
4. Cassell, E.: 1991a, 'Recognizing suffering', *Hastings Center Report* 21, 24-31.
5. Cassell, E.: 1991b, *The Nature of Suffering and the Goals of Medicine,* Oxford University Press, New York, NY.
6. Cassell, E.: 1995, 'Pain and suffering', W.T. Reich (editor-in-chief) *Encyclopedia of Bioethics,* 2nd ed., Simon & Schuster-Macmillan, New York, NY.
7. Hauerwas, S.: 1990, *Naming the Silences,* Wm. B. Eerdmans, Grand Rapids, MI.
8. Kliever, L. D.: 1989, *Dax's Case: Essays in Medical Ethics and Human Meaning,* Southern Methodist University Press, Dallas, TX.
9. Lewis, C. S.: 1962, *The Problem of Pain,* Macmillan, New York, NY.
10. Lewis, C. S.: 1988, *A Grief Observed,* Bantam, New York, NY.
11. Loewy, E. H.: 1989, *Textbook of Medical Ethics,* Plenum, New York, NY.
12. Macquarrie, J.: 1986, 'Suffering', in J. F. Childress, J. Macquarrie (eds.), *The Westminster Dictionary of Christian Ethics,* Westminster Press, Philadelphia, PA.
13. May, W. F.: 1991, *The Patient's Ordeal,* Indiana University Press, Bloomington, IN.
14. Metz, J. B.: 1994, 'Suffering unto God', *Critical Inquiry* 20, 611-622.
15. Miles, S. H.: 1991, 'Informed demand for non-beneficial treatment', *New England Journal of Medicine* 325, 512-515.
16. Nelson, J. L.: 1994, 'Duties of patients to their caregivers', in C. Campbell, B. Lustig (eds.) *Duties to Others,* Kluwer, Dordrecht, The Netherlands.
17. Rawlinson, M. C.: 1986, 'The sense of suffering', *The Journal of Medicine and Philosophy* 11, 39-62.
18. Reich, W. T.: 1989, 'Speaking of suffering', *Soundings* 72, 83-108.
19. Scarry, E.: 1985, *The Body in Pain,* Oxford University Press, New York, NY.
20. Smith, D.: 1987, 'Suffering, medicine, and Christian theology', in A. Verhey, S. Lammers (eds.), *On Moral Medicine: Theological Perspectives in Medical Ethics,* Wm. B. Eerdmans, Grand Rapids, MI.
21. Tolstoy, L.: 1987, *The Death of Ivan Ilyich,* Bantam, New York, NY.
22. Toombs, S. K.: 1992, 'The body in multiple sclerosis: A patient's perspective', in D. Leder (ed.), *The Body in Medical Thought and Practice,* Kluwer, Dordrecht, The Netherlands.

DAVID BARNARD

DOCTORS AND THEIR SUFFERING PATIENTS: COMMENTARY ON CAMPBELL

> But we can see immediately that the recognition of a mystery demands an approach which is quite contrary to that demanded by the solution of a problem. . . . The inroads made in our time by techniques cannot fail to imply for man the obliteration, the progressive effacement, of this world of mystery which is at one a world of presences and of hope; it is not sufficient merely to say that, at the level of mystery, man's desires and fears, which lie behind his technical achievements, are lifted up beyond any assignable limit; we must add that human nature is tending to become more and more incapable of raising itself above desire and fear in their ordinary state, and of reaching in prayer or contemplation a state that transcends all earthly vicissitudes.
>
> —Gabriel Marcel, *Man Against Mass Society*

> It has been said that physicians can relieve symptoms but that patients must find their own peace. Physicians can, however, do much to help by providing the conditions for finding that peace. And these efforts are often successful.
>
> —Ivan Lichter and Esther Hunt, "The Last 48 Hours of Life"

I wish that Courtney Campbell had had the opportunity to meet Milly Jolley. A graduate in pharmacy from Manchester University in England, Milly Jolley developed cancer in the mid-1970s. She established a self-help group for cancer

R. A. Carson and C. R. Burns (eds.), Philosophy of Medicine and Bioethics, 265-272.

patients in 1980. Her experiences led her to certain expectations for medical care, and an overall vision of what is important to patients near the end of life. She wrote,

> As patients we want to be able to trust our doctors. Most of us do not want legislation for active euthanasia. We want informed clinical judgement about when to stop treatment. We want to remain clear-headed to be able to attend to our affairs and relate to family and friends. We don't want to pretend pain is controlled if it is not—in order not to offend. We want to use our gifts and talents to the end. We want to be treated lovingly, knowing we may become unlovable. Most of us would like to die at home in our own surroundings, and hope we will still have friends there when we do ([7], p. 190).

Milly Jolley could have helped Courtney Campbell by imparting to him her rootedness in the everyday reality of life with critical illness, and her appreciation that medical expertise is at once a crucial and a partial resource for someone facing death. Friends, family, and the person herself are the central actors in the drama. At the same time, though her expectations of medicine are modest, they are clear and uncompromising. Because Campbell's essay lacks many of these qualities, I fear that it will mislead readers on some very important points, even though I agree with the thrust of Campbell's overall argument.

Campbell states his main position early on:

> The imperative of medicine must . . . be revised so that it encompasses relief of pain, but recognizes that much suffering is medically intractable. How we respond to suffering is much more a problem of human presence than a lack of medical presence.

This point is unexceptionable. And Campbell follows it with subtly nuanced arguments concerning the tensions between medical epistemology and the personal nature of suffering. He is perceptive in his critique of theological responses to suffering that trivialize the experience of the individual in order to justify suffering on the level of the cosmos. And he writes probingly about the healing power of empathy. Where, then, does he lead us astray?

Trouble is brewing at the very start. Within the scope of his essay's first three paragraphs, Campbell makes the following assertions:

1. "One of contemporary medicine's central purposes is to relieve pain and suffering."

2. "Substantial" effort and funds have been directed to pain research and palliative care.

3. Medicine's broad aim is a "pain-free existence" and enhanced control and quality of life for patients.

4. "In the ideology of contemporary medicine, suffering is an unmitigated evil, worse than death."

5. It is contemporary medicine's goal—abetted by patients—"to banish suffering from the human condition."

6. It "follows" that "chronic, protracted pain and unremitting suffering experienced by patients is a symbol of medical failure."

7. Medicine (erroneously) conflates pain and suffering, thereby treating suffering as a "remediable medical condition."

Each of these assertions is highly dubious and misleading. Yet they form the basis for Campbell's conclusion, namely that "the grasp of contemporary medicine exceeds its reach. It seeks to transform basic features of the human condition into a medical problem and thus inevitably sets itself up for failure."

Campbell is echoing Gabriel Marcel's [11] warning not to transform mysteries into problems, not to impose technicist "solutions" to aspects of human existence that call, instead, for witness, solidarity, and hope. Campbell clearly worries that a technicist approach to suffering will hurt patients, though he never explicitly says how. One possibility might be that medicine will raise patients' expectations for relief of suffering so high that when they are (inevitably) disappointed euthanasia will seem to be the next logical step. (Ironically, the more zealous proponents of hospice care are more likely to bring about this result than are more conventional medical practitioners.) Or, medicine will assault patients with procedures when they need, instead, to be soothed by our empathetic presence. Or, having alleviated physical pain but helpless before other dimensions of suffering, medicine will simply walk away and leave the patient alone.

These are serious dangers. What, then, should medicine do instead?

Campbell is vague about this. He says, as I have quoted him above, that medicine's imperative should include relief of pain, but before the intractability of suffering, medicine should—should what? Here Campbell leaves off, and leaves out what his essay makes us desire most strongly: a framework within which doctors may *integrate* their medical expertise and power with the non-medical aspects of the care of the critically ill and dying. Part of the problem is that Campbell's essay is highly theoretical and abstract. His analyses of concepts such as suffering, theodicy, and empathy move progressively farther away from the day-to-day world of people like Milly Jolley and her doctors. A simple but relevant illustration of Campbell's preference for abstraction is his use of the term "medicine" when I take it he means the activities and perspectives of doctors. By the end of his essay, he has dropped even this relatively specific term in favor of the completely generic "caregiver." As a result, when Campbell elaborates his view of empathetic presence as the right response to intractable suffering, it is unclear whose presence he has in mind, or how that presence ought to be integrated, if at all, with the activities of "medicine."

More importantly, I believe, Campbell is unable to conclude with a convincing statement of the doctor's role in the face of patients' suffering because he has begun by seriously misconstruing medicine's current achievements and failures in this regard. Let me counterpose to Campbell's seven assertions the following statements, which I believe more accurately describe the current state of contemporary medical practice:

1. The primary focus of modern medicine remains the biology and pathophysiology of disease; the patient's experience of illness and the many forms of pain and suffering are of secondary importance [3].
2. The fundamental imperative of modern technological medicine is to prolong life; medicine regards death, not suffering, as the enemy, as may be observed in any intensive care unit or chemotherapy suite ([1], 1993).
3. Despite much rhetoric about quality of life and compassionate care of the dying, the actual needs of the dying and their families remain at the periphery of health-care policy and professional education; while billions are spent to cure cancer, for example, financially strapped hospice organizations and palliative care programs throughout the world are hanging on by a thread [10]; [12]; [14]; [15]; [17].
4. While techniques for pain and symptom control have developed tremendously over the past twenty years, these advances have diffused very slowly and unevenly into general practice; undertreatment of pain is among the most widely documented findings in studies of medical care of the critically ill [4]; [6]; [16].

In summary, and quite contrary to Campbell's assertions, the scandal of modern medicine in the face of critical illness is not its *overreaching* in claiming to eliminate all suffering, but its *underachievement* in failing to address the concrete problems of patients that medicine is capable of addressing effectively. We are far from a reality in which patients' pain and other symptoms are so effectively controlled—or even recognized as important—that the end of suffering seems (mistakenly) to be at hand; for large numbers of patients, undetected or undertreated symptoms *is* reality. And in the face of wildly out-of-control physical pain, nausea, or breathlessness, for example, a patient is unlikely to notice, much less derive spiritual benefit from, our empathetic presence [5]. As Ivan Lichter and Esther Hunt [9] have stated, the physician is in a position to enable a great deal of positive interpersonal and spiritual experience for patients and families—precisely because of his or her ability to manage the physical aspects of pain and suffering. If medicine is to be faulted on this subject, it is for its failure to accomplish the possible, not for its arrogance in pursuing the impossible.

It is certainly true, as Campbell argues, that for many patients suffering has dimensions that elude medical remedy. And for some, complaints of physical distress may be the only available or safe language for expressing loneliness, helplessness, or fear. In these situations misplaced therapeutic zeal can blind doctors and their patients to other important responses to suffering—shared witness, empathically received lament, reformulated goals and transformations of hope, a wordless touch. The doctor's own self—in his or her subjectivity and vulnerability— becomes the primary diagnostic and therapeutic instrument. The doctor, however, does not and cannot act alone. We must ask, then, how the critically ill patient's total environment—doctors, nurses, therapists, loved ones, and the patient himself or herself—can unite and integrate all these different levels of response. And what attitudes and skills does the doctor need to play a constructive and harmonious role within that environment?

Milly Jolley points us in the right directions [7]. Her experience led her to value doctors who spoke the truth to her, who stayed the course with her, who managed her physical symptoms expertly so that she could go about her daily life, and whose care empowered and did not displace her loved ones. These expectations, rooted as they are in simple, day-to-day experience, will disappoint those who idealize the deathbed as the scene of grand spiritual achievement. Yet the modesty and specificity of Milly Jolley's desires caution us against overheated expectations at the end of life, whether formulated medically or theologically.

Robert Kastenbaum's research for the National Hospice Study in the 1980s reinforces this point. Kastenbaum [8] asked a group of hospice patients to "describe the last three days of your life as you would like them to be. Include whatever aspects of the situation seem to be of greatest importance." The most frequent answers were not about grand syntheses and achievements related to suffering and the meaning of life. Rather, they included comments such as:

"I want certain people to be here with me."
"I want to be physically able to do things."
"I want to feel at peace."
"I want to be free from pain."
"I want the last three days of my life to be like any other days."

Modesty of goals needs to be accompanied by modesty with respect to their achievement. On this point Campbell and I, and most experienced clinicians, would surely agree. Lichter and Hunt, in fact, continue their comment about the doctor's potential for assisting patients to find their own peace with the further observation,

> Such a result may not be achieved by all, especially when the situation has been of long standing, though these cases may provide the most rewarding outcomes. A major impediment to success may be the individual's personality. While most patients die peacefully, not all will have died at peace with their relatives and friends, their conscience, or their spirit ([9], p. 13).

There are limits to our skill, imagination, and stamina. Some suffering remains beyond our help—though never beyond the reach of our concern.

There is a tendency in theological and philosophical reflections on medicine to etherealize care. By this I mean subtly devaluing the physical, fleshly aspects of medical and nursing care by viewing them in the "ultimate" contexts of human finitude and spirituality. From this point of view, the care of the body and alleviation of its pain are *only* or *merely* the care of the body, whereas attention to the spirit—to the realms of community, destiny, and meaning—is somehow closer to the true nature of care. No one who has actually accompanied dying persons in the last hours of their lives could possibly adopt this point of view. There are times when the gentle and skillful touch of a doctor or nurse is the only medium we have to express community and love, and it often sets an example for friends and family who may be too frightened to caress the dying person's cheek or hand.

Cicely Saunders, the founder of the modern hospice movement, makes this point well. Referring to the importance of the spiritual dimensions of suffering and empathic dialogue, she observes:

> But many of us have little or no opportunity for the time these meetings may call for—our patients are too ill and we are sometimes too busy. That, too, we often have to accept, but we can always persevere with the practical. Care for the physical needs, the time taken to elucidate a symptom, the quiet acceptance

of a family's angry demands, the way nursing care is given, can carry it all and reach the most hidden places. This may be all we can offer to inarticulate total pain—it may well be enough as our patients finally face the truth on the other side of death ([13], p. 12).

Department of Humanities
Penn State University College of Medicine
Hershey, Pennsylvania, U.S.A.

REFERENCES

1. Callahan, D.: 1993, *The Troubled Dream of Life: Living with Mortality*, Simon and Schuster, New York, NY.
2. Campbell, C.: 1996, 'Must patients suffer?' in this volume, pp. 247-263.
3. Cassell, E. J.: 1991, *The Nature of Suffering and the Goals of Medicine*, Oxford University Press, New York, NY.
4. Cleeland, C., *et al.*: 1994, 'Pain and its treatment in outpatients with metastatic cancer,' *New England Journal of Medicine* 330, 592-596.
5. Doyle, D.: 1992, 'Have we looked beyond the physical and psychosocial?' *Journal of Pain and Symptom Management* 7, 302-311.
6. Foley, K.: 1991, 'The relationship of pain and symptom management to patient requests for physician-assisted suicide,' *Journal of Pain and Symptom Management* 6, 289-297.
7. Jolley, M.: 1988, 'The ethics of cancer management from the patient's perspective,' *Journal of Medical Ethics* 14, 188-190.
8. Kastenbaum, R.: 1986, *Death, Society, and Human Experience*, 3rd ed., Merrill, Columbus, OH.
9. Lichter, I., Hunt, E.: 1990, 'The last 48 hours of life,' *Journal of Palliative Care* 6 (4), 7-15.
10. Magno, J.: 1992, 'USA hospice care in the 1990s,' *Palliative Medicine* 6,158-165.
11. Marcel, G.: 1962, *Man Against Mass Society*, Henry Regnery, Chicago, IL.
12. Rhymes, J.: 1990, 'Hospice care in America,' *JAMA* 264, 369-372.
13. Saunders, C.: 1993, 'Introduction—History and challenge,' in C. Saunders, N. Sykes (eds.), *The Management of Terminal Malignant Disease*, 3rd ed., Edward Arnold, London, U. K.
14. Scott, J.: 1994, 'More money for palliative care? The economics of denial,' *Journal of Palliative Care* 10(3),35-38.
15. Stjernswärd, J.: 1993, 'Palliative medicine: A global perspective,' in D. Doyle, G. Hanks, N. MacDonald (eds.), *Oxford Textbook of Palliative Medicine*, Oxford University Press, Oxford, U. K.

16. SUPPORT Principal Investigators: 1995, 'A controlled trial to improve care for seriously ill hospitalized patients,' *JAMA* , p. 274, 1591-1598.
17. World Health Organization Expert Committee: 1990, *Cancer Pain Relief and Palliative Care*, Technical Report Series 804, World Health Organization, Geneva.

SECTION III

POLICY

BARUCH BRODY

WHATEVER HAPPENED TO RESEARCH ETHICS?

In 1974, the Institute for the Medical Humanities at the University of Texas Medical Branch, in Galveston, sponsored the first of a long series of conferences on bioethics and the philosophy of medicine. In the same year, the United States Congress authorized the creation of the National Commission for the Protection of Human Subjects as an advisory body charged with recommending ethical guidelines for the conduct of research involving human subjects and with identifying the basic ethical principles that should underlie such research. Both of the enterprises were successes. The success of the former can be judged by the high quality of material found in the many volumes of the *Philosophy and Medicine* series that incorporate the papers presented at those many conferences. The success of the latter can be judged by the highly influential Belmont Report [21] and the Department of Health and Human Services (DHHS) regulations on human research issued in the early 1980s [10].

There is, however, a surprise resulting from these parallel successes, one suggested by looking at the content of the series. If one looks at the titles of the first forty-six volumes in the series, one finds only one [25] that explicitly refers to issues in research ethics. While there are a few relevant papers in other volumes, there are not many. I take this to be a symbolic representation of an important fact. During the last ten to fifteen years, the bioethics community has paid little attention to an important set of issues in research ethics that have been much discussed in the research community as well as in the policy/regulatory community. Having succeeded in advocating in the United States and elsewhere a general framework for protecting human subjects in research (independent review of research protocols, appropriate risk-benefit ratios, informed consent), it has turned to other areas, virtually ignoring research ethics. To be sure, there are areas of research ethics that still attract attention (especially reproductive, fetal, and genetic research), and there are journals and individuals (e.g., *IRB* and

R. A. Carson and C. R. Burns (eds.), Philosophy of Medicine and Bioethics, 275-286.

its editor, Robert Levine) that have continued to focus on these issues. But important and fundamental questions in research ethics are not receiving the attention from bioethicists that they deserve.

What sorts of questions? A partial list includes ethical questions raised by fraud and misconduct in research, conflicts of interest in research, the ownership of the results of research (including patents and authorship), clinical trials methodology, and the regulatory approval of the results of research.

It is obviously impossible to demonstrate here the richness of the ethical content of all of these issues and the meagerness of the bioethical contribution to their resolution. What I shall do instead is to focus on two areas: issues of clinical trials methodology and of regulatory approval. In each of these areas, I will identify the rich ethical content of a set of issues and the potential for a major bioethical contribution to their resolution. I will then try to document the modesty of the actual contribution. In the final section, I will offer a hypothesis to explain this gap between potentiality and actuality and a program for bridging it.

I. UNADDRESSED ETHICAL ISSUES IN CLINICAL TRIALS METHODOLOGY

One of the most important characteristics of research in recent years has been the emphasis on controlled trials to validate new interventions before they are widely employed in the clinical setting. Some of this emphasis is due to regulatory pressures (e.g., the need for drug approvals by such agencies as the FDA), while some of it is due to the increasing sophistication of the science of clinical trials. These clinical trials are prospective trials of a well-defined intervention in a population carefully defined by inclusion and exclusion rules. They involve subjects randomly assigned to control groups as well as intervention groups. They are designed to blind both the subjects and the treating professionals to the assignment. They are designed to be of a size that is adequate to determine whether the appropriate endpoints are found more often in one group than in the other. Finally, they are carefully monitored by an independent data board that is authorized to stop the trial if conclusive positive data about efficacy or negative data about safety emerges.

All of these features generate a wide variety of issues that have attracted much attention in the clinical trials community and some, but so far not enough, response in the regulatory community. These issues relate specifically to clinical trials, as opposed to the issues that arise in connection with all human subjects research. Some of the most prominent of these issues are the following:

(1) When is it legitimate to commence a clinical trial? Obviously, there must be enough data supporting the use of the new intervention to justify the expense of running the trial. At the same time, there must not be so much data supporting its use that it would be unethical to deny it to all patients without waiting for the results of the trial. How can the window of legitimacy be defined?

(2) How shall the study group be defined? It would seem safer and therefore more ethical to exclude from the trial patients with co-morbidities who are more vulnerable to the risks of the new intervention. At the same time, including only optimal patients may undermine the clinical relevance of the study to the care of future patients, many of whom have the co-morbidities in question. How shall these two concerns be balanced?

(3) What type of control group shall be used? A clinical trial can involve an historical control group, a concurrent active control group, or a concurrent placebo control group. For a variety of scientific reasons, concurrent placebo control groups offer the most valid data at the least cost. But it is precisely those trials that deny during the trial the potentially valuable intervention to the control group. How can this loss to the placebo control group be minimized? When must some other type of control group be used?

(4) Can alternatives to randomization after full informed consent be employed? Many clinical trials have found it very difficult to enroll subjects because the clinicians and/or the subjects find the idea of randomization to one or another treatment group, about which they are informed and to which they must consent, disturbing. Is there any alternative to this traditional approach? If not, how can we better deal with randomization in the consent process?

(5) What burdens can be imposed on patients to maintain the blinded nature of trials? Blinding both the subjects and the treating clinicians to the subject's assignment in the clinical trial clearly improves the validity of the trial, but may impose special burdens on the subjects. This is particularly true for trials of new surgical interventions (consider what would be involved in doing sham surgery), but it may also be true for trials of medical interventions (consider the burdens posed by taking many placebo shots or pills). Is any additional burden to which the patients would consent legitimate, or are there upper limits on the legitimate burdens that can be imposed to maintain the blind?

(6) What endpoints should be chosen for the clinical trial? Each clinical trial must assess the new intervention in terms of positive endpoints (the benefits of the intervention) and in terms of negative endpoints (the burdens of the intervention). But the importance of various endpoints may vary considerably among potential future patients. Given that well-designed trials for statistical reasons

must either avoid studying a large number of endpoints or must study at great expense a larger number of patients, what then is the basis for choosing the endpoints to be studied?

(7) When shall clinical trials be stopped? Statisticians have developed a variety of rules for stopping clinical trials at an interim analysis if that analysis reveals overwhelming support for the efficacy of the new intervention being tested. But when should such trials be stopped because of concerns about safety revealed by the interim analysis? How should decisions to continue and/or to stop trials be influenced by data from other trials? These questions are not answered by the statistical stopping rules.

These seven important issues have major ethical components. They explore the ways in which we should balance scientific ends with concerns for research subjects and the ways in which trials should be defined to meet their goal of improving future treatment without excessively burdening current patients. It is of interest that there is an increasing recognition of the ethical complexities raised by these issues and that this recognition has led to the appointment of ethicists to serve on the Data Safety and Monitoring Boards (DSMB) that monitor clinical trials. I have served on four such boards at the National Heart, Lungs, and Blood Institute and currently sit with two other ethicists on the board that monitors all of the clinical trials run by the AIDS Clinical Trial Group.

Is there much in the bioethics literature that addresses these issues? Negative existence claims are hard to demonstrate. In any case, I certainly would not claim that there is nothing. Benjamin Freedman [14] has made a number of important contributions, especially his theory of clinical equipoise (it is permissible to begin a clinical trial as long as there continues to be disagreement in the relevant clinical community about the value of the new intervention). The bioethics community [1] provided some useful critiques of Zelen's proposals for dealing with problems of enrollment raised by randomization. Levine's valuable text [20] contains some useful discussion of some of these issues. But there is not that much.

Let me illustrate this paucity by reference to the issue of the early stopping of clinical trials. Levine's textbook, one of whose strengths is its comprehensive treatment of the ethical literature, has three pages (199-202) devoted to the issue of preliminary data, but nearly all of it deals with informing patients and clinicians; only one paragraph (on p. 199) deals with the issue of when to stop the trial. Ruse's discussion [23] addresses these issues without even mentioning the variety of stopping rules created by biostatisticians to guide these decisions. The *Bibliography of Bioethics* discusses issues of human research under a number of headings, including AIDS (human experimentation subheading), Behavioral Research, Biomedical Research, and Human Experimentation. A review of the entries for the last three volumes (18-20) reveals very little. The most important citation is to Stuart Pocock's article [22] in the *British Medical Journal.* Nothing by bioethicists approaches its importance, although many bioethicists would certainly be very concerned with his emphasis on collective ethics in contrast to individual ethics. Of equal importance is the 1993 article by the VA Cooperative Study Group [24] addressing the problem of data from other trials, and exploring the role of reconsenting subjects (obtaining a second consent to continue in the trial despite the data from the other trial), an issue which has attracted almost no attention in the bioethics community.

This issue of early stopping has in recent years produced tremendous controversy in the clinical trials community. One example that has attracted much attention is the decision by the ISIS 2 investigators [18] to not stop their trial despite their own impressive interim data supporting the use of streptokinase and despite other supporting data. Another example that has attracted much attention is the decision by the AIDS Clinical Trials Group [13] to stop their trials of the early use of AZT in light of their own impressive interim data about CD4 counts and the early onset of AIDS-related infections. Both decisions have been criticized and defended, and it is clear that the issue of early stopping (both on the basis of interim data and of data from other trials) requires much further analysis. The silence of the bioethics literature during the period in question is therefore very disappointing.

This is, then, my first example of the problem I have identified. Clinical trials are becoming central to the scientific/medical enterprise, and there are many issues with significant ethical content raised by such trials. While important examples of these issues have led the clinical trials community to debate them, and while the bioethics community has been invited to participate in the discussion of them, the extant bioethics literature has said little about them. What have we contributed to the discussion, for example, of the interim stopping of clinical trials? There is an obvious need to explain this failure. I shall return to that explanation in the final section. Before doing so, however, I want to discuss another example.

II. UNADDRESSED ETHICAL ISSUES IN REGULATORY APPROVAL

While all clinical studies are designed to answer scientific and clinical questions, some are also designed to play a role in securing permission from a national regulatory agency to produce and sell a new drug. This often means that research is driven by the relevant regulations and their requirements as to what must be proven in order to obtain regulatory approval. As a result, the content of these regulations is an important issue for researchers, and the ethical component of that issue is an important aspect of research ethics.

In order to understand this issue and its ethical component, one must first understand the history of these regulatory agencies and the regulatory scheme that has emerged. The U. S. history is not atypical, so I will focus on it and on its resulting regulatory scheme, although I will say something about developments in other countries.

The U. S. history of drug regulation began with a 1906 law prohibiting the adulteration or mislabeling of drugs. In 1938, after a scandal in which more than one hundred children died, Congress mandated that no drug could be sold without the producer demonstrating to the Food and Drug Administration (FDA) in a new drug application (NDA) that the drug was safe when used as intended. This gave rise to the need for research that would satisfy the FDA about safety. In response to the thalidomide tragedy, Congress, in 1962, mandated that clinical trials of new drugs could not commence until the researchers and/or their sponsors secured approval of an investigational new drug (IND) application from the FDA. This gave rise to the need for preclinical research, including animal research, that would satisfy the FDA that it was reasonable to begin clinical research. Moreover, Congress also mandated, in 1962, that no drug could be sold without the sponsor demonstrating to the FDA that the drug was effective when used as intended. This gave rise to the need for research that would satisfy the FDA about effectiveness. In more recent years, in response both to long-standing concerns that its demands for firm evidence were delaying access for Americans to valuable drugs and to AIDS-generated concerns that it was insensitive to the needs of those facing life-threatening illnesses with no available options, the FDA has relaxed its requirements for evidence (relying, for example, on surrogate endpoints), and this has modified the type of research being undertaken.

The resulting scheme in the United States is that after running the appropriate preclinical studies, the sponsor of a new drug must submit an IND application to the FDA. If that application is approved, clinical trials designed to demonstrate safety and effectiveness are conducted. At the FDA's insistence, these are clinical trials of the type described in the previous section. They are, in other words, randomized controlled (usually, placebo controlled) trials, involving well-defined inclusion and exclusion criteria and maximal blinding, and analyzed according to strict statistical criteria. The results of these trials form the basis for the sponsor's application for approval.

The regulatory schemes in most European countries are similar, but some only require notification (as opposed to approval) before clinical trials commence, some emphasize primarily the demonstration of safety, and some, rather than mandating the use of evidence from strict clinical trials, are satisfied by whatever clinical evidence convinces experts. These less strict standards are responsible for earlier drug approvals, with the benefit of earlier access but with the risk of more mistaken approvals. There is a movement now, because of the internationalization of the drug industry, to produce uniform standards for drug approval. It remains to be seen whether that will occur, and if it does occur, whether the results will resemble the stricter U. S. scheme or the less strict European scheme.

If one goes beyond the details of the regulatory scheme, it becomes clear that there really are two fundamental questions that every scheme needs to answer. One is the question of how the goals of effectiveness and safety should be balanced in the drug-approval process. We want drugs that are effective. We want drugs that are safe. Sometimes, however, the more effective drugs carry greater safety risks while the safer drugs are less effective. So what is the appropriate safety-effectiveness ratio for drug approval? This question relates to the content of what must be established before drugs are approved for marketing by the regulatory agency. The second question is how the demand for adequate evidence and the demand for quicker approval and access should be balanced in the drug-approval process. We want firm evidence of a satisfactory safety-effectiveness ratio before a drug is approved. We want quick access to useful drugs. Sometimes, however, obtaining firm evidence requires time that delays access, while speeding up access means making do with less evidence (e.g., data about surrogate endpoints). What is the appropriate balance between these two legitimate demands? This question relates to the quality of the evidence that must be presented before drugs are approved for marketing by the regulatory agency. These two questions, the content question and the epistemic question, are obviously related to each other, but they are also different questions. One subtle way of understanding the differences between the regulatory schemes in different countries is to see them as offering different answers to these two questions. These are the two fundamental issues about drug regulatory schemes.

I want to suggest that these issues involve a rich ethical component and are not merely technical regulatory issues. I will do so by presenting an argument that I call the drug-approval paradox, an argument that claims that the adoption of any drug regulatory scheme is incompatible with the fundamental moral commitments of a liberal society. The argument runs as follows:

(1) any scheme for regulating drug approvals necessarily must answer both the content question and the epistemic question, for it is the answer to these two questions that structures (implicitly or explicitly) the regulatory scheme;

(2) each of these questions can be answered only by balancing goals, the goals of safety and effectiveness and the goals of access and firm evidence. Different members of society will, in light of their different values, balance those goals differently;

(3) any society-wide drug regulatory scheme will involve a balancing of those goals that is quite consonant with the balancing that some citizens would make in light of their values but is quite dissonant with the balancing that other citizens would make in light of their values;

(4) therefore, any society-wide scheme will impose the values of some citizens on other citizens who do not share those values;

(5) this is incompatible with the fundamental moral commitment of liberal societies to tolerance of diverging values, and so any society-wide drug regulatory scheme is morally illegitimate.

There are, of course, many possible responses to this paradox. Some will say that this reveals a need to move from liberalism to some form of communitarianism. Others will say that drug regulations may fall under one of the exceptions to the liberal principle of tolerance. Still others will say that a scheme might be developed that is compatible with the differing values of different citizens. My point here is not to adjudicate between the paradox and these responses; it is simply to conclude that the issues raised by drug regulatory schemes have a rich ethical content.

These issues have been extensively debated in the last few years. There is a growing recognition in the various calls for general regulatory reform that these fundamental issues must be addressed, rather than focussing on narrow details of the regulatory scheme. While these general debates have been going on, some clinicians, some members of the clinical trials community, and some regulatory officials have usefully debated these fundamental questions in specific cases. An excellent example is the Angell-Kessler debate [2], [19] about the FDA's withdrawal of general approval for the use of breast implants filled with silicone gel. Another excellent example is the current debate about the use of surrogate endpoints in AIDS trials [13]. A few thinkers, mostly economists [15] and political theorists [16] of a libertarian persuasion, have even argued for terminating drug regulatory schemes as we know them today, using arguments that resemble the paradox.

I know of no discussions that have fully articulated a bioethical response that leads to a structuring of a regulatory scheme that can be justified morally other than one that I have recently provided [5]. A review of the *Bibliography of*

Bioethics for these last few years reveals that there has been almost no contribution to this debate in the bioethics literature, with the exception of some material related to AIDS. This silence is particularly distressing in light of the fact that the issues have been discussed so fully in recent years without the full understanding that a bioethical analysis of the paradox could contribute.

This, then, is my second example of the paradox about research ethics. There has been a growing recognition in the regulatory community of the need to address the fundamental issues about drug regulatory schemes. Clinicians have debated these issues in particular cases. The extant bioethics literature has failed, however, to offer deeper understandings and resolutions. There is a need to explain this failure.

III. A DIAGNOSIS AND A MODEST PROPOSAL

By the early 1980s, a consensus had been reached about the basic ethics of research involving human subjects, and that consensus had been incorporated into parallel regulations from the National Institutes of Health (NIH) and the FDA. Since that time, as we have seen, other important issues have arisen and have been discussed without a significant contribution from the bioethics literature. We need to explain that failure and to develop a program for dealing with it.

It is useful, I believe, to compare research ethics and clinical ethics. By the early 1980s, a consensus had also emerged about some of the most fundamental issues in clinical ethics, the definition of death and the criteria for limiting life-prolonging interventions in critically ill patients, and that consensus was clearly enunciated in various reports from the President's Commission [6]. But in those areas, the contributions of bioethics to the resolution of new issues has not ceased. Let me briefly demonstrate that point.

Important recent bioethical literature has greatly enriched our understanding of the issues that arose after the basic consensus in clinical ethics was reached. In the area of brain death, one can cite such diverse contributions as the literature concerning anencephalics as potential organ donors [9], the Pittsburgh protocol for procuring organs from non-heart-beating donors [3], and the continued existence of some brain functioning after the usual clinical criteria of brain death are met [7]. In the area of limiting life-prolonging interventions, one can cite such diverse contributions as the literature concerning artificial nutrition and hydration for PVS patients [4], the provision of futile life-prolonging interventions [7], active euthanasia [11], the reliability of surrogate decision making [12], and ethnic differences about the value of life-prolonging interventions [8].

Many more examples could be provided. There is no question about the continuing contribution of bioethics to the new issues about brain death and about limiting life-prolonging interventions. All of this makes the failure of bioethicists to address ethical issues in research more puzzling and troublesome, but suggests to me the beginnings of an explanation.

We have seen in the last decade tremendous growth in clinical bioethics. Some of this work has been done by Ph.D. or J.D. bioethicists who have immersed themselves in the clinical realities of these issues and who are now dealing with those issues at a clinical level. Some of this work has been done by M.D.s who have received considerable training in bioethics and who are now using that training to deal with these issues clinically. Sometimes, this work has been organized around the activities of health-care ethics committees. Sometimes, this work has been organized around ethics consultants. We have today many regional networks of such committees and a national society of such consultants, the Society for Bioethics Consultations. All of this concrete work has stimulated the development of the above-cited literature, published either in major clinical journals or in the bioethics journals.

Nothing similar has happened in the area of research ethics. There are not a significant number of basic or clinical researchers who have received considerable training in bioethics and who are using that training to deal with the new issues of research ethics. Nor are there a large number of Ph.D. or J.D. bioethicists who have immersed themselves in the day-by-day realities of research and who are providing bioethical insight to those realities. There is no national society of people doing research ethics. There are, of course, the Institutional Review Boards (IRBs) that administer at the institutional level the NIH-FDA regulations, but it is interesting to note that, unlike the case of the ethics committees, there has not emerged networks of such IRBs. As a result, I suggest, we have not seen the emergence of a powerful bioethics literature on research ethics, leaving aside, as noted at the beginning of the paper, the special areas of genetic, fetal, and animal research.

My diagnosis, in short, is that research ethics has suffered because of the non-emergence of practicing research ethicists comparable to the large number of practicing clinical ethicists. The obvious solution to that problem is to encourage their emergence. Let me make some modest proposals along those lines.

Most research institutions, in response to recent federal regulations, have developed committees to oversee issues of conflicts of interest, issues of scientific misconduct, and issues of technology transfer. All of these issues have major ethical components. Bioethicists at such institutions need to secure appointment to these committees and to use those appointments as an occasion to immerse themselves into the realities of research practice.

Clinical trials are increasingly being monitored by outside DSMBs, and some portions of the NIH have already set the precedent of appointing bioethicists to those DSMBs. Bioethicists need to lobby to be appointed to more of them. Similarly, they need to lobby to be appointed to FDA Advisory Boards, where major decisions are being made about regulatory approvals. This will be harder, for I am not aware of any precedents. Finally, the new federal mandate for the teaching of research ethics to trainees offers another splendid opportunity for bioethicists. They need to become actively involved in doing that teaching, preferably together with research colleagues, but they also need to make sure that what they teach relates to the actual practice of research. All of these will offer the opportunity for immersion into the realities of research practice.

Bioethicists who actively engage in all of these activities, thereby becoming research ethicists, will soon find themselves identifying new issues of research ethics and creatively resolving them. In addition to making this type of institutional contribution, they will find this service work leading to important publications in the bioethics literature, akin to the contributions made by the clinical ethicists.

There is another side to this proposal. Clinical ethics has benefitted considerably from the emergence of clinicians who received substantial bioethics training, as well as from the clinical activities of bioethicists. Research ethics similarly needs researchers who receive substantial bioethics training. Training programs, funded privately or publicly, are needed to provide that training.

I believe that this program holds out the hope for a revitalization of research ethics. What it requires is a frank understanding of what has gone wrong and a commitment to set it straight. If this happens, we can realistically expect to see many more volumes in the *Philosophy and Medicine* series devoted to research ethics.

Center for Medical Ethics and Health Policy
Baylor College of Medicine
Houston, Texas, U.S.A.

REFERENCES

1. Angell, M.: 1984, 'Patients' preferences in randomized clinical trials', *New England Journal of Medicine* 310, 1385-1387.
2. Angell, M.: 1992, 'Breast implants—protection or paternalism?' *New England Journal of Medicine* 326, 1695-1696.

3. Arnold, R. and Youngner, S.: 1993, 'The dead-donor rule: Should we stretch it, bend it, or abandon it?', *Kennedy Institute of Ethics Journal* 3, 263-278.
4. Brody, B.: 1992, 'Special ethical issues in the management of PVS patients', *Law Medicine and Health Care* 20, 104-115.
5. Brody, B.: 1994, *Ethical Issues in Drug Testing, Approval, and Pricing,* Oxford University Press, New York, NY.
6. Brody, B.: 1995, 'Limiting life-prolonging medical treatment' (forthcoming).
7. Brody, B. and Halevy, A.: 1995, 'Is futility a futile concept?' *Journal of Medicine and Philosophy* 20, 123-144.
8. Caralis, P. V. *et al.*: 1993, 'The influence of ethnicity and race on attitudes towards advance directives, life-prolonging treatments, and euthanasia', *Journal of Clinical Ethics* 4, 155-165.
9. Council on Ethical and Judicial Affairs: 1994, *Code of Medical Ethics: Current Opinions with Annotations,* AMA, Chicago, IL.
10. Department of Health and Human Services (DHHS): 1983, 'Protection of human subjects', 45 CFR 46.
11. Emanuel, E.: 1994, 'Euthanasia: Historical, ethical, and empiric perspectives', *Archives of Internal Medicine* 154, 1890-1901.
12. Emanuel, E. and Emanuel, L.: 1992, 'Proxy decision making for incompetent patients', *JAMA* 267, 2067-2071.
13. Fleming, T.: 1992, 'Evaluating therapeutic interventions', *Statistical Science* 7, 428-456.
14. Freedman, B.: 1987, 'Equipoise and the ethics of clinical research', *New England Journal of Medicine* 317, 141-145.
15. Grabowski, H. and Vernon, J.: 1983, *The Regulation of Pharmaceuticals,* American Enterprise Institute, Washington, D. C.
16. Green, D.: 1987, *Medicines in the Marketplace,* IEA Health Unit, London, U. K.
17. Halevy, A. and Brody, B.: 1993, 'Brain death: reconciling definitions, criteria, and tests', *Annals of Internal Medicine* 119, 519-525.
18. ISIS Investigators: 1987, 'Intravenous streptokinase given within 4 hours of onset of myocardial infarction', *The Lancet,* 397-401.
19. Kessler, D.: 1992, 'The basis of the FDA's decision on breast implants', *New England Journal of Medicine* 326, 1713-1715.
20. Levine, R.: 1988, *Ethics and Regulation of Clinical Research,* 2nd ed., Yale University Press, New Haven, CT.
21. National Commission: 1979, 'The Belmont Report', *Federal Register* 44.
22. Pocock, S.: 1992, 'When to stop a clinical trial', *British Medical Journal* 305, 235-240.
23. Ruse, M.: 1988, 'Statistical certainty and trial interruption', in S. F. Spicker (ed.), *The Use of Human Beings in Research,* Kluwer, Dordrecht, The Netherlands.
24. Simberkoff, M. S., et.al.: 1993, 'Ethical dilemmas in continuing a zidovudine trial after early termination of similar trials', *Controlled Clinical Trials* 14, 6-18.
25. Spicker, S. F., et al. (eds.): 1988, *The Use of Human Beings in Research,* Kluwer, Dordrecht, The Netherlands.

HAROLD Y. VANDERPOOL

WHAT'S HAPPENING IN RESEARCH ETHICS? COMMENTARY ON BRODY

I. INTRODUCTION

The title of Baruch Brody's essay, "Whatever Happened to Research Ethics?," suggests that research ethics, having once held a visible and powerful place in bioethics, has now disappeared, or virtually disappeared, from public view. Brody's discussion centers on identifying significant and weighty topics in the ethics of research that are *not* being scrutinized and on proposing why bioethicists have neglected these issues. After commenting on the issues Brody develops, I will present a supplementary and contrasting perspective in the form of a brief survey of what *is* being scrutinized in the research ethics literature.

Brody argues that, after emerging as a major topic of public concern and discussion in the late 1970s and early 1980s, research ethics began to be virtually ignored by "the bioethics community." He adds the proviso that certain areas ("especially reproductive, fetal, and genetic research"), individuals (e.g., Robert J. Levine) and journals (namely, *IRB: A Review of Human Subjects Research* that Levine edits) continue to focus on the ethics of research, but that "important and fundamental questions in research ethics are not receiving the attention from bioethicists which they deserve."

II. AREAS OF NEGLECT AND NEED

The "partial list" of important and neglected issues in research ethics identified by Brody includes questions over fraud, misconduct, and conflicts of interest in research, the ownership of the results of research, clinical trials methodology, and regulatory approval of research findings. Although issues of fraud and misconduct have been receiving considerable attention in the literature, Brody dwells on the last two of these issues that have indeed been seriously neglected: questions regarding the initiation, conducting, and ending of clinical trials; and

R. A. Carson and C. R. Burns (eds.), Philosophy of Medicine and Bioethics, 287-297.

questions regarding the goals and standards of new drug and medical-device approvals at the national level.

Brody's identification and analysis of ethical questions inherent to these two sets of issues consumes two thirds of his essay. The questions he develops are insightful and convincing with respect to their number, import (including their import for public policy), complexity, and the degrees to which they have received little attention from bioethicists. His exposition of important questions that beg for answers should further awaken bioethicists from their not-so-dogmatic slumbers—from believing that rectifying root problems behind the abuses of research brought to light by Henry K. Beecher (1966, pp. 1354-60) and by shocking revelations over the U. S. Public Health Service Syphilis Study (commonly called the Tuskegee Syphilis Study) would prevent future abuses, or from assuming that the National Commission's reports fifteen years ago and the ensuing Department of Health and Human Services' regulations of research thereafter resolved most of the ethical issues pertaining to the conducting of clinical trials and the drug approval process.

The body of Brody's essay takes us well beyond the common and reigning assumption that research ethics should be equated with newly-emerging medical and scientific developments (such as genetics or the many issues surrounding human reproduction), with ways to involve new subject-patient populations in research (such as those receiving emergency treatment), and with certain newly discovered "horrors."[1]

Brody goes back to basics, to the ever-present practices of designing and conducting clinical trials, and then shepherding new drugs and medical devices through a nation's approval process. He demonstrates how these areas bristle with ethical issues that bioethicists need to explore in tandem with the designers of clinical trials, researchers, biostatisticians, and members of oversight and approval committees at institutions. Over the last fifteen years medical ethicists have successfully demonstrated that their investigations of the everyday, "low profile" issues in clinical medicine rival or exceed the importance of their analyses of new and news-making "high profile" issues. Brody charts an equally impressive course for future consideration of ubiquitous "low profile" issues in the ethics of research involving human subjects.

Brody recognizes that a few bioethicists are now exploring some of the issues associated with the questions he raises. Indeed, his own book-length study (1994) addresses several of the issues he outlines in this essay both with respect to the conducting of clinical trials and with respect to the ethics of testing and approving therapeutic agents.[2] Other authors are also exploring some of the issues Brody identifies in this essay. For example, two recently published studies address problems pertaining to one of the seven questions Brody poses about clinical trials—the question of "blinding" research subjects with respect to their being enrolled in one of several segments or "arms" of a trial.[3] Additional authors are dealing with contentious issues that are being raised over the use of placebo control groups.[4] Beyond infrequent publications such as these and the work of the authors Brody mentions, the "silence," "paucity," and "failure" he laments are real.

The issues Brody identifies with respect to clinical trials and the process of new drug approval are critically important, but not exhaustive. In clinical trials, for example, Benjamin Freedman proposes in a soon-to-be-published study that Institutional Review Boards (IRBs) should consider how clinical trials ought to be designed in order to increase the possibility that these trials will effect beneficial and avoid harmful changes in clinical practice. Freedman views a research regimen's effects on clinical practice as an essential, but often neglected, ingredient in an IRB's risk-benefit, beneficence-rooted analysis (Freedman, 1996). How clinical trials should be reported is also replete with greater ethical import than is commonly assumed.[5]

The notable contribution Brody makes in this essay centers on his critique of and reformist charge to bioethicists as a community of investigators: return to basic concerns that were initially, but by no means fully, explored when bioethics emerged as disciplined inquiry regarding clinical medicine and research. Importantly, Brody links his charge to specific and significant areas of neglect and need.

III. DIAGNOSIS AND THERAPY FOR BIOETHICISTS

In the last section of his essay Brody compares the short stature and ill health of research ethics to the tremendous growth and health of clinical bioethics. He accounts for the latter by observing that Ph.D. and J.D. bioethicists have immersed themselves in "the clinical realities" of contemporary medical practice at the same time that M.D.s have received considerable training in bioethics. Both groups of clinical ethicists profit from training programs, national professional associations, several ethics journals, and notable medical journals that are receptive to the contributions of clinical ethicists.

In contrast, Brody believes that research ethics languishes in its lack of these intellectual and organizational supports. To revitalize research ethics, those who wish to explore its ever-present and emerging problems need to immerse themselves "into the realities of research practice" by securing footholds in any number of local and national committees and boards that are now largely staffed by non-bioethicists. Furthermore, research ethics needs to be nurtured and supported by bioethics training programs for researchers and IRB members, national network organizations, and the development of courses in research ethics.

The discussion that follows will indicate why I view research ethics as less sickly than does Brody. His metaphors suggest reanimation in the face of a failure to thrive. Mine suggest reorientation in light of Brody's insights and suggestions. At the same time, I may be more skeptical about reorienting research ethics along the lines Brody suggests than he is about its reanimation.

Some of the measures Brody advocates can and should be readily followed. Ethicists can and should become far more actively involved on research design and monitoring committees. They can and probably should establish a national society for research ethics, which should, however, be linked to presently existing networks. And training fellowships in bioethics should be developed for worthy and interested researchers.

Other suggestions made by Brody face daunting impediments at the present time. Consider two of the points Brody borrows from existing programs in the ethics of clinical research—the possibilities of developing training programs in research ethics and the challenge of becoming immersed in "the realities" of research ethics. While these sound exciting, how would they be put into effect?

Due to the organization and visibility of hospital-based medical practice and training and to twenty-five years of groundbreaking trial and error, clinical practice is relatively open and explorable by ethicists who can gain first and secondhand experience and understanding by making ward rounds, attending grand rounds, functioning as ethics consultants, and teaching in required and elective medical ethics courses. In spite of the relatively visible and porous nature of clinical practice in academic medical centers, its accessibility to non-M.D. bioethicists was arduously achieved over the course of many years.[6]

In contrast, knowledge and experience with research ethics is far less subject to "hands on" observation and participation, especially by trainee outsiders. Important levels of knowledge can be gained within the framework of committee and sub- committee involvement, and occasionally by serving as the primary reviewer of research protocols for one's IRB. But for a host of reasons, these research-related committees do not lend themselves to intricate and innovative discussions of ethics; and outside the confines of these committees, medical researchers give no grand rounds akin to the give and take within clinical departments and divisions. In addition, much of the information about the extent and types of an institution's research—including the number, types, and process of recruitment of research subjects—is either limited or off limits. IRB members by and large review protocols on paper. Rarely, if ever, are committee members or outside ethics observers able to be present in person as research is being conducted.

While these points call into question the ease with which contemporary research ethics might be reoriented or reinvigorated—which Brody deals with only indirectly—they admit to Brody's diagnosis of the problem. To rectify current deficits, some of his suggested remedies can be employed now. Others need to be thoroughly aired and debated by experienced researchers, IRB chairs and administrators, and experienced research ethicists in light of the fundamental dynamics of the research enterprise and the experiences of, at best, a handful of interdisciplinary programs that are exploring the design and management of clinical trials. Symposia and conferences would do well to address the critical challenges raised in this brief last section of Brody's essay.

IV. WHAT'S HAPPENING IN RESEARCH ETHICS

In order to begin to understand the content and extent of current writing about research ethics and to assess the accuracy of Brody's assertions about the degrees to which bioethicists are "virtually ignoring research ethics," I surveyed the literature on the ethics of research published during 1994 and 1995. While the following results are less than complete, they do include the articles and annotated listings of articles and books in *IRB, Kennedy Institute of Ethics Journal, Journal of Medicine and Philosophy, Medical Humanities Review, Hastings Center Report, Cambridge Quarterly of Healthcare Ethics, Perspectives in Biology and Medicine,* and *Journal of Clinical Ethics.* This survey does not include publications on the ethics of animal research or on researcher fraud and misconduct, and does not include book chapters, encyclopedia articles,[7] and collections of readings.

During 1994 and 1995, 116 articles and twelve books were published on the ethics and regulation of research involving human subjects. Because these articles were published in no less than thirty-five different journals (from the *American Journal of Pediatrics,* through *JAMA* and the *New England Journal of Medicine (NEJM),* to the *Saint Louis University Law Journal* and *Transfusion*), they are far from readily accessible to most readers. If, for example, an ethicist subscribed to and read the *Hastings Center Report, JAMA, NEJM,* and *Journal of Medicine and Philosophy* during these years, she or he would have discovered only 23 (20 percent) of these articles—actually, a *high* percentage in light of the fact that three of these journals were among the five that contained four or more articles on research ethics over this two-year period. If he or she subscribed to or regularly read *IRB,* 27 (21 percent) of these articles would have been discovered, and another 31 (27 percent) would have been discoverable in *IRB*'s annotated listings.

Overall, this survey of two years of research ethics literature indicates that it is unwarranted for Brody to derive "an important fact" about the paucity of literature in research ethics from the unquestionable fact that only one of the forty-six volumes of the *Philosophy of Medicine* series explicitly deals with the ethics of research. This survey *and* earlier bibliographical searches and summaries show that the lack of attention to research ethics in the *Philosophy of Medicine* series raises a question about this series, rather than the reverse.[8]

The large body of literature published during 1994 and 1995 can be divided into two general categories. The first category consists of seventy-one articles and nine books that contain "ethics content" in that they discuss ethical problems, ethical theory, and/or are written by one or more persons with a "bioethics" title or who work with bioethics-related organizations. A number of these books and articles were authored by well-known Ph.D. or M.D. bioethicists, including Terrence F. Ackerman (1995, pp. 1-5), Michael A. Grodin and Leonard H. Glanz (1994), Ruth Macklin (1994, pp. 209-226), Jay Katz (1994, 7-54), Sissela Bok (1995, pp. 1-18), Benjamin Freedman (1994, pp. 1-6), and Stuart E. Lind (1994, pp. 828-834). As a corrective to Brody's statements that most of the continuing work in the ethics of research deals with issues surrounding genetics and reproduction, only 12 (15 percent) of the articles and books in this category centered on genetics, and another 10 (13 percent) with embryo, cloning, and fetal tissue research. Seven (9 percent) focused on clinical trials, 11 (14 percent) on children and adolescents, 16 (31 percent) on autonomy and informed consent, and 5 (6 percent) on the inclusion and recruitment of women as research subjects.

The second general category consists of forty-five articles and three books that deal with background and supplemental information pertaining to the ethics of clinical research involving human subjects. This literature includes social scientific studies of informed consent (Harth and Thong, 1995, pp. 1647-1651), studies about and experience within IRBs (Hayes, Hayes, and Dykstra, 1995, pp. 1-6), the development of new ethical and regulatory guidelines (Green, 1994, pp. 345-356), and thirteen books and articles on the history of research ethics and/or the abuses of research subjects (Advisory Committee, 1995; Lasagna, 1995, pp. 96-98). This category of publications is itself supplemented by the journal *Human Research Report: Protecting Researchers and Research Subjects,* which summarize and reviews developments and trends at the national and international levels, including new initiatives and regulatory guidelines from the National Institutes of Health and the Food and Drug Administration. *Human Research Report* also regularly reviews noteworthy court cases regarding research with human and animal subjects.

Finally, in order to further assess the degrees to which research ethics is more or less vital or moribund and more or less nationally organized, I developed a list of conferences on research ethics and regulations regarding human subjects that were held in the U. S. during 1994 and 1995. This list is surely not exhaustive, and it does not include a number of conferences each year that focus on the ethics of research with animals on the one hand and on the ethics and regulation of scientific/researcher misconduct on the other.

My search revealed that twenty-four conferences were held on the ethics and regulation of human subject research during these two years—eleven in 1994 and thirteen in 1995. Six of these conferences were developed and put on by the organization that for the last twenty-two years has served as a network for discussing research ethics and regulation at the national level—Public Responsibility in Medicine and Research (PRIM&R). The majority of attendees of PRIM&R conferences are members and administrators of IRBs,[9] and approximately fifty nationally known ethicists, medical school and government administrators, and medical journal editors serve as faculty at each of these gatherings. The PRIM&R conferences last for two days, and those who attend usually buy a topically arranged, book-size set of readings that includes recently published articles, book chapters, and news stories on the ethics and regulation of research involving human subjects.

According to their descriptions, a majority of the twenty-four conferences held during 1994 and 1995 dealt with the regulation, ethics, and oversight of several generally recognized "current," "contemporary," or "emerging" issues in research ethics—the inclusion and recruitment of women and minorities into clinical trials, genetics, AIDS, informed consent for prisoners or mentally ill subjects, fetal research, and so on. But a number of them were more targeted: research with children as subjects at Texas Medical Center in Houston; neurobiological research in Baltimore; teaching research ethics in Bloomington, Indiana; research involving subjects in emergency circumstances at Bethesda; "lessons" from the history of human experimentation in New York (Columbia University); and the research relationships between universities and industry at the University of Maryland in Baltimore. Many of these conferences listed the National Institutes of Health (NIH), the Food and Drug Administration (FDA), and the NIH's Office for Protection from Research Risks (OPRR) as joint sponsors.

By definition, this brief survey of the literature and conferences on research ethics and regulations during 1994 and 1995 does not deal with the degrees to which the topics identified here are or are not being explored innovatively and in depth. Nor does it begin to account for what I take to be a widely shared assumption: that "new contributions in research ethics" primarily entail applying the ethical verities of *The Belmont Report* to new medical and scientific procedures and to expanding populations of research subjects. This overview nevertheless indicates that even if research in the ethics of genetics and human reproduction is excluded, research ethics has hardly disappeared from public view or as a multifaceted subject of continued scholarly concern. The agenda of issues that Brody charts should both redirect and intensify existing and ever-changing levels of interest, analysis, and continuing controversy in research ethics.

Institute for the Medical Humanities
The University of Texas Medical Branch
Galveston, Texas, U. S. A.

NOTES

1. The last section of this essay will explore current interest in research ethics. With respect to newly discovered "horrors," compare the alarmist reporting of John Schwartz ([33], pp. 10-11) with the thorough-going analysis of the Institute of Medicine [17] regarding the Fialuridine (FIAU) clinical trials.
2. A number of Brody's [5] key points seem to have been lost on some of his reviewers. See Thomas Preston ([29], pp. 42-43).
3. The fifth question Brody raises with respect to the conducting of clinical trials asks, "What burdens can be imposed on patients to maintain the blinded nature of trials?" This "blinding" involves keeping both researchers and subjects from knowing which segment, arm or "wing" of a trial (experimental arm, presently-standard therapy arm or placebo arm) subjects are placed in. Blinding is practiced in order to decrease subjective bias on the part of researchers (who may be tempted to attribute greater effects to the experimental drug) and patient-subjects (whose psychological-placebo responses could skew a trial's results). Two articles that deal with "blinding" desperate patients are by Gerald Logue and Stephen Wear ([26], pp. 57-64) and Brendan P. Minogue, *et al.* ([28], pp. 43-56).
4. The third question Brody poses regarding the ethics of clinical trials involves the inclusion of control groups—in particular, administering a placebo to one of the trial's groups of subjects. Controversy over the ethics of placebo controls is evident in the articles by Kenneth J. Rothman and Karin B. Michel ([32], pp. 394-398) and by Gunnel Elander and Goran Hermeren ([8], pp. 171-182).
5. See. e.g., Kathleen Cranley Glass ([12], pp. 327-338) and *Lancet*, Editorial ([22], pp. 347-348).
6. Compare the perspectives of David J. Rothman [31], and Albert R. Jonsen ([18], pp. 42-47) and Daniel M. Fox ([9], pp. 47-50).

[7] The 1995 revised edition of the *Encyclopedia of Bioethics,* Warren T. Reich, editor-in-chief, contains at least seven entries on the ethics and history of research involving human subjects, including articles by Loretta Kopleman, Robert J. Levine and Nancy M. P. King [30].

[8] In our survey of the cancer and ethics literature between 1966 and 1983, Gary B. Weiss and I [35] that some 102 articles and letters had discussed ethical issues pertaining to randomized clinical trials, some thirty-six with questions over stopping clinical trials, forty-eight with IRBs, eighty-three with the ethics of Phase I, II and III trials, and so on. See also, e.g., the eighty-two entries listed in *Cambridge Quarterly of Healthcare Ethics* ([7], pp. 574-577).

[9] The names and positions of the early registrants to these conferences are listed each year in the PRIM&R conference proceedings. These listings call into question Brody's assertion that networks of IRBs have not yet emerged. While entire IRBs are not meeting with one another, their representative members do. Freeman has discussed one of the reasons why IRBs should consult with each other more frequently ([10], pp. 1-6).

REFERENCES

1. Ackerman, T. F.: 1995, 'The ethics of Phase I pediatric oncology trials,' *IRB* 17, 1-5.
2. Advisory Committee on Human Radiation Experiments: 1995, *Final Report on Human Radiation Experiments,* U.S. Government Printing Office, Washington, D.C.
3. Beecher, H. K.: 1966, 'Ethics and clinical research,' *New England Journal of Medicine* 274, 1354-1360.
4. Bok, S.: 1995, 'Shading the truth in seeking informed consent for research purposes,' *Kennedy Institute of Ethics Journal* 5, 1-18.
5. Brody, B.: 1994, *Ethical Issues in Drug Testing, Approval, and Pricing,* Oxford University Press, New York, NY.
6. Brody, B.: 1996, 'Whatever happened to research ethics?' in this volume, pp. 275-286.
7. *Cambridge Quarterly of Healthcare Ethics:* 1994, Special Section: 'Research ethics: Ethics at the borders of medical research,' 3, 574-577.
8. Elander, G., and Hermeren, G.: 1995, 'Placebo effect and randomized clinical trials,' *Theoretical Medicine* 16, 171-182.
9. Fox, D. M.: 1991, 'Negotiating the ascendancy of bioethics: Two views,' *Medical Humanities Review* 5, 47-50.
10. Freedman, B.: 1994, 'Multicenter trials and subject eligibility: Should local IRBs play a role?' *IRB* 16, 1-6.
11. Freedman, B.: 1996, 'The ethical analysis of clinical trials: New lessons for and from cancer research,' in H.Y. Vanderpool (ed.), *The Ethics of Research Involving*

Human Subjects: Facing the 21st Century, University Publishing Group, Frederick, MD.

12. Glass, K. C.: 1994, 'Toward a duty to report clinical trials accurately: The clinical alert and beyond,' *Journal of Law, Medicine and Ethics* 22, 327-338.
13. Green, R. M.: 1994, 'At the vortex of controversy: Developing guidelines for human embryo research,' *Kennedy Institute of Ethics Journal* 4, 345-356.
14. Grodin, M. A., and Glanz, L. H. (eds): 1994, *Children as Research Subjects: Science, Ethics and Law,* Oxford University Press, New York, NY.
15. Harth, S. C., and Thong, Y. H.: 1995, 'Parental perceptions and attitudes about informed consent in clinical research involving children,' *Social Science and Medicine* 41, 1647-1651.
16. Hayes, G. J., Hayes, S. C., and Dykstra, T.: 1995, 'A survey of university institutional review boards: Characteristics, policies, and procedures,' *IRB* 17, 1-6.
17. Institute of Medicine: 1995, *Review of the Fialuridine (FIAU) Clinical Trials,* National Academy Press, Washington, D.C.
18. Jonsen, A. R.: 1991, 'Negotiating the ascendancy of bioethics: Two views,' *Medical Humanities Review* 5, 42-47.
19. Katz, J.: 1994, 'Human experimentation and human rights,' *Saint Louis University Law Journal* 38, 7-54.
20. King, N. M. P.: 1995, 'Privacy and confidentiality in research,' in W.T. Reich (ed.), *Encyclopedia of Bioethics,* Simon and Schuster-Macmillan, New York, NY.
21. Kopelman, L.: 1995, 'Research methodolgy: II. Controlled clinical trials,' in W.T. Reich (ed.), *Encyclopedia of Bioethics,* Simon and Schuster-MacMillan, New York, NY.
22. *Lancet:* 1994, Editorial, 'Reporting clinical trials: Message and medium,' 344, 347-348.
23. Lasagna, L.: 1995, 'The Helsinki Declaration: Timeless guide or irrelevant anachronism?' *Journal of Clinical Psychopharmacology* 15, 96-98.
24. Levine, R. J. : 1995, 'Informed consent: III. Consent issues in human research,' in W.T. Reich (ed.), *Encyclopedia of Bioethics,* Simon and Schuster-MacMillan, New York, NY.
25. Lind, S. E.: 1994, 'Ethical considerations related to the collection and distribution of cord blood stem cells for transplantation to reconstitute hematopoietic function,' *Transfusion* 54, 828-834.
26. Logue, G., and Wear, S.: 1995, 'A desperate solution; Individual autonomy and the double-blind controlled experiment,' *Journal of Medicine and Philosophy* 29, 57-64.
27. Macklin, R.: 1994, 'Splitting embryos on the slippery slope: Ethics and public policy,' *Kennedy Institute of Ethics Journal* 4, 209-226.
28. Minogue, B. P., et al.: 'Individual autonomy and the double-blind controlled experiment: The case of desperate volunteers,' *Journal of Medicine and Philosophy* 29, 43-56.

29. Preston, T.: 1996, 'Protection or paternalism,' *Hastings Center Report* 26, 42-43.
30. Reich, W. T. (editor-in-chief): 1995 (revised ed.), *Encyclopedia of Bioethics,* Simon and Schuster-MacMillan, New York, NY.
31. Rothman, D. J.: 1991, *Strangers at the Bedside: A History of How Law and Bioethics Transformed Medical Decision Making,* Basic Books, New York, NY.
32. Rothman, K. J., and Michel, K. B.: 1994, ' The continuing use of placebo controls,' *New England Journal of Medicine* 331, 394-398.
33. Schwartz, J.: 1993, 'Nightmare at NIH: Warning signs in drug trial went unheeded, and then the patients started dying,' *Washington Post National Weekly Edition,* September 13-19, 10-11.
34. Vanderpool, H. Y.: 1996, *Ethics of Research Involving Human Subjects: Facing the 21st Century,* University Publishing Group, Frederick, MD.
35. Weiss, G. B., and Vanderpool, H. Y.: 1984, *Ethics and Cancer: An Annotated Bibliography,* University of Texas Medical Branch, Galveston, TX.

E. HAAVI MORREIM

AT THE INTERSECTION OF MEDICINE, LAW, ECONOMICS, AND ETHICS: BIOETHICS AND THE ART OF INTELLECTUAL CROSS-DRESSING

I. INTRODUCTION

When I first started doing bioethics in the clinical setting, I was struck by the moral importance of factual complexity. It's all very well to study principles in their purity, I thought, but solving real moral problems in medicine requires us to attend to political clashes and turf battles, emotional attachments and personality conflicts, communication glitches, legal threats both real and fictional, practical obstacles, fiscal limitations, and a host of other details that constitute the actual and often highly complex substance of a moral problem ([46] [44]). Solving these problems requires something more than purely philosophical contemplation.

> Our moral lives are comprised, not of terrible hypotheticals from which there is no escape, but of complex situations whose constituent elements are often amenable to considerable alteration. And our moral aim should be, not to make dramatic choices which honor one value at a terrible sacrifice to some competing value, but to create a resolution which honors all important values maximally . . . to be as thoughtful, even inventive, as possible. . . . The philosopher may be right in thinking that sometimes we must make terrible priority choices between important values. But with a bit of ingenuity, such occasions can be far rarer than we are inclined to think ([46], p. 49).

Nowhere is this need to appreciate details and to negotiate with practicalities more true than the area in which I've done much of my writing over the past several years: the intersection of medicine, law, economics, and ethics. As outlined below, economic relationships in medicine have, in the past few years, posed moral and legal challenges that have become so complicated and so fast-changing, it is difficult to keep pace. We cannot even understand, let alone re-

R. A. Carson and C. R. Burns (eds.), Philosophy of Medicine and Bioethics, 299-325.

solve, these challenges without crossing disciplinary lines freely and comfortably. Those who work in this area of bioethics must be "intellectual cross-dressers," as comfortable talking about "exclusive provider organizations" and "*stare decisis*"[1] as they are about beneficence and autonomy.

This article will explore this claim by focusing particularly on one area: the standard of care that physicians owe their patients. On one level it is a legal issue concerning the kind of care that physicians must deliver, lest they be found negligent and thereby civilly liable for any injuries their substandard conduct causes. As I will argue, the traditional appeal to customary or prevailing practice is profoundly challenged by a variety of powerful economic factors pressuring physicians to change their practices. Because many of these pressures point in opposing directions, there is far less consensus about appropriate practice and, therefore, is not clear on what bases the law's standard should now be set.

On another level are moral questions concerning the level of care that physicians owe their patients. Traditional moral expectations of virtually unlimited resources and unstinting loyalty are impugned by newer realities, as it is no longer possible for physicians to produce resources with just a signature. If "ought" implies "can," it may now be unfair to expect physicians to commandeer resources belonging to third parties who refuse to furnish them, or to expect physicians to place their own well-being endlessly in jeopardy in that quest. Reappraising physicians' duties raises anew the question what it means for the physician to be a fiduciary of the patient, and how far the duties of advocacy should be pressed.

I will not attempt definitive analyses here. Rather, the purpose of this essay is much simpler: to show that any acceptable answers must be intensely interdisciplinary. Good ethics and good law begin with good facts. Only when the full complexity of the situation is appreciated can plausible resolutions be constructed. Otherwise we will badly miss the mark.

II. THE NEW, THICKER ALPHABET SOUP

A bit of alphabet soup will highlight some of the economic changes.[2] A decade ago we learned about DRGs (diagnosis related groups), PPSs (prospective payment systems), HMOs (health maintenance organizations) and their variants, the IPAs (independent practice associations). These new arrangements emerged from the realization that traditional FFS (fee-for-service) reimbursement had the economically perverse incentive of encouraging providers to maximize the number of services they provided and to "unbundle" their care into as many individually billable services as possible. By prospectively setting the payment for services, whether by hospital episode as in DRGs, or per patient per year as in HMOs and IPAs, the various governments, businesses, and insurers who paid for care hoped to reverse the perverse. Traditional indemnity insurers added their own cost containment in the form of PPOs (preferred provider organizations), comprised of providers who agreed to discount their fees in exchange for higher volumes of business. At the same time, all kinds of payers intensified their UR (utilization review) to ensure that only "necessary" services were delivered in "appropriate" settings.

As the alphabet soup has grown thicker, we now see MCOs (managed care organizations) diversifying to include PHOs (physician-hospital organizations), EPOs (exclusive provider organizations), GPWWs (group practices without walls), MSOs (management services organizations), IDSs (integrated delivery systems) or IDNs (integrated delivery networks), and a staggering variety of other arrangements.[3] "Consumers" (a term occasionally referring to patients who consume services, but more often to businesses who purchase health plans) are becoming more concerned to ensure that cost containment has not inordinately compromised quality of care, and so traditional QA (quality assurance) is making way for CQI (continuous quality improvement) and TQM (total quality management).

Other economic changes will be highlighted below. For now, a few observations will illuminate their overall character. First, money flows much less freely than it did a decade or two ago, when health care was financed from a seemingly bottomless Artesian well of money [53] in which providers could deliver whatever services they deemed appropriate, assured of payment for virtually whatever fee they asked. Now, providers' earnings are being bargained downward or placed on fee schedules, and resource utilization is aggressively monitored, prospectively and concurrently as well as retrospectively, with easier denial of payment for unauthorized services.

Second, physicians have considerably less control over health-care resources than they once did. Besides utilization guidelines, limits on pharmacy formularies and other rules curb what physicians can order. And where physicians actually retain the clinical authority to use health-care resources as they see fit, they are usually constrained by an incentive system ensuring that profligate spenders pay a personal price for imprudence. Even outside of managed care, physicians whose patients overspend their welcome may lose their hospital staff privileges through "economic credentialing"—or they may be "deselected" (fired) from their PPO or MCO.[4]

These economic changes have in turn spawned a major jurisprudential challenge that I began to discuss in prior writings [52] [47] [49] [48]. In tort law (the branch of civil law enabling citizens to hold each other accountable when they harm one another), plaintiffs must prove that their injuries were caused by the defendant's negligence. This, in turn, requires the law to determine a standard of conduct, below which lies negligence and above which lies unfortunate but nonculpable harm. Ordinarily the standard is that people must conduct themselves in a reasonable and prudent way. In tort law that standard is defined by appeal to a shared notion of what reasonable, prudent people would do in a given situation, and determined in practice by the consensus of a jury of ordinary people. In professional negligence cases the professional is held to a higher standard than the ordinary person. He or she is expected to provide, not the care that a layman would, but the care that a reasonable and prudent member of the same profession would, under the same or similar circumstances. Traditionally, that level is established for medicine empirically, by appealing to "prevailing practice"—that is, by asking what physicians generally in fact do.[5]

Two features in this way of setting physicians' standard of care have been especially significant. First, over the years that standard has come to require that physicians not only be knowledgeable and skillful, but that they deliver certain potentially costly interventions, such as to keep a patient hospitalized as long as medically necessary or to take X-rays if indicated. Second, our tort system entitles every citizen to the same level of reasonable consideration from fellow citizens. In medicine, this has come to mean that every patient should receive the same basic quality of care, regardless of ability to pay [72]. Because medical technologies are often owned or financed by third parties, the legal system therefore has expected physicians to appropriate other people's money and property for their patients' benefit.

In the past, both these requirements were easy enough to satisfy, because Artesian economics permitted a remarkable degree of fiscal and thereby medical and legal parity. Emerging costly technologies were almost immediately accepted by payers,[6] and virtually all patients could be assured a roughly equal level of care by a cost-shifting in which those who could pay for care subsidized those who couldn't. Hence, virtually anyone who had access to care at all enjoyed roughly the same technologically high standards.

However, payers' more recent refusal any longer to subsidize the uninsured has given rise to a "stratified scarcity" in which the uninsured have had significantly less access to health care than the well-insured. It has become implausible to expect physicians to deliver even the roughly uniform standard of care they once could, because current economic conditions essentially preclude it.

Accordingly, the jurisprudential dilemma is to determine whether physicians should still be held potentially liable for inadequate care when the inadequacies stem from resource constraints beyond their control,[7] or whether alternatively we should reconceive the moral and legal expectations we place on physicians.

A few years ago this problem was fairly straightforward: stratified scarcity sometimes precluded physicians from delivering to their poorer patients the same level of care that they delivered to their insured patients [45]. There were mainly two levels of care—well insured versus uninsured—and the main question was whether to permit the physician to rebut the expectation of equality in cases where it could not be achieved [47].

The situation has changed dramatically since then. A simple stratified scarcity no longer begins to capture the numerous, radically diverse, and often conflicting sources of medical and economic standards by which physicians are expected to deliver care. If we are to understand the scope and magnitude this jurisprudential challenge has now assumed, we must array some of the varying ways in which those who control the resources expect physicians to practice. These economic limits and expectations, in turn, powerfully shape the "prevailing" practices to which courts look in determining to what standard the physician should be held legally accountable. Ultimately it will be evident that there is virtually no such thing as prevailing practice any more, and the judiciary needs to look elsewhere for medical standards.

It will also be important to outline how the standard of care figures into physicians' moral commitment to patients, and how it is affected by these economic changes. While our legal expectations of each other hardly equate to a moral code, nevertheless many of our most important legal precepts embody a shared social morality. In criminal law we condemn murderers and thieves, not just because a rulebook says we are supposed to, but because we deem these offenses morally so abhorrent that we bring the power of the state to bear against them. Similarly, civil law marks out the minimum ways in which decent members of society must act toward one another. Tort law in particular embodies the moral principle that those who carelessly or deliberately cause injury should restore to wholeness those they have harmed. And in so decreeing, the law must set standards of behavior that help us to define the boundaries of acceptable conduct.

Here, as we inquire what physicians owe their patients, moral issues of fairness arise: fairness to patients, as they depend on physicians for help and bring certain expectations to the relationship; fairness to physicians, because traditional expectations may no longer be reasonable; and fairness to all the others who contribute to and depend on the same health-care resources, because as costs become more constrained, their needs and interests compete more keenly with the traditional simple picture of physicians serving only their patients. In the next section we will examine these economic forces and, in the ensuing two sections, the way they change the legal and moral issues.

III. THE STANDARDS OF CARE

One can identify at least nine different sources that purport to identify the kind and level of care physicians should provide to patients. Note that, within each source, different parties can construct differing medical guidelines, while on other occasions different sources may in fact produce similar guidelines. The nine sources can be distinguished, however, by their varying economic perspectives and agendas. Most important for current purposes, they can and sometimes do place physicians under widely differing, even opposing commands.

1. Artesian standards: The era of affluent insurance promoted certain values: it is better to do too much than too little; if an intervention might possibly help the patient, and is unlikely to harm, it should be used; if a new drug or device has been proven safe and effective, it should immediately be approved for payment; no one should be denied potentially beneficial care on account of inability to pay; high-tech is better than low-tech is better than no-tech [53]. Because most patients were covered either by Artesian financing or by cost-shifting, physicians' prevailing practices were medically thorough and technologically rich, and courts have used them to define the standard of care that physicians owe their patients for insured and uninsured alike.

This lavish standard is alive and remarkably well. Even if relatively few people retain the first-dollar coverage that so many once enjoyed, generous insurance still occupies a strong sector of the market. And although insurers are exercising considerably closer scrutiny over their expenditures, the fact remains that denials of payment or prospective authorization are still relatively rare.[8] Further, many physicians fervently espouse Artesian values as part of their moral commitment to their patient.[9] And the power of collective habits and customs, combined with early medical training, likewise reinforces the notion that physicians owe their patient every potentially helpful intervention [8] [63] [28]. Hence, a spare-no-expense approach to care still flourishes in many sectors of medicine that, under the "prevailing practices" approach to setting legal standards, retains considerable dominance.

2. Managed standards: In sharp contrast, many insurers and MCOs have developed their own concepts of appropriate care, with accompanying guidelines to determine which interventions are medically necessary. These standards are based on the beliefs that more is not always better; that physicians' customs do not always ensure good outcomes; that scientific evidence, not the anecdotes or personal wisdom of local department heads, should determine which interventions are medically warranted; and that in cases of doubt the burden of proof should rest on those who want to intervene.

Accordingly, managed care expects medical standards to be lean, not lavish. Strict controls on utilization contrast with the "if it might work, do it" ethic of Artesian financing ([34], p. 81; [57], p. 44-48; [77], p. 30; [9]). A payer may expect physicians to discharge patients with a particular diagnosis from the hospital within a limited number of days, for instance, or might decline to authorize epidural anesthesia for uncomplicated obstetric cases ([34], p. 93; [5]; [11]).

The case of Nellene Fox highlights this clash between Artesian and managed standards. A thirty-eight-year-old woman with advanced breast cancer, Nellene Fox, asked HealthNet, her HMO, to pay for high-dose chemotherapy with autologous bone marrow transplant. It was the most advanced treatment for her condition and probably her only, even if slim, hope for survival. HealthNet denied coverage on the ground that it was experimental, while Fox demanded it on the premise that it was widely practiced [43].

In fact, both sides were correct. The treatment has become quite widely available. Partly this is because many physicians, acting on Artesian values, have been willing to undertake this treatment that offers promise to a group of patients with no other hope. And partly it is because many patients have managed to secure court injunctions mandating insurance coverage. Others secured it merely by threatening legal action [60]. Where a costly treatment is fully paid for, it is not difficult to find a provider willing to say that it is medically indicated.

At the same time, insurers and MCOs can rightly point to the lack of controlled scientific studies documenting that this treatment is safe and effective for this particular condition. Although the National Institutes of Health has a study underway, results are not yet available and may never be. Because so many women have direct access to the treatment through their insurers, it has become very difficult to recruit enough women willing to enter controlled trials in which only half the women actually receive the treatment [38].

Thus, a clear conflict emerges between Artesian standards ("It's the patient's only hope, and it's paid for, so it would be wrong not to provide it") and managed standards ("Absent clear scientific evidence documenting its value, it should not be paid for as a routine form of care"). Nellene Fox and HealthNet were both caught in the conflict. And so are physicians whose actions can be appraised from these opposing perspectives at the same time.[10]

3. Corporate standards: Many large business corporations have chosen not to purchase outside health insurance for their workers, but rather to pay the cost of their workers' health care directly through self-insurance. The option is attractive because federal law permits these corporations to save money by avoiding costly state insurance mandates.[11] In addition, many companies are coming to feel that they can do a better job of controlling rising health-care expenditures themselves, without paying third parties to do it.

Accordingly, many business corporations are directly creating their own healthcare systems. Some companies directly recruit their own networks of physicians, hospitals, and other providers; some engage in case management of all workers whose care exceeds a specified amount; and some specifically commission the creation of medical guidelines to govern their workers' care [65]. These may be adopted from medical specialty groups, or purchased from entrepreneurs whose sole purpose is to create and sell such guidelines, or adapted from MCO protocols, or even created ad hoc, not based on readily identifiable guidelines from elsewhere in the health-care scene.

Compared with MCOs, these in-house corporate guidelines may have a distinctive time-frame from which to evaluate the value of various medical interventions. Although it is common to say that MCOs save money by emphasizing preventive care, the fact remains that most preventive care saves money, if at all, only in the long run. Only if an MCO can keep a patient enrolled for many years, for instance, will promoting exercise and healthy eating actually save that MCO money later on. Where people are free to change their health plans every year, there is no assurance that an MCO will reap such long-term rewards. And so some MCOs may actually de-emphasize some preventive care as a result.[12]

In contrast, some business corporations can expect to keep many of their employees for a considerably longer period of time, so that its views about what health care is cost-effective may in fact be quite different from those of an MCO.

4. Government standards: Federal and state governments pay about forty percent of the nation's health-care bill. Like other payers, they are well aware that wide variations in medical practice are not always justified by severity of patients' illness or other medically obvious factors. In the interest of promoting more prudent purchasing, the federal government has launched several initiatives to study which medical interventions are most appropriate for which conditions. The Agency for Health Policy and Research, for instance, is conducting outcomes studies and creating detailed guidelines for a range of conditions such as cataracts, bed sores, depression, and pain management ([67], p. 67; [25], p. 3032; [26], p. 1664). The U.S. Preventive Services Task Force plans to issue recommendations for preventive services ([25], p. 3032; [66]).

Not all of these government suggestions are welcomed by clinicians. After the National Institutes of Health issued advisory guidelines on the treatment of high blood pressure, a number of cardiologists argued that the protocols tended to be simplistic and potentially harmful to patients. Defenders retorted that the guides left considerable room for flexibility, and that they were based on evidence rather than on less well-substantiated, even if widely practiced, prevailing approaches [2].

In many cases, these government guidelines may lie somewhere between Artesian standards, on the one hand, and managed or corporate standards, on the other. To the extent that they are created by leaders in academic medicine, they may call for technologically rich practices, reflecting the economically insulated views still fairly common in academic settings. To the extent that they are created by people interested in saving the government money, they may be leaner, more like managed or corporate standards.

5. *Subspecialty medical organization standards:* Many medical subspecialty organizations are developing their own protocols. The American College of Physicians, for instance, has an extensive program in which academic leaders review medical literature and other factors such as costs [18]. Other specialty organizations have wide-ranging efforts of their own.

Although specialist guidelines may often be motivated by concern for patients, some are also accused of being vehicles for preserving turf in an increasingly competitive marketplace, and for ensuring high utilization of revenue-generating procedures ([77], p. 2547); [67], p. 73; [1]). Further, when they are created by physicians in academic environments, they may more closely reflect economically insulated values than the leaner realities of the marketplace.

6. *Group practice standards:* According to the American Medical Association, there are about 1500 guidelines currently in print ([67], p. 67). As is obvious from the foregoing, these are created by a wide variety of organizations, with widely differing agendas. And yet even these do not complete the picture. Over the past decade, physicians' alliances with one another have been changing dramatically. Many are joining into large, multispecialty group practices that can offer economically attractive health-care packages to businesses and insurers. Many of these groups develop their own in-house standards to guide members' delivery of care. Some draw mainly on their own members, while others use outside consultants ([18], p. 315; [59]). Perhaps more important for

current purposes, these standards can be very different from any of the above, depending on whether the group is trying to contain its costs or enhance its billable revenue, and they can vary also according to their creators' differing training, values, and conscientiousness.

7. Malpractice insurers' standards: An entirely different agenda is captured in the guidelines that malpractice insurers establish for their physician clients. Here, the objective is not to save money for the hospital, HMO, or health insurer, nor is it to protect any particular physician's turf or income. Rather, these protocols aim to minimize the risk of lawsuit or, in the event of suit, to minimize damage awards.

Accordingly, many insurers have developed their own guidelines in such high-risk areas as breast cancer diagnosis, anesthesia monitoring, heart attacks, head trauma, antibiotics, and pulmonary embolism. Some offer discounts to physicians who agree to follow them. And at least one insurer actually mandates adherence, on pain of dropping noncompliant physicians.[13]

These standards aim to reduce litigation partly by ensuring that necessary interventions are taken, and partly by improving the chart documentation that helps physicians and patients alike. Sometimes, however, they may require a level of care that is economically or logistically unsupportable in a given physician's circumstances. In a rural area, for instance, it may not be possible to sustain these insurers' ideal level of personnel and facilities.[14]

8. Patients' expectations: Patients and their actual and perceived expectations introduce one of the most unpredictable sources of the standards that physicians are expected to fulfill. On the one hand, many patients, economically insulated from the costs of care, bring to the medical encounter an entitlement mentality demanding the best of care, spare no expense ([30], p. 1785; [61], p. 141; [62]; [70]). They have become accustomed to Artesian financing and its implicit values. Physicians fearful of displeasing a potential litigant may feel impelled to deliver every intervention the patient explicitly demands or presumably wants.

On the other hand, recent evidence indicates that patients do not necessarily sue out of displeasure with the technical quality or quantity of their care. The great majority of negligent medical injuries do not precipitate any kind of claim [7] [39]. And often the most salient predictor of whether a physician will be sued is the extent to which patients feel that they are being treated with honesty, respect, and personal interest [32] [33]. Indeed, there appears to be rather little correlation between physicians' technical quality of care and their likelihood of being sued [17].

However, patients' expectations still can conflict with other demands on physicians. Patients who want their physician to take plenty of time, explain everything carefully, and invite lots of questions will be disappointed if the physician's heavy caseload—the MCO's or group practice's "productivity" expectation—does not permit such leisurely conversation.

9. Individual clinicians' ad hoc standards: Physicians' responses to these opposing pressures to contain costs, preserve turf, avoid liability, and gain market share are not always well organized. The various written guidelines discussed above may not be carefully constructed, let alone scientifically grounded, rigorously tested, and regularly updated [77]. However, even less carefully formulated are the ad hoc informal practice changes that sometimes spring from economic panic. Previously routine medical practices may be suddenly changed, perhaps in hasty response to economic changes and not for any better medical reason than a passing realization that perhaps old routines weren't so well justified after all.

When specialists like cardiologists, urologists, or dermatologists are paid on a fee-for-service basis, for instance, they have little economic reward for identifying ways to cut back on the number of procedures they do, while conversely they may be well rewarded for hanging on to all the "turf" they can, as other specialists compete to do some of the same procedures. However, traditional ways of paying specialists are rapidly changing. Although many specialists are still paid fee-for-service, many MCOs and large group practices are now placing specialists under capitation in which each one is paid a set amount every month for every patient on his or her list, regardless of the level of service these patients need.

Suddenly, many of these specialists must reconsider longstanding medical habits. Capitated cardiologists may forego costly nuclear treadmill tests for cheaper stress echocardiograms ([68], p. 124), for instance, and do fewer ordinary treadmill tests ([69], p. 26E). Capitated obstetricians may perform fewer cesarean deliveries ([69], p. 26E), while gastroenterologists may decide they don't need a biopsy from every colonoscopy ([68], p. 131).

Turf issues are literally reversed. Urologists may teach family physicians to perform vasectomies in their offices, while dermatologists may teach primary care internists how to freeze off warts and do simple skin biopsies ([68], p.131-32; [69], p. 26F). The fewer such procedures the specialists do, after all, the more time they have left to expand their panel of patients and thereby their capitated revenue.

Illustratively, one study showed that referral patterns and routines of care in one MCO changed dramatically every time reimbursement methods changed. When the plan paid primary physicians fee-for-service, with strong financial and bureaucratic incentives to "keep the gates" against excessive use of specialists, specialists got very few referrals. When the financial arrangements changed so that all the physicians were capitated and referrals to specialists required almost no paperwork, specialists were deluged with patients, many of whose ailments were only marginally if at all related to the specialist's field [78].

These widely varying standards and shifts in practice patterns might be viewed cynically, of course, as being inspired more by crass self-interest than by a concern for patients' welfare. But on a deeper level they reflect the formidable if not impossible pressures that physicians now face. Medicine's rapidly changing economic arrangements have spawned an incredible array of payment and practice arrangements that, in turn, demand different kinds of health-care delivery. One can argue, in theory, that physicians ought to consider only their patient's best interests. But if in fact the insurer will not pay for what the physician has ordered, or if reimbursement changes threaten bankruptcy to the physician who does not streamline his practices, prevailing Artesian practices will be rapidly trimmed by realities.

As a result of these economic changes, the jurisprudential challenge identified at the outset in this article is vastly more complicated, and even the moral challenges have changed. We will explore each in the next two sections.

IV. MULTIPLE STANDARDS: LEGAL CHALLENGES

As noted, Artesian funding permitted a remarkably uniform standard of care. Most patients had direct or indirect access to the same level of funding, and physicians throughout medicine could find considerable agreement in their practices.[15] It was genuinely possible to describe certain practices as "prevailing," even if not unanimous.

Clearly from the foregoing discussion, it is now impossible to identify any one style of practice as customary or prevailing. What prevails is an incredible array of practices. Neither can we simply point to two or three stratified tiers of care and ask to which one of them the physician should be held. We cannot even identify any single most credible source of guidelines. After all, physicians as members of a group practice may not produce the same standards that they would endorse as members of a subspecialty organization. Managed care organizations, corporations, and other payers likewise can have very different views about what kinds of care are warranted, when. Malpractice insurers have their own agendas. And patients can hold a still different set of expectations. There are too many guidelines with too many variations, both gross and subtle, pointing in too many opposing directions, to identify any of them as dominant or even as distinctly more credible than any of the others.

Accordingly, it is now untenable to argue that physicians should be judged according to any one particular standard or to expect uniformity among physicians' practices. Anything in medicine that requires resources, from major surgery to physicians' time, is subject to widely discrepant views about how resources ought to be used. Physicians retain little power to decide who will receive what, and the economic agents who do have the power represent widely varying agendas. As a result, the legal system must reconceive the way in which it establishes physicians' obligations to their patients.

Concern for the way in which law sets the standard of care is not new. Previous commentators also questioned setting the standard of care empirically via prevailing routines. Some twenty years ago Joseph King pointed out that the actual patterns that emerge in the hectic conditions of actual daily practice are not nearly as good a source of standards as the style of care that physicians judge to be most appropriate after study and contemplation ("accepted practice") [36]. Concurrently Bovbjerg, recognizing that HMOs are designed to provide a very different approach to care than indemnity-insured providers, argued that HMOs should be held to their own separate standard of care [6].

Such views were on the right track, but are no longer directly applicable. Because costly resources play such a major role in medical care, and because there are so many approaches to resource use—each with its own set of values—King's concept of "accepted practice" would presuppose a level of agreement that can no longer be found even within the physician community, let alone among MCOs, corporations, governments, malpractice insurers, and patients. And Bovbjerg's separation of HMO standards from the rest of medicine seems to presuppose a degree of uniformity within these two sectors that likewise can no longer be found.

Neither can we resolve the problem by appealing to the law's traditional permission for diversity through "respectable minorities." Medical practice has become a cacophony of minorities. Virtually every published or informal in-house guideline, every *ad hoc* local routine, represents such a minority. And it has become essentially impossible to determine which ones are "respectable" and which are not, because to do so would require agreement about basic criteria by which to judge which guidelines and practices are acceptable. Unfortunately, it is unlikely that agreement could ever be found on such criteria. Various guidelines embody different sets of values and they rank priorities differently: the high-cost needs of desperate individuals versus the more mundane needs of the wider group, short-term versus the long-term results, cost-effectiveness versus innovation, and so forth. Because all these values are important, and because a wide number of different priorities can be defensible, any attempt to choose one set of criteria over another would assuredly amount to a political battle in which victory comes through clout rather than rational discourse.

One major option remains. In earlier work on this topic, I described the growing gap between what the legal system says physicians owe their patients and what physicians actually can deliver, given the fast-changing economic scene [52] [49] [51]. I proposed that we divide the standard of care into two distinct components: the professional expertise that physicians owe the patient, versus the resources that others owe.

Physicians can only owe what is theirs to give—their knowledge, skill, and effort and, in the realm of resources and patients' access to care, their diligent advocacy. This I called the Standard of Medical Expertise, or SME. Tort law should govern physicians' duties under the SME. Like the traditional malpractice standard of care, the SME would be a relatively uniform standard asking physicians to render the same level of professional skill, knowledge, and effort to all whom they accept for care.

On the other side are insurers, MCOs, governments, businesses, and other parties with specific obligations to furnish patients with financial or medical resources. Theirs is the Standard of Resource Use, or SRU. The level of resources to which someone is entitled should generally be a function of what level he actually purchased, or that was purchased on his behalf. Hence, the SRU would mainly be conceived as a matter of contract, not tort as in the SME. Tort liability might also apply, if for instance an insurer or MCO maliciously or recklessly harms a patient by denying benefits it clearly owes.[16]

There is much to recommend this approach. It is difficult to justify holding physicians legally liable for others' resource decisions, or expecting them to commandeer others' money and property in pursuit of an Artesian ideal of medicine that no longer can claim moral exclusivity as the obviously best form of practice. Particularly now, given the plethora of guidelines under which physicians are expected to practice, it makes far more sense to hold the creators and administrators of the various funding packages and guidelines, rather than physicians, accountable for the level of resources they authorize or provide for each patient.

Arguably, this divided standard of care is the only approach that is now tenable. Current economics have led to such a chaotic diversity of practice patterns that any claim to describe prevailing practices has lost credibility except in the most obvious situations, as where a physician reads an X-ray backwards, fails to ask the patient about allergies, or the like. Notably, these obvious cases typically concern medical expertise—the knowledge and skills that any credible physician should possess. Some cases, of course, could concern resources, if for instance a plan refused to approve basic antibiotics for a bacterial infection. But overall, any such obvious cases stand in stark contrast to the rapid shifts in resource rules that preclude most generalities about appropriate or expected levels of resources. Only with a divided standard of care can we, on the one hand, continue to expect physicians to devote careful thought to every patient and, on the other, acknowledge that no longer can all patients be legally entitled to receive the same level of resources and that physicians are not the ones who provide or control those resources.

Even here, however, economic factors introduce further complications. First, not everyone has legal entitlement to a morally adequate level of care. Standards of adequacy differ, as we have seen. But the fact remains that some citizens have access only to medical stabilization in an emergency room, and few would argue that emergency care alone represents adequate care. This poses a problem. On the one hand, it is difficult to insist that private health-care providers somehow are legally obligated to deliver costly care for which they will not be paid; on the other hand, surely the uninsured patient is at least morally, even if not legally, entitled to receive better. The problem that was once covered by cost-shifting is now an open sore, as payers increasingly refuse to pay higher charges to reimburse the care for people outside their sphere of strict obligation.

Further, in many cases it is morally difficult to argue that patients should always be strictly held to the terms of their health-plan contracts. Many people have no choice over their health plan. Nearly eighty-four percent of firms that provide health coverage offer only one health plan ([3], p. 243). And many other employers offer only a couple of choices—one insurance plan and one MCO, for instance. There is usually little opportunity for beneficiaries to negotiate the terms of such health plans, as they are mostly offered on a take-it-or-leave-it basis. Accordingly, it is difficult to argue that patients should be held to the contractual terms of health plans that others have chosen. Fairly enforcing the SRU becomes more complicated than originally envisioned.[17]

The problem is compounded if we suppose that health plans ought to bear tort liability when an illegitimate denial of medical resources results in harm to a patient. In principle, there is no obstacle to such liability, and insurers and MCOs have been held liable for such torts as wrongful death, intentional infliction of emotional distress, fraud, and bad faith breach of contract.[18] However, there is in fact a major obstacle. The Employee Retirement Income Security Act (ERISA) is a federal law designed to protect employee benefits such as retirement pensions by bringing them under uniform federal regulations. As federal law, ERISA preempts state laws and, particularly pertinent here, it preempts all state-based tort litigation against health plans that are acquired as employee benefits—including claims such as wrongful death, bad faith breach of contract, or negligent infliction of emotional distress. Therefore, even if a health plan commits numerous and serious offenses that would clearly be torts under state law, such claims are preempted so long as that plan is an employee benefit. At most, the patient can sue for limited federal damages, such as requiring the payer to pay the cost of care it initially refused to reimburse.[19] Because most nonelderly citizens' health insurance comes through the workplace, this means that most people cannot hold their health plans liable in tort for causing injury through negligent utilization review or claims administration.

These challenges do not defeat the notion that the medical standard of care should now separate matters of medical expertise from resource issues. Whatever the current difficulties for holding health plans accountable in contract or tort, the basic point is that resource issues should be separated from questions of professional expertise: physicians should be liable for their own conduct, not for others' resource decisions. And plans should be held, not to what someone thinks they should ideally cover, but according to what they have agreed and been paid to cover. What these challenges do mean, however, is that further adjustments will have to be made before we can fully define and enforce the SRU.[20]

V. MULTIPLE STANDARDS: ETHICAL CHALLENGES

Artesian funding kept the moral question of what the physician owes the patient relatively simple, economically and morally. Issues of justice do not arise where there is no scarcity. And where physicians are paid to do virtually whatever they want, issues of divided loyalty are minimal. There has always, of course, been the well-known conflict of interest implicit in fee-for-service remuneration. But so long as money was no object, the physician could usually be unequivocally devoted to each patient's interests, only rarely worrying about resources [51] [53]. The current economic upheaval utterly precludes such moral simplicity, as physicians now have entered into new affiliations that bring widely diverging and sometimes incompatible duties toward numerous parties.

Membership in an MCO, or partnership in a group practice, for instance, brings contractual promises and thereby moral obligations to honor the rules and promote the interests of the organization. At the same time, one's malpractice insurance may require a commitment to abide by a highly specific set of guidelines that can, on occasion, directly conflict with those that the physician has promised to follow as part of his employment contract. And either set of standards can conflict with patients' interests.[21]

And one patient's interests can conflict with the needs of a greater population that a physician is obligated to serve. Although it is traditional to suppose that physicians should be devoted exclusively to their own patients' interests, David Eddy argues that physicians should see themselves as part of a team whose goal is the health of an entire defined population. On this view, physicians must understand that their own patients have no intrinsic priority over other members of the group, and that justice may require that the best interests of their own patient must sometimes give way to broader concerns.[22]

Such conflicts among moral obligations to individual patients, populations of patients, financial agents, and medical groups are also accompanied by conflicts of interest that have grown in both size and scope. The familiar conflicts embedded in fee-for-service medicine have become stronger than ever, and added to them are some powerful newer conflicts, as incentive schemes reward physicians financially for containing costs by reducing care.

As these changes first emerged, it became evident that part of the resolution is to recognize that physicians have lost much of the control they once had over resources, and that this alters their moral obligations. Instead of conscripting others' resources, they should take on the role of intercessor and advocate, helping patients to receive the care to which they are entitled in an increasingly complex health-care system [51].

However, as the economic upheaval becomes increasingly turbulent, even the mandate of advocacy now requires further qualification. Ardent advocacy can carry a heavy price, and it has become essential to recognize reasonable limits on this duty. Institutional "hassle factors" can pose prohibitive costs to physicians' time and energy. Physicians who protest utilization review decisions too vigorously or too frequently may find themselves deselected from managed care organizations, and those who are deselected may not be accepted into other MCOs. In areas of the country that are heavily penetrated by managed care, those physicians may find themselves unemployable. Indeed, this problem was perceived to be of such severity in the state of California that the legislature enacted a statute providing some protection for physicians who can show that their deselection was occasioned by their efforts to advocate on their patients' behalf.[23] And in Connecticut, patients have actually joined their physicians in suing an MCO over physician deselection.[24] In other settings, physicians who protest funding denials too frequently may even find themselves *persona non grata* with their own institutions. For instance, if a physician succeeds in securing extra funding for his inpatients too often, he may give an MCO sufficient reason to deselect his entire hospital as one of its preferred providers.[25]

One cannot resolve this challenge simplistically, for example, by proposing that the physician need only ensure that his contract with the MCO or group practice allows him to be a vigorous advocate and fully ethical physician. In many cases, physicians have little flexibility to negotiate the terms of those contracts. In some areas of the country a few large, powerful organizations control most, if not all, health care in the region. If virtually all patients' employers or third-party payers direct them to these providers, individual physicians may have little leverage to negotiate the terms of their contracts. And if physicians band together to insist on more favorable contractual arrangements they could, under some circumstances, offend antitrust law [19].

Further hazards emerge from the fact that many plans do not openly disclose to patients their utilization limits or the incentives they place on physicians. Hence, patients expecting the traditionally lavish Artesian level of care may be annoyed when told that they cannot have the antibiotic or diagnostic test they expected. Additionally, since health plans that are acquired as a workplace benefit are mostly immune from state-based tort suits,[26] the physician may be the only one available to sue if the patient is unhappy with his care.[27]

In sum, greater economic complexity has brought increased moral complexity. Thirty years ago, unlimited funding permitted relatively simple obligations for physicians to serve their own patients and refrain from exploiting them for profit. Now, however, constrained funding has given rise to a plethora of conflicting obligations to patients, colleagues, payers, and populations, and to increasingly pervasive and difficult conflicts of interest.

It even prompts reassessment of what medicine is all about. What is medically "indicated," what a patient "needs," is very much a function of who is asked, what goals of care have been established, and what particular cost/benefit values have been embraced. And no one set of values is obviously superior. It is good to seek the best health of a population, but when resources are limited this goal will inevitably deprive some patients of some care from which they would benefit, and it will require individual physicians sometimes to do less than they could for their own patients. Similarly, it is not obviously wrong for an MCO to devote its resources to interventions of demonstrated value for ordinary illnesses, rather than to spend large amounts on promising but unproven treatments for the most desperately ill. But again, the physician who goes along with this priority will sometimes forego the best interest of his own patients in favor of others' benefit.[28]

We may likewise need to reassess what it means for physicians to be professionals. Standard definitions cite independence of judgment and action, and point to esoteric knowledge, applied with judgment to varying individual cases, as hallmarks of professionalism. A moral commitment to service and a priority on the needs of the patient over one's own interests are further defining characteristics. However, such independence and individuality of care are steadily eroded by the guidelines of insurers, MCOs, malpractice insurers, and group practices, in favor of more generalized approaches that can serve the group better at the expense of some individuals. And with the rise of incentive systems and threats of deselection, it is virtually impossible for physicians to give patients' interests unswerving priority.

Undoubtedly first steps toward redefining medical professionalism will require recognition that Artesian financing spawned unrealistic expectations in the first place. Other professions, whether law, ministry, or architecture, have not required their members to commandeer virtually limitless amounts of others' resources, partly because only in medicine has there even temporarily been such lavish funding for nearly all citizens. Still, careful assessment of professionalism will be required in the case of medicine, again because of its economic peculiarities. Other professions do not rely on massive third-party financing that inevitably leads to the sharing of authority and decision making we find in medicine today. Hence, other professions do not find their members placed between the enormous conflicting pressures created by such arrangements. Our description of what is obligatory for physicians as professionals must rely on what is possible, not just on what is ideal.

VI. CONCLUSIONS

At this point several things should be clear. First, it is virtually impossible to provide any single, credible answer to the question what standard of care physicians legally owe their patients. Old answers have been displaced by economic changes. Courts may not yet recognize the problem, but it is nonetheless real.

Second, we cannot adequately determine what physicians' legal obligations ought to be until we get a clearer idea how their moral obligations should be conceived. Much of our law, after all, is a reflection and enforcement of our most basic moral concepts. Unfortunately, it is now terribly unclear what physicians morally owe their patients. Physicians must have a significant role in suggesting appropriate resource use, because they are uniquely qualified to evaluate the likely consequences of providing, versus not providing, particular interventions under specified conditions. On the other hand, such recommendations also involve important decisions about values and priorities that are the product, not of medical expertise, but of decisions about what kinds of medical outcomes are most important for what range of people, and how medical goals ought to be weighed against other possible uses of funds. Even the simpler claim that physicians ought to be advocates to help patients obtain needed resources has become clouded by the sometimes very large personal price that physicians may pay for such efforts.

Third and perhaps most important, simplistic answers will not do. Good ethics begins with good facts, and good moral problem-solving will not happen unless we fully understand the problem. In this area, the problem has become strikingly complicated. The bioethicist cannot offer morally credible resolutions unless they are also practically workable. And, as the foregoing account should make obvious, this cannot be done without a broad understanding of the many factors that compose the situation. One must understand the changes on a global level, but must also appreciate the specific details of each situation. Truly, it requires an intellectual "cross-dressing" that spans medicine, law, economics, and ethics.

Department of Human Values and Ethics
University of Tennessee, College of Medicine
Memphis, Tennessee, U.S.A.

Acknowledgement: The author is grateful for the very helpful comments provided on earlier drafts of this manuscript by William Winslade, Ronald Carson, Chester Burns, and the participants in the William Bennett Bean Symposium on Philosophy and Medicine.

NOTES

1. *Stare decisis*, or "let the decision stand," is the legal precept that appellate level judicial decisions set precedents that must be followed in subsequent similar cases within a given jurisdiction.

2. For a more detailed discussion of medicine's economic changes and their ethical implications, see [51]. For an excellent summary of basic elements and variations in managed care, see [71].

3. For further discussion of these entities see, e.g., ([64], pp. 370-71).

4. Another term, allegedly preferred by MCOs, is "decertification." See [12].

5. Some room for flexibility is permitted. Deviation from prevailing practice is seen to be below the standard of care, so long as the physician's care is consistent with the practices

of some "respectable minority" viewpoint within the profession. For further discussion, see ([37], p. 65).

6. Technologies were often incorporated before their most effective or cost-effective use had been determined. See, e.g., [25] ([26], p. 1638); [75] [8] [63].

7. Proponents of this view argue that physicians should remain under legal pressure to deliver the best possible care to their patients, so that they will aggressively seek the resources their patients need. An easing of liability could invite a less diligent advocacy. See, e.g., [21].

8. As of 1989, utilization-review panels refused to authorize payment for proposed medical interventions only one to two percent of the time. See ([24], p. 77).

9. For a discussion of some of these views, see [27].

10. Note, these "managed standards" are not confined exclusively to managed care organizations. Traditional indemnity insurers likewise can establish guidelines they expect physicians to follow. In 1993, for example, Blue Cross and Blue Shield of Illinois made adherence to a set of clinical practice guidelines part of its contract with physicians. The organization did not specify sanctions for physicians who fail to comply, but it did indicate that those who consistently deviate despite the company's educational and persuasive efforts may be dropped from the Blues' network. See [4].

11. The Employee Retirement Income Security Act (ERISA) of 1974 generally preempts state laws governing employee benefits, including health-care benefits. Some of those state laws include mandates that health insurers doing business in that state cover specified services—from in vitro fertilization in Massachusetts and many other states, to wigs for alopecia in Minnesota. Because these insurance mandates are costly, and because corporations can avoid them by self-insuring rather than buying commercial insurance, businesses have a powerful incentive to cover employees' health costs directly.

12. As one commentator observed, high-quality routine care for diabetic patients is much costlier in the short term, but can save considerably in the long term, if patients have fewer episodes of diabetic ketoacidosis, renal failure, blindness, nonhealing sores, amputations, and the like. However, if the patient is only likely to be in the MCO a year or two, then the higher immediate costs of good routine care may result in savings, not for this MCO, but for some other payer. It thus may not be in the MCO's financial interests to deliver the higher quality preventive care. See [58].

13. See [56]. Some states are establishing medical standards in hopes of reducing malpractice litigation. Maine is in the midst of a five-year Medical Liability Demonstration Project. In the areas of gynecology, emergency medicine, anesthesiology, and radiology, physicians who volunteer for the program and follow established guidelines will be immune from liability. See ([67], p. 83; [10], p. 18-19). Similar programs are underway in Minnesota and Florida. See ([56], p. 43).

14. An obstetrics protocol, for instance, may ask the physician to watch and wait for a certain length of time before attempting a cesarean delivery. However, if the region's only anesthesiologist is about to go home for the night, it may make considerably more sense to curtail the protocol a bit. See ([67], p. 74).

One glimmer of relief has appeared recently from the Colorado Supreme Court. For medically sound reasons, a physician had declined to follow the guidelines of his malpractice insurer, COPIC, in diagnostic evaluation of a breast mass. The mass turned out to be malignant, but in subsequent litigation the Colorado court ruled that malpractice insurers' guidelines were not admissible as evidence. The guides were promulgated by a private insurance company and did not reflect any general standard of care. Further, the jury might have been prejudiced by any mention of the guidelines, since that would disclose that the doctor had insurance coverage. See [56] [10].

[15.] As noted above, these values included: it is better to do too much than too little; if an intervention might possibly help the patient, and is unlikely to harm, it should be used; no one should be denied a potentially beneficial intervention on account of inability to pay; high-tech is better than low-tech is better than no-tech.

[16.] Where health coverage is obtained through the workplace, such suits are generally preempted by federal ERISA law, a point to be discussed just below.

[17.] It can be argued that one important approach may be to provide patients with considerably more choice over their health plans than most people currently enjoy. Various commentators have proposed an array of ways in which to accomplish this. In the medical savings account (MSA), for example, patients could purchase high-deductible catastrophic policies and concurrently establish tax-free MSAs to cover the costs of that deductible; on this approach, because the patient directly pays for routine medical expenses, he would enjoy considerably more control over his own money and medical care. See [22] [23].

On a different, but not incompatible, approach, people might be able to select among tiers of coverage: a basic package covering only care that is demonstrated to be safe, effective and cost effective; a more generous package covering care that is safe and effective, but not necessarily maximally cost effective; and a high-end package that would also add the most innovative technologies not yet proven to be effective. See [35] [26]. Other approaches are also available for empowering patients to make more of their own resource decisions. We need not choose among them here. The important point is that, the more choice the patient has exercised over his health plan, the more plausible it is to hold him to the terms of his SRU.

[18.] See [72] [73] [74]. For further discussion of bad faith breach of contract, see [50].

[19.] For further discussion, see [55].

[20.] One way of bringing greater fairness and accountability to both patients and health plans is to place greater economic control, and consequences, in patients' hands. Some of these approaches are described in [22] [31] [54].

[21.] Guidelines can rarely define optimal care with certainty, due to poor science, imperfect analytic processes, and differences in patients. Recommendations are often worded in highly specific language that achieves clarity at the expense of scientific validity. Rigid enforcement of such guidelines could harm patients, interfere with individualization of care, increase costs, and promote unfair judgments against clinicians who deviate from them for good reasons ([77], p. 2646). See also ([67], p. 72 ff).

Guidelines can also be constructed with little interest in patients' welfare. In some cases, for instance, payers or business corporations hire outside utilization review agencies, who are so sure that their unpublished, proprietary criteria will save money that they are willing to perform their services on a contingency basis, garnering a percentage of whatever they save. See ([67], p. 68).

[22] More specifically, in his "Principles for making difficult decisions in difficult times," Eddy recommends in his principle #5 that: "The objective of health care is to maximize the health of the population served, subject to the available resources" ([16], p. 1793). In his principle #6 he proposes: "The priority a treatment should receive should not depend on whether the particular individuals who would receive the treatment are our personal patients" ([16], p. 1796). See also [14] ([64], p. 375).
[23] See [40]. Such legislation admittedly provides rather little protection for physicians. Since many MCO contracts are "at will" contracts in which either party can end the relationship without cause, it may be difficult for the physician to establish that his advocacy efforts were the reason he was deselected by the MCO.
[24] See [42]. Although similar suits in Texas have not been successful, the addition of patients to the list of plaintiffs adds a distinctive element to the new litigation.
[25] "Practitioners who persuade third party payers to extend benefits not only antagonize these companies, they often become enemies of the institutions for which they work. The hospital, clinic, or agency needs managed care dollars to survive and outspoken patient care advocates jeopardize relationships with the companies that control these funds" ([76], p. 393).
[26] Federal ERISA law preempts state-based common law suits where health plans or other benefits are derived through the workplace. Precluded, for instances, are suits for wrongful death, bad faith breach of conflict, fraud, intentional or negligent infliction of emotional distress, and the like. See [55] [54]. See below for further discussion of ERISA.
[27] See *Harrell v. Total Health Care, Inc.*, [29]. On page 62 of the decision the court notes that the ERISA statute "does not deny the plaintiff a remedy for the wrong done to her. She has her right of action against the negligent surgeon. The statute simply limits her access to an additional pocket. In so doing, it serves a proper legislative purpose. We know of no case holding that the legislature may not determine which persons and corporations are liable for the consequences of medical malpractice."

Physicians can also encounter other legal hazards in the course of moral advocacy. Recent evidence indicates that published guidelines, such as those promulgated by subspecialty organizations, can be used as a litigation weapon against the physician who deviates from them [20] [10]. Unfortunately, the court testimony rarely reveals that the physician was practicing under multiple competing, often incompatible, guidelines [41].
[28] For further discussion of the value trade-offs implicit in competing guidelines, and the mixed allegiances that physicians face as they consider when to honor guidelines and when to pursue patients' best interests, see [13] [14] [15].

REFERENCES

1. Ayres, W. H. : 1994, 'Dilemmas and challenges: A clinician's perspective', *Journal of the American Academy of Child and Adolescent Psychiatry* 33, 153-157.
2. Bishop, J. E.: 1994, 'Some doctors denounce U. S. guidelines on drugs to treat high blood pressure', *Wall Street Journal*, 11/17, p. B-12.

3. Blendon, R. J., Brodie, M., and Benson, J.: 1995, 'What should be done now that national health system reform is dead?', *Journal of the American Medical Association* 273, 243-244.
4. Borzo, G.: 1993, 'Illinois blues: If you sign a contract, you follow guidelines', *American Medical News,* 12/6, p. 8.
5. Borzo, G.: 1994, 'Rhode Island doctors face "absurd" inpatient limits', *American Medical News*, 3/21, pp. 1, 9.
6. Bovbjerg, R.: 1975, 'The medical malpractice standard of care: HMOs and customary practice', *Duke Law Journal*, 1375-1414.
7. Brennan, T. A., Leape, L. L., and Laird, N. M.: 1991, 'Incidence of adverse events and negligence in hospitalized patients: Results of the Harvard Medical Practice Study I', *New England Journal of Medicine* 324, 370-376.
8. Burnum, J. F.: 1987, 'Medical practice a la mode', *New England Journal of Medicine* 317, 1220-1222.
9. Council on Ethical and Judicial Affairs, American Medical Association.: 1995, 'Ethical issues in managed care', *Journal of the American Medical Association* 273, 330-335.
10. Crane, M.: 1994, 'When doctors are caught between dueling clinical guidelines', *Medical Economics* (general surgery edition), 13(9), 16-20.
11. Crittenden, D.: 1994, 'Don't give birth up here', *Wall Street Journal,* 3/31, p. A-14.
12. Dunn, L. F.: 1994, 'Beware: Managed care is gaining the upper hand in law', *American Medical News,* Dec 5, pp. 45, 48.
13. Eddy, D. M.: 1991, 'The individual vs. society: Is there a conflict?', *Journal of the American Medical Association* 265, 1446, 1449-1450.
14. Eddy, D. M.: 1993, 'Broadening the responsibilities of practitioners: The team approach', *Journal of the American Medical Association* 269, 1849-1855.
15. Eddy, D. M.: 1993, 'Three battles to watch in the 1990s', *Journal of the American Medical Association* 270, 520-526.
16. Eddy, D. M.: 1994, 'Principles for making difficult decisions in difficult times', *Journal of the American Medical Association* 271, 1792-1798.
17. Entman, S. S., Glass, C. A., Hickson, G. B., Githens, P. B., Whetten-Goldstein, K., and Sloan, F. A.: 1994, 'The relationship between malpractice claims history and subsequent obstetric care', *Journal of the American Medical Association* 272, 1588-1591.
18. Farmer, A. L: 1993, 'Medical practice guidelines: Lessons from the United States', *British Medical Journal* 397, 313-317.
19. Felsenthal, E.: 1994, 'Doctors seek right to join forces to negotiate with health plans', *Wall Street Journal,* 1/3, p. B-10.
20. Felsenthal, E.: 1994, 'Doctors' own guidelines hurt them in court', *Wall Street Journal,* 10/19, pp. B-1, B-9.
21. Furrow, B. R.: 1985-86, 'Medical malpractice and cost containment: Tightening the screws', *Case Western Reserve Law Review* 36, 985-1032.
22. Goodman, J., Musgrave, G.: 1994, *Patient Power*, Cato Institute, Washington, D.C.

23. Gramm, P.: 1994, 'Why we need medical saving accounts', *New England Journal of Medicine* 330, 1752-1753.
24. Gray, B. H., and Field, M. J.: 1989 (eds.), *Controlling Costs and Changing Patient Care? The Role of Utilization Management*, National Academy Press, Washington, D.C.
25. Grimes, D. A: 1993, 'Technology follies: The uncritical acceptance of medical innovation', *Journal of the American Medical Association* 269, 3030-3033.
26. Hall, M. A., and Anderson, G. F.: 1992, 'Health insurers' assessment of medical necessity', *University of Pennsylvania Law Review* 140, 1637-1712.
27. Hall, M. A.: 1994, 'The ethics of health care rationing', *Public Affairs Quarterly* 8, 33-50.
28. Hardison, J. E.: 1979, 'To be complete', *The New England Journal of Medicine* 300, 193-194.
29. *Harrell v. Total Health Care, Inc.*, 781 S.W.2d 58 (Mo. 1989) (en banc).
30. Havighurst, C.: 1992, 'Prospective self-denial: Can consumers contract today to accept health care rationing tomorrow?', *University of Pennsylvania Law Review* 140, 1755-1808.
31. Havighurst, C. C.: 1995, *Health Care Choices: Private Contracts as Instruments of Health Reform*, American Enterprise Institute, Washington, D.C.
32. Hickson, G. B., Clayton, E. W., Entman, S. S., Miller, C. S., Githens, P. B., Whetten-Goldstein, K., and Sloan, F. A.: 1994, 'Obstetricians' prior malpractice experience and patients' satisfaction with care', *Journal of the American Medical Association* 272, 1583-1587.
33. Hickson, G. B., Clayton, E. W., Githens, P. B., and Sloan, F. A.: 1992, 'Factors that prompted families to file medical malpractice claims following perinatal injuries', *Journal of the American Medical Association* 267, 1359-1363.
34. Hirshfeld, E.: 1994, 'The case for physician direction in health plans', *Annals of Health Law* 3, 81-102.
35. Kalb, P. E.: 1990, 'Controlling health care costs by controlling technology: A private contractual approach', *Yale Law Journal* 99, 1109-1126.
36. King, J. H.: 1975, 'In search of a standard of care for the medical profession: The "accepted practice" formula', *Vanderbilt Law Review* 28, 1214-1276.
37. King, J. H.: 1986, *The Law of Medical Malpractice*, West, St. Paul, MN.
38. Kolata, G.: 1995, 'Women resist trials of marrow transplants', *New York Times*, Feb. 2, pp. A-1, B-7.
39. Leape, L. L., Brennan, T. A., and Laird, N. M.: 1991, 'Incidence of adverse events and negligence in hospitalized patients: Results of the Harvard Medical Practice Study II', *New England Journal of Medicine* 324, 377-384.
40. McCormick, B.: 1994, 'What price patient advocacy?', *American Medical News*, 3/28, pp. 1, 13, 14.
41. McCormick, B.: 1994, 'Managed care posing new liability risks, insurers warn', *American Medical News*, 6/13, pp. 3, 54, 55, 57.
42. McCormick, B.: 1994, 'Patients, doctors sue CIGNA in deselection flap', *American Medical News*, 9/16, pp. 3, 36.

43. Meyer, M., Murr, A.: 1994, 'Not my health care', *Newsweek*, 123(2), 36-38.
44. Morreim, E. H.: 1983, 'The philosopher in the clinical setting', *The Pharos* 46(1), 2-6.
45. Morreim, E. H.: 1985-86, 'Stratified scarcity and unfair liability', *Case Western Reserve Law Review* 36(4), 1033-1057.
46. Morreim, E. H.: 1986, 'Philosophy lessons from the clinical setting: Seven sayings that used to annoy me', *Theoretical Medicine* 7, 47-63.
47. Morreim, E. H.: 1987, 'Cost containment and the standard of medical care', *California Law Review* 75(5), 1719-1763.
48. Morreim, E. H.: 1988, 'Cost constraints as a malpractice defense', *Hastings Center Report* 18(1), 5-10.
49. Morreim, E. H.: 1989, 'Stratified scarcity: Redefining the standard of care', *Law, Medicine and Health Care* 17, 356-367.
50. Morreim, E. H.: 1990, 'Physician investment and self-referral: Philosophical analysis of a contentious debate', *The Journal of Medicine and Philosophy* 15(4), 425-448.
51. Morreim, E. H.: 1991, *Balancing Act: The New Medical Ethics of Medicine's New Economics*, Kluwer, Dordrecht, The Netherlands.
52. Morreim, E. H.: 1992, 'Rationing and the law', in Strosberg, M. A., Wiener, J. M., Baker, R., and Fein, I. A. (eds.), *Rationing America's Medical Care: The Oregon Plan and Beyond*, Brookings Institution, Washington, D.C.
53. Morreim, E. H.: 1994, 'Redefining quality by reassigning responsibility', *American Journal of Law and Medicine* 20, 79-104.
54. Morreim, E. H.: 1996, 'Diverse and perverse incentives in managed care', *Widener Law Symposium Journal* 1, 89-139.
55. Morreim, E. H.: 1995, 'Moral justice and legal justice in managed care', *Journal of Law, Medicine, and Ethics* 23(3), 247-265.
56. Oberman, L.: 1994, 'Risk management strategy: Liability insurers stress practice guidelines', *American Medical News*, 9/5, pp. 1, 42, 43.
57. Orient, J.: 1994, *Your Doctor Is Not In: Healthy Skepticism about National Health Care,* Crown, New York, NY.
58. Page, L.: 1995, 'Can plans manage the preventive care diabetics need?', *American Medical News*, March 20, pp. 4-5.
59. Perry, K.: 1994, 'Where salaried practice feels like private practice', *Medical Economics* 71(17), 65-75.
60. Peters, W. P., and Rogers, M. C.: 1994, 'Variation in approval by insurance companies of coverage for autologous bone marrow transplantation for breast cancer', *The New England Journal of Medicine* 330, 473-477.
61. Reinhardt, U.: 1992, 'American values: Are they blocking health-system reform?', *Medical Economics* 69(21), 126-148.
62. Reinhardt, U.: 1992, 'You pay when business bankrolls health care', *Wall Street Journal,* Dec 2, p. A-14.
63. Reuben, D. B.: 1984, 'Learning diagnostic restraint', *The New England Journal of Medicine* 310, 591-593.

64. Sederer, L. I.: 1994, 'Managed mental health care and professional compensation', *Behavioral Sciences and the Law* 12, 367-378.
65. Slomski, A. J.: 1994, 'How business is flattening health costs', *Medical Economics* 71(13), 87-100.
66. Sox, H. C., and Woolf, S. H.: 1993, 'Evidence-based practice guidelines from the US preventive services task force', *Journal of the American Medical Association* 269, 2678.
67. Stevens, C.: 1993, 'Guidelines spread, but how much impact will they have?', *Medical Economics* 70(13), 66-89.
68. Terry, K.: 1994, 'Look who's guarding the gate to specialty care', *Medical Economics* 71(16), 124-132.
69. Terry, K.: 1994, 'Is this the best way to divide HMO income?', *Medical Economics* 71(19), 26B-26F.
70. Weaver, J.: 1992, 'The best care other people's money can buy', *Wall Street Journal*, Nov 19, p. A-14.
71. Weiner J. P., and de Lissovoy, G.: 1993, 'Razing a tower of babel: A taxonomy for managed care and health insurance plans', *Journal of Health Politics, Policy and Law* 18(1), 75-103.
72. *Wickline v. State of California*, 228 Cal. Rptr. 661 (Cal. App. 2 Dist. 1986).
73. *Wickline v. State of California*, 192 Cal. App. 3d 1630 (1987).
74. *Wilson v. Blue Cross of Southern Cal.*, 271 Cal. Rptr. 876 (Cal. App. 2 Dist. 1990).
75. Wong, E. T., and Lincoln, T. L.: 1983, 'Ready! Fire!. . . Aim!', *Journal of the American Medical Association* 250, 2510-2513.
76. Wooley, S. C.: 1993, 'Managed care and mental health: The silencing of a profession', *International Journal of Eating Disorders* 14, 387-401.
77. Woolf, S. H.: 1993, 'Practice guidelines: A new reality in medicine. III. Impact on patient care', *Archives of Internal Medicine* 153, 2646-2655.
78. Zoloth-Dorfman, L.: 1994, 'Standing at the gate: Managed care and daily ethical choices', *Managed Care Medicine* 1(6), 26-38.

WILLIAM J. WINSLADE

INTELLECTUAL CROSS-DRESSING: AN ECCENTRICITY OR A PRACTICAL NECESSITY? COMMENTARY ON MORREIM

Haavi Morreim's provocative title and, more importantly, her probing analysis of the legal and ethical challenges to physicians as a result primarily of changes in health-care financing in the United States, raise several important questions [1]. What is required to understand current changes in health-care services and delivery? How do these changes affect the legal standards of care relevant to medical malpractice? What impact do economic changes in the health-care system have on the ethical standards regulating the physician-patient relationship? Morreim offers plausible but incomplete answers to these questions. After critiquing them, I will take her conclusions as a point of departure for an exploration of additional implications for bioethicists, physicians, and patients. In particular, I will offer some suggestions about the education of patients and physicians as "intellectual crossdressers."

I. UNDERSTANDING MEDICAL PRACTICE

The complexity of medical practice is not, as Morreim points out, reducible to principles or formulas. Much more than many other human interactions, medical practice includes scientific complexity as well as uncertainty, psychodynamics and power relationships, barriers to communication about injury, illness, and especially death. Legal regulations, professional standards, and fiscal limitations result in changes, if not chaos, in health-care financing. Legal and ethical issues must be analyzed in the context of these multiple factors, thus making the field of bioethics inherently interdisciplinary. Bioethicists must know how to find their way among several professions—medicine, law, ethics, and economics. But bioethicists must also know the factual details of current policies and practices.

R. A. Carson and C. R. Burns (eds.), Philosophy of Medicine and Bioethics, 327-334.

II. MEDICAL MALPRACTICE STANDARDS

Morreim's thesis about medical malpractice standards is that continuing economic and organizational changes in health-care delivery undermine the stability of prevailing practices in medicine. This, in turn, makes it difficult, if not impossible, to define legal standards to determine whether a particular course of conduct by physicians falls short of acceptable practice. "Ultimately it will be evident," Morreim writes, "that there is virtually no such thing as prevailing practice anymore, and the judiciary needs to look elsewhere for medical standards."

She goes on to show how standards of care are shaped by "varying economic perspectives and agendas" for the organization and delivery of health care. She distinguishes nine "standards," ranging from the lavish artesian standard of providing any potentially beneficial treatment regardless of cost through more circumscribed standards based on managed care, corporate, government, subspecialty, group practice, malpractice insurers,' patients' expectations, or individual clinicians' ad hoc standards. The multiplicity of standards results n decreasing control over the economic resources available to physicians in establishing practice patterns. Her proposal is that physicians can only reasonably be held legally accountable for their professional knowledge and expertise within the limits of the resources available to them. Others—insurers, managed-care organizations, businesses, or other payers—should be held accountable for allocation of resources.

Morreim's proposal would provide welcome legal relief for physicians who feel beleaguered by the tumultuous changes in the economics of health-care delivery. The elegance of her proposal lies in its sharp distinction between professional expertise and resource issues. But is this bifurcation of responsibility realistic? It sounds plausible in theory, but it becomes problematic in practice when Morreim turns her attention to ethical challenges for physicians.

III. ETHICAL CHALLENGES TO THE PHYSICIAN-PATIENT RELATIONSHIP

Occasionally, but rarely, patients and physicians stand in a simple dyadic relationship. A patient consults a physician for diagnosis and treatment. A physician makes a recommendation and is paid a cash fee for services. More often the patient and physician interact in the context of polyadic relationships with families, other health professionals, health-care institutions, third-party payers, government regulators, professional organizations, researchers, and educators. The relationships are not only polyadic, but also polyvalent. Physicians may experience multiple forces that exert varying degrees of influence over their loyalties and duties. Morreim reminds us that not only are the multiple relationships of physicians with payers, institutions, government, and other professionals in flux, but also the roles of physicians in their relationships with patients and families are undergoing revisions. Are physicians fiduciaries, advocates, friends, advisors, paternalists, or information sources for their patients? Are physicians independent professionals, employees, contractors, intermediaries, or mere subordinates of payers, institutions, or the government? In the midst of this uncertainty, what are patients legally and morally entitled to expect from physicians?

Morreim realizes that the "current economic upheaval" creates new and conflicting moral obligations for physicians. Physicians no longer can act solely as advocates for their patients. Nor are physicians able to serve as fiduciaries for their patients. Physicians must perform their professional services subject to the economic arrangements they or their patients have accepted by contract, regulation, or default. This forces physicians to rethink the nature and scope of their professional authority and even their identity as professionals.

Some commentators have argued for a new model for the physician-patient relationship, one that brings physicians as advocates and educators into a closer alliance with their patients to counter the social forces—primarily economic—that threaten to rupture their relationship [1]. If Morreim is correct—and I think she is—economic forces have already transformed the physician-patient relationship. One might yearn for family physicians who are friends, teachers, and advocates for their patients. But health care is already dominated by a business model in which patients are customers and consumers. Not long ago the fiduciary metaphor characterized the physician-patient relationship. Physicians' overriding loyalty to their patients was supposed to justify patients' trust and reliance on their physicians' recommendations. The fiduciary metaphor, as Marc Rodwin has pointed out, is at best strained and at worst obsolete [3]. Morreim's thesis implies that the fiduciary metaphor is not merely strained, but dead. As she puts it, "constrained funding has give rise to a plethora of conflicting obligations to patients, colleagues, payers, and populations, and to increasingly pervasive and difficult conflicts of interest." Furthermore, "with the rise of incentive systems and threats of deselection, it is virtually impossible for physicians to give patients' interests unswerving priority."

Morreim's brief discussion of ethical challenges to the traditional physician-patient relationship shows that the price of insulation from legal liability is indeed high. If physicians become subordinated to economic forces and mere functionaries of those who allocate resources, they not only lose control over the provision of medical services, they also lose their professional identity. Physicians become technicians who provide only services paid for by the customers' contracts or the economic arrangements that determine their level of care. This profound change in the authority of physicians to exercise their professional knowledge, skill, and clinical judgement is an unwelcome and undesirable implication of the economic alterations of the delivery of health care.

Morreim diagnosed ethical challenges, but she does not offer us much help in rethinking how to respond to them. She does say that "[p]hysicians must have a significant role in suggesting appropriate resource use, because they are uniquely qualified to evaluate the likely consequences of providing, versus not providing, particular interventions under specified conditions." This implies that physicians, not just bioethicists, must become intellectual cross-dressers to properly understand ethical obligations. In fact, although Morreim does not say this, I believe it is precisely because of the economic changes she so artfully describes that the education of bioethicists, physicians, and patients require a broader interdisciplinary vision.

IV. EDUCATING INTELLECTUAL CROSS-DRESSERS

When we become patients, most of us are woefully ignorant of our true medical needs. We may know that we have been injured or fallen ill, but we rarely know for sure what to do or even whether we need professional health care. When the problems don't go away or seem potentially serious we may seek advice, but not necessarily paid professional consultation. It is not only parsimoniousness that motivates us. We also fear bad news—and inconvenient procedures, or lengthy rehabilitation, or pain and suffering. So we often procrastinate or simply avoid health care until reason prevails or, more likely, discomfort drives us to seek help.

When some patients with sufficient insurance or economic reserves consult health professionals about a potential, feared, or actual ailment, they tend to focus on their health problem rather than economics. But even here economic considerations are not irrelevant or entirely ignored. Patients lacking economic resources may fail to seek health care or try to use intermittent emergency services as a substitute for an ongoing health-care relationship. These patterns are familiar to most of us who study or work in health-care settings. But most new or prospective patients lack this familiarity. Their ignorance makes them vulnerable. And vulnerable patients, even those who are generally well educated but specifically ignorant about health care, may feel confused, manipulated, exploited, or harmed.

One important but inadequate solution to this problem is, of course, patient education. But a major difficulty is that by the time persons become patients they are much less receptive to education. They just want their medical problems to be fixed and to go away. Health professionals and other educators face a daunting task of reforming health education at all levels. When persons become patients they need to know much more about what they are getting into, what to expect, and how to promote their needs and protect their interests. Health professionals need to collaborate with other educators, especially bioethicists, to help make patient education a reality rather than empty rhetoric. For example, the educational imperatives of the federal Patient Self-Determination Act are rarely realized, even if they are sincerely and seriously attempted. Imagine how difficult patient education becomes when health-care financing education is seriously undertaken. Morreim has ably demonstrated that bioethicists must be intellectual cross-dressers. So should health professionals. They must appreciate legal, ethical, and economic issues as well as understand the facts about patterns and practices in health-care financing. This knowledge must then be conveyed to others. Otherwise we serve neither ourselves, our students, nor our patients.

The educational mission must begin long before persons become patients. In theory, health education occurs in schools. In practice, most health education, like drivers' training, is inadequate, uninteresting, and sometimes misleading. The health-care community, in collaboration with educators, should revisit the aims and practices of health education as an essential phase of normal development, not merely a reaction to medical crisis or end-of-life decisions. If health education becomes an integral rather than a peripheral component of the education of children, the ignorance and vulnerability of persons who become patients may be decreased.

Bioethicists and other intellectual cross-dressers should begin to talk less only among themselves and more to others in health care as well as to the general public. For example, medical journalists, insurance executives, government policy makers, administrators, educators, and students need to know more about the goals of health care as well as legal, ethical, and economic issues, and other aspects of bioethics. It is important that issues such as confidentiality, informed consent, or end-of-life decisions be addressed prior to a personal crisis. Emotional reactions may overwhelm reflective and rational responses, such as when a family is unexpectedly confronted with end-of-life decisions after a life-threatening accident. Especially because the roles and relationships among health professionals and patients are rapidly evolving, thinking prior to a crisis (what John Dewey called "dramatic rehearsal") may help all parties cope more effectively with decisions in the midst of changing patterns of health care. When speaking to public groups I am reminded how interested but uninformed so many persons are about even the most elementary issues in health care and bioethics. Those of us who spend much of our professional lives thinking about health care forget that most people prefer not to think about it unless they can't avoid it. One of our public responsibilities in this era of health-care reform is to continually try to inform and educate others about options and implications of changes, not only in health-care financing but also in health-care services.

V. HEALTH PROFESSIONALS: THE NEED FOR NEW LIAISONS

The provision of health care in late twentieth-century America is not and should not be dominated by physicians. Physicians occupy a role in society that has transformed them from the apocryphal country doctor into a extensively trained technician with psychological as well as professional authority and power. But health care—preventive, diagnostic, primary, and various specialty services—requires the cooperation and collaboration of other health professionals as well as clerks, secretaries, technicians, consultants, administrators, institutions, and payers. Physicians must cope with an ever-increasing array of complex variables in order to achieve professional competence. Medical education is only slowly and grudgingly responding to the needs of trainees who must practice their profession in the midst of the alphabet soup so aptly described by Morreim.

The role of physician, like that of the bioethicist, today requires at least intellectual cross-dressing. It may also demand that physicians rethink their fundamental loyalties—to patients, to research, to institutions, to careers, to society. Different physicians may emphasize different professional roles. But, like patients, ignorance of the complexity of health care makes them vulnerable to confusion, manipulation, exploitation, and harm. Unlike patients, however, physicians have a professional duty to learn about the patterns as well as the pitfalls of health-care financing. Their professional integrity demands no less.

Physicians today must continue to forge new relationships with other health professionals, not merely for economic but also for moral and pragmatic reasons. Provisions of health-care services requires systematic efficiency that was neglected during the artesian era. This does not mean that physicians must abandon personal care of their patients, but it must be achieved in creative ways with the collaboration of other health professionals.

Physicians have a duty to educate their patients. It is also critical that they educate payers that health care is not only a technical manipulation but also personal care that requires attention to patients as persons—their bodies, their psyches, their families, and their lives. This takes time and it costs money. More importantly, it requires dedication to the goals of promoting, preserving, and enhancing the quality of life. Health care—like any other caring relationship—is a process that is only partially constituted by its products and outcomes. Unless physicians and other health professionals educate each other as well as patients and payers, we all suffer.

I want to divert some attention away from Morreim's legitimate concerns about malpractice standards to the more aspirational goals of health-care practitioners as educators, advocates, and fiduciaries of their patients' best interests. At times Morreim seems to teeter on the brink of acquiescing in the domination of health care by economic forces. At other times she points toward the critical importance of rethinking how to integrate inevitable economic factors with the maintenance of a health-care system that serves the health needs of the many rather than the avarice of the few. This cannot be achieved unless physicians—individually and collectively—resist the temptation to convert health care into a corporate model driven primarily, if not exclusively, by a profit motive.

Physicians and other health professionals are confronted with a formidable challenge. They must overcome their own longstanding rivalries to forge new professional collaborations to achieve the worthy but elusive goals of health care: cure of disease, care of the ill, restoration of function, and comfort for the dying. If health professionals bow to and are dominated by economic forces, dangerous liaisons indeed will ensue. This is an important theme in Morreim's paper that I have tried to articulate more fully.

The main reason that so much economic power has shifted away from health professionals to payers is that the costs of health care have increased much faster than demonstrated benefits. The fear that excessive attention to costs and profits may overcorrect for escalating costs at the expense of appropriate services is, however, a realistic concern. For this reason, it is important for health professionals and bioethicists to become informed about and responsive to health-care policies, changing patterns of health-care financing, and public and political discussion about such issues. Health professionals, administrators, and planners have a common cause with patients—and all of us are potential patients—to rethink health-care policies, financing, and reform with less concern for professional or personal selfinterest and more concern about public responsibility. This is more likely to occur if intellectual cross-dressing is not the exclusive province of bioethicists or a few health professionals. The issues are significant enough for us all that the public at large must be invited to learn about and participate in the discussions and debates that shape future health policy. Morreim has, through her insightful observations in this paper and other writings, shown that intellectual cross-dressing is not an eccentricity but a practical necessity. We all share the responsibility to bring this to the attention of public audiences.

Institute for the Medical Humanities
The University of Texas Medical Branch
Galveston, Texas, U.S.A.

REFERENCES

1. Balint, J., and Wayne, S.: 1996, 'Regaining the initiative: Forging a new model of the patient-physician relationship', *Journal of the American Medical Association* 275, 887-891.
2. Morreim, E. H.: 1996, 'At the intersection of medicine, law, economics, and ethics: Bioethics and the art of intellectual cross-dressing', in this volume, pp. 299-325.
3. Rodwin, M.: 1995, 'Strains in the fiduciary metaphor: Divided physician loyalties and obligations in a changing health care system', *American Journal of Law and Medicine* XXI, 241-257.

NOTES ON CONTRIBUTORS

Judith Andre, Ph.D., is Professor at The Center for Ethics & Humanities in the Life Sciences, Michigan State University, East Lansing, Michigan.

David Barnard, Ph.D., is University Professor and Chairman of the Department of Humanities at The Pennsylvania State University College of Medicine in Hershey, Pennsylvania.

Baruch Brody, Ph.D., is the Leon Jaworski Professor of Biomedical Ethics and Director of the Center for Medical Ethics and Health Policy at Baylor College of Medicine in Houston, Texas.

Courtney Campbell, Ph.D., is Director of the Program for Ethics, Science, and the Environment, and Associate Professor in the Department of Philosophy at Oregon State University at Corvallis, Oregon.

Ronald A. Carson, Ph.D., is Kempner Distinguished Professor of Humanities and Director of the Institute for the Medical Humanities at The University of Texas Medical Branch at Galveston, Texas.

Larry R. Churchill, Ph.D., is Chairman of the Department of Social Medicine at the University of North Carolina School of Medicine at Chapel Hill, North Carolina.

K. Danner Clouser, Ph.D., is University Professor of Humanities at the Pennsylvania State University College of Medicine in Hershey, Pennsylvania.

Thomas R. Cole, Ph.D., is Professor and Graduate Program Director at the Institute for the Medical Humanities at The University of Texas Medical Branch at Galveston, Texas.

Carl Elliott, M.D., Ph.D., is Assistant Professor at the McGill Centre for Medicine, Ethics, and Law at McGill University and Clinical Ethicist at Montreal Children's Hospital in Montreal, Canada.

H. Tristram Engelhardt, Jr., Ph.D., M.D., is Professor in the Departments of Medicine, Community Medicine, and Obstetrics and Gynecology, and a member of the Center for Medical Ethics and Health Policy at Baylor College of Medicine; Professor in the Department of Philosophy at Rice

University; and an Adjunct Research Fellow at the Institute of Religion in Houston, Texas.

Grant Gillett, FRACS, D.Phil. (Oxon), is Professor of Neurosurgery and a member of the Bioethics Centre at the University of Otago Medical School in New Zealand.

Anne Hudson Jones, Ph.D., is Professor of Literature and Medicine at the Institute for the Medical Humanities and Professor in the Department of Preventive Medicine and Community Health at The University of Texas Medical Branch at Galveston, Texas.

Loretta M. Kopelman, Ph.D., is Professor and Chairman of the Department of Medical Humanities at East Carolina University School of Medicine in Greenville, North Carolina.

Gerald P. McKenny, Ph.D., is Associate Professor in the Department of Religious Studies at Rice University in Houston, Texas.

E. Haavi Morreim, Ph.D., is Professor in the Department of Human Values and Ethics at The University of Tennessee College of Medicine in Memphis, Tennessee.

Edmund D. Pellegrino, M.D., is Director of the Center for Clinical Bioethics and John Carroll Professor of Medicine and Medical Ethics at the Georgetown University Medical Center

Stuart F. Spicker, Ph.D., is Professor at the Center for Medical Ethics and Health Policy (and the Department of Community Medicine) at Baylor College of Medicine in Houston, Texas, and Emeritus Professor at the University of Connecticut School of Medicine in Farmington, Connecticut.

Henk ten Have, M.D., Ph.D., is Chairman of the Department of Ethics, Philosophy and History of Medicine at the Catholic University of Nijmegen, The Netherlands.

Stephen Toulmin, Ph.D., is Henry Luce Professor in the Center for Multiethnic & Transnational Studies at the University of Southern California.

Harold Y. Vanderpool, Ph.D., Th.M., is Professor in the History and Philosophy of Medicine at the Institute for the Medical Humanities at The University of Texas Medical Branch at Galveston, Texas.

William J. Winslade, Ph.D., J.D., is James Wade Rockwell Professor of Philosophy of Medicine at the Institute for the Medical Humanities, and Professor in the Departments of Preventive Medicine and Community Health and Psychiatry and Behavioral Sciences at The University of Texas Medical Branch at Galveston, Texas.

Marx Wartofsky, Ph.D., was Distinguised Professor in the Department of Philosophy at Baruch College, City University of New York.

INDEX